KU-686-906

Sound and Recording:
An Introduction

Fourth edition

Francis Rumsey BMus (Tonmeister), PhD
Tim McCormick BA, LTCL

1478485

CHESTER COLLEGE

ACC No. 01091804	DEPT
CLASS No. 621.389 Rum	

LIBRARY

Focal Press

OXFORD AMSTERDAM BOSTON LONDON NEW YORK PARIS
SAN DIEGO SAN FRANCISCO SINGAPORE SYDNEY TOKYO

Focal Press
An imprint of Elsevier Science
Linacre House, Jordan Hill, Oxford OX2 8DP
225 Wildwood Avenue, Woburn, MA 01801-2041

First published 1992
Reprinted 1994
Second edition 1994
Reprinted 1995, 1996
Third edition 1997
Reprinted 1998 (twice), 1999, 2000, 2001
Fourth edition 2002

Copyright © 1992, 1994, 1997, 2002, Francis Rumsey and Tim McCormick.
All rights reserved

The right of Francis Rumsey and Tim McCormick to be identified as the
authors of this work has been asserted in accordance with the Copyright,
Designs and Patents Act 1988

No part of this publication may be reproduced in any material form (including
photocopying or storing in any medium by electronic means and whether
or not transiently or incidentally to some other use of this publication) without
the written permission of the copyright holder except in accordance with the
provisions of the Copyright, Designs and Patents Act 1988 or under the terms of
a licence issued by the Copyright Licensing Agency Ltd, 90 Tottenham Court Road,
London, England W1T 4LP. Applications for the copyright holder's written
permission to reproduce any part of this publication should be addressed
to the publisher.

British Library Cataloguing in Publication Data
A catalogue record for this book is available from the British Library

Library of Congress Cataloguing in Publication Data
A catalogue record for this book is available from the Library of Congress

ISBN 0 240 51680 X

For information on all Focal Press publications visit our
website at www.focalpress.com

Printed and bound in Great Britain

Contents

Fact File Directory

Series Introduction

The Focal Press Music Technology Series is intended to fill a growing need for authoritative books to support college and university courses in music technology, sound recording, multi-media and their related fields. The books will also be of value to professionals already working in these areas and who want either to update their knowledge or to familiarise themselves with topics that have not been part of their mainstream occupations.

Information technology and digital systems are now widely used in the production of sound and in the composition of music for a wide range of end uses. Those working in these fields need to understand the principles of sound, musical acoustics, sound synthesis, digital audio, video and computer systems. This is a tall order, but people with this breadth of knowledge are increasingly sought after by employers. The series will explain the technology and techniques in a manner which is both readable and factually concise, avoiding the chattiness, informality and technical woolliness of many books on music technology. The authors are all experts in their fields and many come from teaching and research backgrounds.

Dr Francis Rumsey
Series Consultant

Preface to the Second Edition

One of the greatest dangers in writing a book at an introductory level is to sacrifice technical accuracy for the sake of simplicity. In writing *Sound and Recording: An Introduction* we have gone to great lengths not to fall into this trap, and have produced a comprehensive introduction to the field of audio, intended principally for the newcomer to the subject, which is both easy to understand and technically precise. We have written the book that we would have valued when we first entered the industry, and as such it represents a readable reference, packed with information. Many books stop after a vague overview, just when the reader wants some clear facts about a subject, or perhaps assume too much knowledge on the reader's behalf. Books by contributed authors often suffer from a lack of consistency in style, coverage and technical level. Furthermore, there is a tendency for books on audio to be either too technical for the beginner or, alternatively, subjectively biased towards specific products or operations. There are also quite a number of American books on sound recording which, although good, tend to ignore European trends and practices. We hope that we have steered a balanced course between these extremes, and have deliberately avoided any attempt to dictate operational practice.

Sound and Recording: An Introduction is definitely biased towards an understanding of 'how it works', as opposed to 'how to work it', although technology is never discussed in an abstract manner but related to operational reality. Although we have included a basic introduction to acoustics and the nature of sound perception, this is not a book on acoustics or musical acoustics (there are plenty of those around). It is concerned with the principles of audio recording and reproduction, and has a distinct bias towards the professional rather than the consumer end of the market. The coverage of subject matter is broad, including chapters on digital audio, timecode synchronisation and MIDI, amongst other more conventional subjects, and there is comprehensive coverage of commonly misunderstood subjects such as the decibel, balanced lines, reference levels and metering systems.

This second edition of the book has been published only two years after the first, and the subject matter has not changed significantly enough in the interim to warrant major modifications to the existing chapters. The key difference between the second and first editions is the addition of a long chapter on stereo recording and reproduction. This important topic is covered in considerable detail, including historical developments, principles of stereo reproduction, surround sound and stereo microphone techniques. Virtually every recording or broadcast happening today is made in stereo, and although surround sound has had a number of notable 'flops' in the past it is likely to become considerably more important in the next ten years. Stereo and surround sound are used extensively in film, video and television production, and any new audio engineer should be familiar with the principles.

Since this is an introductory book, it will be of greatest value to the student of sound recording or music technology, and to the person starting out on a career in sound engineering or broadcasting. The technical level has deliberately been kept reasonably low for this reason, and those who find this frustrating probably do not need the book! Nonetheless, it is often valuable for the seasoned audio engineer to go back to basics. Further reading suggestions have been made in order that the reader may go on to a more in-depth coverage of the fields introduced here, and some of the references are considerably more technical than this book. Students will find these suggestions valuable when planning a course of study.

Francis Rumsey
Tim McCormick

Preface to the Third Edition

Since the first edition of *Sound and Recording* some of the topics have advanced quite considerably, particularly the areas dependent on digital and computer technology. Consequently I have rewritten the chapters on digital recording and MIDI (Chapters 10 and 15), and have added a larger section on mixer automation in Chapter 7. Whereas the first edition of the book was quite 'analogue', I think that there is now a more appropriate balance between analogue and digital topics. Although analogue audio is by no means dead (sound will remain analogue for ever!), most technological developments are now digital.

I make no apologies for leaving in the chapter on record players, although some readers have commented that they think it is a waste of space. People still use record players, and there is a vast store of valuable material on LP record. I see no problem with keeping a bit of history in the book – you never know, it might come in useful one day when everyone has forgotten (and some may never have known) what to do with vinyl disks. It might even appease the faction of our industry that continues to insist that vinyl records are the highest fidelity storage medium ever invented.

Francis Rumsey
Guildford

Preface to the Fourth Edition

The fourth edition is published ten years after *Sound and Recording* was first published, which is hard to believe. The book has been adopted widely by students and tutors on audio courses around the world. In that time audio technology and techniques have changed in some domains but not in others. All the original principles still apply but the emphasis has gradually changed from predominantly analogue to quite strongly digital, although many studios still use analogue mixers and multitrack tape recorders for a range of purposes and we do not feel that the death-knell of analogue recording has yet been sounded. Readers of earlier editions will notice that the chapter on record players has finally been reduced in size and relegated to an appendix. While we continue to believe that information about the LP should remain in the literature as the format lingers on, it is perhaps time to remove it from the main part of the book.

In this edition a new chapter on surround sound has been added, complemented by a reworked chapter preceding it that is now called 'two-channel stereo'. Surround sound was touched upon in the previous edition but a complete chapter reflects the increased activity in this field with the coming of new multichannel consumer replay formats.

The chapter on auditory perception has been reworked to include greater detail on spatial perception and the digital audio chapter has been updated to include DVD-A and SACD, with information about Direct Stream Digital (DSD), the MiniDisc, computer-based editing systems and their operation. Chapter 5 on loudspeakers now includes information about distributed-mode loudspeakers (DML) and a substantial section on directivity and the various techniques used to control it. Finally a glossary of terms has now been provided, with some additional material that supports the main text.

Francis Rumsey
Tim McCormick

Chapter 1

What is sound?

1.1 A vibrating source

Sound is produced when an object (the source) vibrates and causes the air around it to move. Consider the sphere shown in Figure 1.1. It is a pulsating sphere which could be imagined as something like a squash ball, and it is pulsating regularly so that its size oscillates between being slightly larger than normal and then slightly smaller than normal. As it pulsates it will alternately compress and then rarefy the

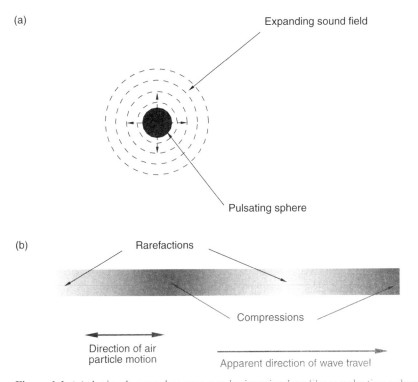

Figure 1.1 (a) A simple sound source can be imagined as like a pulsating sphere radiating spherical waves. (b) The longitudinal wave thus created is a succession of compressions and rarefactions of the air

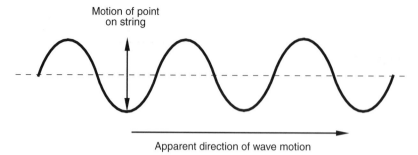

Figure 1.2 In a transverse wave the motion of any point on the wave is at right angles to the apparent direction of motion of the wave

surrounding air, resulting in a series of compressions and rarefactions travelling away from the sphere, rather like a three-dimensional version of the ripples which travel away from a stone dropped into a pond. These are known as *longitudinal waves* since the air particles move in the same dimension as the direction of wave travel. The alternative to longitudinal wave motion is *transverse* wave motion (see Figure 1.2), such as is found in vibrating strings, where the motion of the string is at right angles to the direction of apparent wave travel.

1.2 Characteristics of a sound wave

The rate at which the source oscillates is the *frequency* of the sound wave it produces, and is quoted in *hertz* (Hz) or cycles per second (cps). 1000 hertz is termed 1 *kilohertz* (1 kHz). The amount of compression and rarefaction of the air which results from the sphere's motion is the *amplitude* of the sound wave, and is related to the loudness of the sound when it is finally perceived by the ear (see Chapter 2). The distance between two adjacent peaks of compression or rarefaction as the wave travels through the air is the *wavelength* of the sound wave, and is often represented by the Greek letter *lambda* (λ). The wavelength depends on how fast the sound wave travels, since a fast-travelling wave would result in a greater distance between peaks than a slow-travelling wave, given a fixed time between compression peaks (i.e.: a fixed frequency of oscillation of the source).

As shown in Figure 1.3, the sound wave's characteristics can be represented on a graph, with amplitude plotted on the vertical axis and time plotted on the horizontal axis. It will be seen that both positive and negative ranges are shown on the vertical axis: these represent compressions ($+$) and rarefactions ($-$) of the air. This graph represents the *waveform* of the sound. For a moment, a source vibrating in a very simple and regular manner is assumed, in so-called *simple harmonic motion*, the result of which is a simple sound wave known as a *sine wave*. The most simple vibrating systems oscillate in this way, such as a mass suspended from a spring, or a pendulum swinging on a fulcrum (see also section 1.7 on phase). It will be seen that the frequency (f) is the inverse of the time between peaks or troughs of the wave ($f = 1/t$). So the shorter the time between oscillations of the source, the higher the

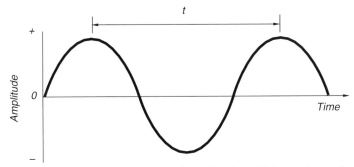

Figure 1.3 A graphical representation of a sinusoidal sound waveform. The period of the wave is represented by t, and its frequency by $1/t$

frequency. The human ear is capable of perceiving sounds with frequencies between approximately 20 Hz and 20 kHz (see section 2.2); this is known as the *audio frequency range* or *audio spectrum*.

1.3 How sound travels in air

Air is made up of gas molecules and has an elastic property (imagine putting a thumb over the end of a bicycle pump and compressing the air inside – the air is springy). Longitudinal sound waves travel in air in somewhat the same fashion as a wave travels down a row of up-ended dominoes after the first one is pushed over. The half-cycle of compression created by the vibrating source causes successive air particles to be moved in a knock-on effect, and this is normally followed by a balancing rarefaction which causes a similar motion of particles in the opposite direction.

It may be appreciated that the net effect of this is that individual air particles do not actually travel – they oscillate about a fixed point – but the result is that a wave is formed which appears to move away from the source. The speed at which it moves away from the source depends on the density and elasticity of the substance through which it passes, and in air the speed is relatively slow compared with the speed at which sound travels through most solids. In air the speed of sound is approximately 340 metres per second ($m\ s^{-1}$), although this depends on the temperature of the air. At freezing point the speed is reduced to nearer 330 m s^{-1}. In steel, to give an example of a solid, the speed of sound is approximately 5100 m s^{-1}.

The frequency and wavelength of a sound wave are related very simply if the speed of the wave (usually denoted by the letter c) is known:

$$c = f\lambda \text{ or } \lambda = c/f$$

To show some examples, the wavelength of sound in air at 20 Hz (the low-frequency or LF end of the audio spectrum), assuming normal room temperature, would be:

$$\lambda = 340/20 = 17 \text{ metres}$$

whereas the wavelength of 20 kHz (at the high-frequency or HF end of the audio spectrum) would be 1.7 cm. Thus it is apparent that the wavelength of sound ranges from being very long in relation to most natural objects at low frequencies, to quite short at high frequencies. This is important when considering how sound behaves when it encounters objects – whether the object acts as a barrier or whether the sound bends around it (see Fact File 1.5).

1.4 Simple and complex sounds

In the foregoing example, the sound had a simple waveform – it was a sine wave or *sinusoidal* waveform – the type which might result from a very simple vibrating system such as a weight suspended on a spring. Sine waves have a very pure sound because they consist of energy at only one frequency, and are often called *pure tones*. They are not heard very commonly in real life (although they can be generated electrically) since most sound sources do not vibrate in such a simple manner. A person whistling or a recorder (a simple wind instrument) produces a sound which approaches a sinusoidal waveform. Most real sounds are made up of a combination of vibration patterns which result in a more complex waveform. The more complex the waveform, the more like noise the sound becomes, and when the waveform has a highly random pattern the sound is said to be *noise* (see section 1.6).

The important characteristic of sounds which have a definite pitch is that they are repetitive: that is, the waveform, no matter how complex, repeats its pattern in the same way at regular intervals. All such waveforms can be broken down into a series of components known as *harmonics*, using a mathematical process called Fourier analysis (after the mathematician Joseph Fourier). Some examples of equivalent line spectra for different waveforms are given in Figure 1.4. This figure shows another way of depicting the characteristics of the sound graphically – that is, by drawing a so-called *line spectrum* which shows frequency along the horizontal axis and ampli-tude up the vertical axis. The line spectrum shows the relative strengths of different frequency components which make up a sound. Where there is a line there is a frequency component. It will be noticed that the more complex the waveform the more complex the corresponding line spectrum.

For every waveform, such as that shown in Figure 1.3, there is a corresponding line spectrum: waveforms and line spectra are simply two different ways of showing the characteristics of the sound. Figure 1.3 is called a *time-domain* plot, whilst the line spectrum is called a *frequency-domain* plot. Unless otherwise stated, such frequency-domain graphs in this book will cover the audio-frequency range, from 20 Hz at the lower end to 20 kHz at the upper end.

In a reversal of the above breaking-down of waveforms into their component frequencies it is also possible to *construct* or *synthesise* waveforms by adding together the relevant components.

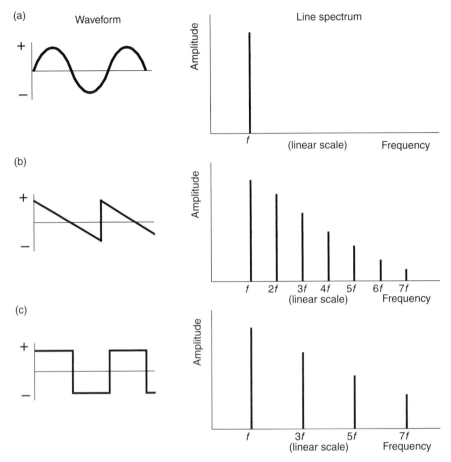

Figure 1.4 Equivalent line spectra for a selection of simple waveforms. (a) The sine wave consists of only one component at the fundamental frequency f. (b) The sawtooth wave consists of components at the fundamental and its integer multiples, with amplitudes steadily decreasing. (c) The square wave consists of components at odd multiples of the fundamental frequency

1.5 Frequency spectra of repetitive sounds

As will be seen in Figure 1.4, the simple sine wave has a line spectrum consisting of only one component at the frequency of the sine wave. This is known as the *fundamental* frequency of oscillation. The other repetitive waveforms, such as the square wave, have a fundamental frequency as well as a number of additional components above the fundamental. These are known as *harmonics*, but may also be referred to as *overtones* or *partials*.

Harmonics are frequency components of a sound which occur at integer multiples of the fundamental frequency, that is at twice, three times, four times and so on. Thus a sound with a fundamental of 100 Hz might also contain harmonics at 200 Hz, 400 Hz and 600 Hz. The reason for the existence of these harmonics is that most simple vibrating sources are capable of vibrating in a number of harmonic *modes* at the same time. Consider a stretched string, as shown in Figure 1.5. It may be made to vibrate in any of a number of modes, corresponding to integer multiples of the fundamental frequency of vibration of the string (the concept of a 'standing wave' is introduced in section 1.13, below). The fundamental corresponds to the mode in which the string moves up and down as a whole, whereas the harmonics correspond to modes in which the vibration pattern is divided into points of maximum and minimum motion along the string (these are called *antinodes* and *nodes*). It will be seen that the second mode involves two peaks of vibration, the third mode three peaks, and so on.

In accepted terminology, the fundamental is also the *first harmonic*, and thus the next component is the *second harmonic*, and so on. Confusingly, the second harmonic is also known as the *first overtone*. For the waveforms shown in Figure 1.4, the fundamental has the highest amplitude, and the amplitudes of the harmonics decrease with increasing frequency, but this will not always be the case with real sounds since many waveforms have line spectra which show the harmonics to be higher in amplitude than the fundamental. It is also quite feasible for there to be harmonics missing in the line spectrum, and this depends entirely on the waveform in question.

It is also possible for there to be overtones in the frequency spectrum of a sound which are *not* related in a simple integer-multiple fashion to the fundamental. These cannot correctly be termed harmonics, and they are more correctly referred to as *overtones* or *inharmonic partials*. They tend to arise in vibrating sources which have a complicated shape, and which do not vibrate in simple harmonic motion but have a number of repetitive modes of vibration. Their patterns of oscillation are often

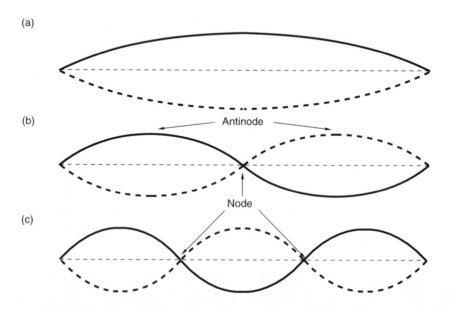

Figure 1.5 Modes of vibration of a stretched string. (a) Fundamental. (b) Second harmonic. (c) Third harmonic

unusual, such as might be observed in a bell or a percussion instrument. It is still possible for such sounds to have a recognisable pitch, but this depends on the strength of the fundamental. In bells and other such sources, one often hears the presence of several strong inharmonic overtones.

1.6 Frequency spectra of non-repetitive sounds

Non-repetitive waveforms do not have a recognisable pitch and sound noise-like. Their frequency spectra are likely to consist of a collection of components at unrelated frequencies, although some frequencies may be more dominant than others. The analysis of such waves to show their frequency spectra is more complicated than with repetitive waves, but is still possible using a mathematical technique called *Fourier transformation*, the result of which is a frequency-domain plot of a time-domain waveform.

Single, short pulses can be shown to have continuous frequency spectra which extend over quite a wide frequency range, and the shorter the pulse the wider its frequency spectrum but usually the lower its total energy (see Figure 1.6). Random waveforms will tend to sound like hiss, and a completely random waveform in which the frequency, amplitude and phase of components are equally probable and constantly varying is called *white noise*. A white noise signal's spectrum is flat, when averaged over a period of time, right across the audio-frequency range (and theoretically above it). White noise has equal energy for a given bandwidth, whereas another type of noise, known as *pink noise*, has equal energy per *octave*.

Figure 1.6 Frequency spectra of non-repetitive waveforms. (a) Pulse. (b) Noise

For this reason white noise sounds subjectively to have more high-frequency energy than pink noise.

1.7 Phase

Two waves of the same frequency are said to be 'in phase' when their compression (positive) and rarefaction (negative) half-cycles coincide exactly in time and space (see Figure 1.7). If two in-phase signals of equal amplitude are added together, or superimposed, they will sum to produce another signal of the same frequency but twice the amplitude. Signals are said to be out of phase when the positive half-cycle of one coincides with the negative half-cycle of the other. If these two signals are added together they will cancel each other out, and the result will be no signal.

Clearly these are two extreme cases, and it is entirely possible to superimpose two sounds of the same frequency which are only partially in phase with each other. The resultant wave in this case will be a partial addition or partial cancellation, and the phase of the resulting wave will lie somewhere between that of the two components (see Figure 1.7(c)).

Phase differences between signals can be the result of time delays between them. If two signals start out at sources equidistant from a listener at the same time as each other then they will be in phase by the time they arrive at the listener (although see section 2.5 on directional perception). If one source is more distant than the other then it will be delayed, and the phase relationship between the two will depend upon the amount of delay (see Figure 1.8). A useful rule-of-thumb is that sound travels about 30 cm (1 foot) per millisecond, so if the second source in the above example were 1 metre (just over 3 ft) more distant than the first it would be delayed by just over 3 ms. The resulting phase relationship between the two signals, it may be appreciated, would depend on the frequency of the sound, since at a frequency of around 330 Hz the 3 ms delay would correspond to one wavelength and thus the delayed signal would be *in* phase with the undelayed signal. If the delay had been half this (1.5 ms) then the two signals would have been out of phase at 330 Hz.

Phase is often quoted as a number of degrees relative to some reference, and this must be related back to the nature of a sine wave. A diagram is the best way to illustrate this point, and looking at Figure 1.9 it will be seen that a sine wave may be considered as a graph of the vertical position of a rotating spot on the outer rim of a disk (the amplitude of the wave), plotted against time. The height of the spot rises and falls regularly as the circle rotates at a constant speed. The sine wave is so called because the spot's height is directly proportional to the mathematical *sine* of the angle of rotation of the disk, with zero degrees occurring at the origin of the graph and at the point shown on the disk's rotation in the diagram. The vertical amplitude scale on the graph goes from minus one (maximum negative amplitude) to plus one (maximum positive amplitude), passing through zero at the halfway point. At 90° of rotation the amplitude of the sine wave is maximum positive (the sine of 90° is +1), and at 180° it is zero (sin 180° = 0). At 270° it is maximum negative (sin 270° = −1),

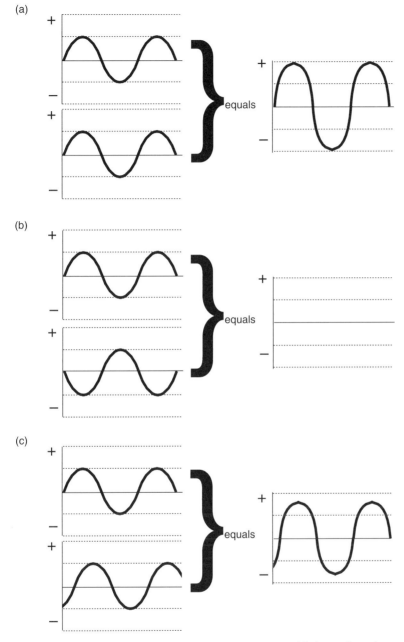

Figure 1.7 (a) When two identical in-phase waves are added together, the result is a wave of the same frequency and phase but twice the amplitude. (b) Two identical out-of-phase waves add to give nothing. (c) Two identical waves partially out of phase add to give a resultant wave with a phase and amplitude which is the point-by-point sum of the two

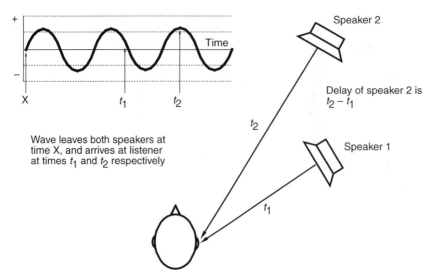

Figure 1.8 If the two loudspeakers in the drawing emit the same wave at the same time, the phase difference between the waves at the listener's ear will be directly related to the delay $t_2 - t_1$

and at 360° it is zero again. Thus in one cycle of the sine wave the circle has passed through 360° of rotation.

It is now possible to go back to the phase relationship between two waves of the same frequency. If each cycle is considered as corresponding to 360°, then one can say just how many degrees one wave is ahead of or behind another by comparing the 0° point on one wave with the 0° point on the other (see Figure 1.10). In the example wave 1 is 90° out of phase with wave 2. It is important to realise that phase is only a relevant concept in the case of continuous repetitive waveforms, and has little meaning in the case of impulsive or transient sounds where time difference is the more relevant quantity. It can be deduced from the foregoing discussion that (a) the higher the frequency, the greater the phase difference which would result from a given time delay between two signals, and (b) it is possible for there to be more than

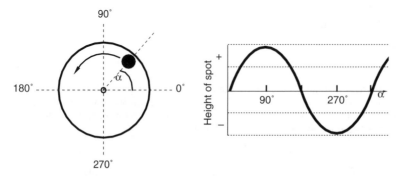

Figure 1.9 The height of the spot varies sinusoidally with the angle of rotation of the wheel. The phase angle of a sine wave can be understood in terms of the number of degrees of rotation of the wheel

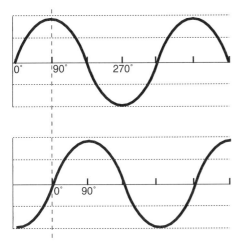

Figure 1.10 The lower wave is 90° out of phase with the upper wave

360° of phase difference between two signals if the delay is great enough to delay the second signal by more than one cycle. In the latter case it becomes difficult to tell how many cycles of delay have elapsed unless a discontinuity arises in the signal, since a phase difference of 360° is indistinguishable from a phase difference of 0°.

1.8 Sound in electrical form

Although the sound that one hears is due to compression and rarefaction of the air, it is often necessary to convert sound into an electrical form in order to perform operations on it such as amplification, recording and mixing. As detailed in Fact File 3.1 and Chapter 4, it is the job of the microphone to convert sound from an *acoustical* form into an *electrical* form. The process of conversion will not be described here, but the result is important because if it can be assumed for a moment that the microphone is perfect then the resulting electrical waveform will be exactly the same shape as the acoustical waveform which caused it.

The equivalent of the amplitude of the acoustical signal in electrical terms is the *voltage* of the electrical signal. If the voltage at the output of a microphone were to be measured whilst the microphone was picking up an acoustical sine wave, one would measure a voltage which changed sinusoidally as well. Figure 1.11 shows this situation, and it may be seen that an acoustical compression of the air corresponds to a positive-going voltage, whilst an acoustical rarefaction of the air corresponds to a negative-going voltage. (This is the norm, although some sound reproduction systems introduce an *absolute phase reversal* in the relationship between acoustical phase and electrical phase, such that an acoustical compression becomes equivalent to a negative voltage. Some people claim to be able to hear the difference.)

The other important quantity in electrical terms is the *current* flowing down the wire from the microphone. Current is the electrical equivalent of the air particle motion discussed in section 1.3. Just as the acoustical sound wave was carried in the

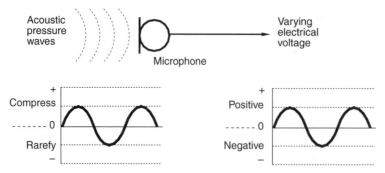

Figure 1.11 A microphone converts variations in acoustical sound pressure into variations in electrical voltage. Normally a compression of the air results in a positive voltage and a rarefaction results in a negative voltage

motion of the air particles, so the electrical sound wave is carried in the motion of tiny charge carriers which reside in the metal of a wire (these are called *electrons*). When the voltage is positive the current moves in one direction, and when it is negative the current moves in the other direction. Since the voltage generated by a microphone is repeatedly alternating between positive and negative, in sympathy with the sound wave's compression and rarefaction cycles, the current similarly changes direction each half cycle. Just as the air particles in section 1.3 did not actually go anywhere in the long term, so the electrons carrying the current do not go anywhere either – they simply oscillate about a fixed point. This is known as *alternating current* or AC.

A useful analogy to the above (both electrical and acoustical) exists in plumbing. If one considers water in a pipe fed from a header tank, as shown in Figure 1.12, the voltage is equivalent to the pressure of water which results from the header tank, and the current is equivalent to the rate of flow of water through the pipe. The only difference is that the diagram is concerned with a *direct current* situation in which the direction of flow is not repeatedly changing. The quantity of *resistance* should be introduced here, and is analogous to the diameter of the pipe. Resistance impedes the flow of water through the pipe, as it does the flow of electrons through a wire

Figure 1.12 There are parallels between the flow of water in a pipe and the flow of electricity in a wire, as shown in this drawing

and the flow of acoustical sound energy through a substance. For a fixed voltage (or water pressure in this analogy), a high resistance (narrow pipe) will result in a small current (a trickle of water), whilst a low resistance (wide pipe) will result in a large current. The relationship between voltage, current and resistance was established by Ohm, in the form of Ohm's law, as described in Fact File 1.1. There is also a relationship between power and voltage, current and resistance.

In AC systems, resistance is replaced by *impedance*, a complex term which contains both resistance and *reactance* components. The reactance part varies with the frequency of the signal; thus the impedance of an electrical device also varies with the frequency of a signal. Capacitors (basically two conductive plates separated by an insulator) are electrical devices which present a high impedance to low-frequency signals and a low impedance to high-frequency signals. They will not pass direct current. Inductors (basically coils of wire) are electrical devices which present a high impedance to high-frequency signals and a low impedance to low-frequency signals. Capacitance is measured in farads, inductance in henrys.

1.9 Displaying the characteristics of a sound wave

Two devices can be introduced at this point which illustrate graphically the various characteristics of sound signals so far described. It would be useful to (a) display the waveform of the sound, and (b) display the frequency spectrum of the sound. In other words (a) the time-domain signal and (b) the frequency-domain signal.

An *oscilloscope* is used for displaying the waveform of a sound, and a *spectrum analyser* is used for showing which frequencies are contained in the signal and their amplitudes. Examples of such devices are pictured in Figure 1.13. Both devices

---

FACT FILE 1.1   Ohm's law

Ohm's law states that there is a fixed and simple relationship between the current flowing through a device (*I*), the voltage across it (*V*), and its resistance (*R*), as shown in the diagram:

$V = IR$

or:

$I = V/R$

or:

$R = V/I$

Thus if the resistance of a device is known, and the voltage dropped across it can be measured, then the current flow may be calculated, for example.

There is also a relationship between the parameters above and the *power* in watts (*W*) dissipated in a device:

$$W = I^2R = V^2/R$$

(a)   (b)

Figure 1.13 (a) An oscilloscope displays the *waveform* of an electric signal by means of a moving spot which is deflected up by a positive signal and down by a negative signal. (b) A spectrum analyser displays the *frequency spectrum* of an electrical waveform in form of lines representing the amplitudes of different spectral components of the signal

accept sound signals in electrical form and display their analyses of the sound on a screen. The oscilloscope displays a moving spot which scans horizontally at one of a number of fixed speeds from left to right and whose vertical deflection is controlled by the voltage of the sound signal (up for positive, down for negative). In this way it plots the waveform of the sound as it varies with time. Many oscilloscopes have two inputs and can plot two waveforms at the same time, and this can be useful for comparing the relative phases of two signals (see section 1.7).

The spectrum analyser works in different ways depending on the method of spectrum analysis. A real-time analyser displays a constantly updating line spectrum, similar to those depicted earlier in this chapter, and shows the frequency components of the input signal on the horizontal scale together with their amplitudes on the vertical scale.

1.10 The decibel

The unit of the decibel is used widely in sound engineering, often in preference to other units such as volts, watts, or other such absolute units, since it is a convenient way of representing the *ratio* of one signal's amplitude to another's. It also results in numbers of a convenient size which approximate more closely to one's subjective impression of changes in the amplitude of a signal, and it helps to compress the range of values between the maximum and minimum sound levels encountered in real signals. For example, the range of sound intensities (see section 1.11) which can be handled by the human ear covers about fourteen powers of ten, from 0.000 000 000 001 W m^{-2} to around 100 W m^{-2}, but the equivalent range in decibels is only from 0 to 140 dB.

Some examples of the use of the decibel are given in Fact File 1.2. The relationship between the decibel and human sound perception is discussed in more detail in Chapter 2. Operating levels in recording equipment are discussed further in sections 3.4, 7.2 and 8.5.

Decibels are not only used to describe the ratio between two signals, or the level of a signal above a reference, but they are also used to describe the *voltage gain* of a device. For example, a microphone amplifier may have a gain of 60 dB, which is the equivalent of multiplying the input voltage by a factor of 1000, as shown in the example below:

$$20 \log 1000/1 = 60 \, dB$$

1.11 Sound power and sound pressure

A simple sound source, such as the pulsating sphere used at the start of this chapter, radiates sound power *omnidirectionally* – that is, equally in all directions, rather like a three-dimensional version of the ripples moving away from a stone dropped in a pond. The sound source generates a certain amount of power, measured in *watts*, which is gradually distributed over an increasingly large area as the wavefront travels further from the source; thus the amount of power per square metre passing through the surface of the imaginary sphere surrounding the source gets smaller with increasing distance (see Fact File 1.3). For practical purposes the intensity of the direct sound from a source drops by 6 dB for every doubling in distance from the source (see Figure 1.14).

The amount of acoustical power generated by real sound sources is surprisingly small, compared with the number of watts of electrical power involved in lighting a light bulb, for example. An acoustical source radiating 20 watts would produce a sound pressure level close to the threshold of pain if a listener was close to the source. Most everyday sources generate fractions of a watt of sound power, and this energy is eventually dissipated into heat by absorption (see below). The amount of heat produced by the dissipation of acoustic energy is relatively insignificant – the chances of increasing the temperature of a room by shouting are slight, at least in the physical sense.

Acoustical power is sometimes confused with the power output of an amplifier used to drive a loudspeaker, and audio engineers will be familiar with power outputs from amplifiers of many hundreds of watts. It is important to realise that loudspeakers are very inefficient devices – that is, they only convert a small proportion of their electrical input power into acoustical power. Thus, even if the input to a loudspeaker was to be, say, 100 watts electrically, the acoustical output power might only be perhaps 1 watt, suggesting a loudspeaker that is only 1 per cent efficient. The remaining power would be dissipated as heat in the voice coil.

Sound *pressure* is the effect of sound power on its surroundings. To use a central heating analogy, sound power is analogous to the heat energy generated by a radiator into a room, whilst sound pressure is analogous to the temperature of the air in the room. The temperature is what a person entering the room would feel, but the heat-generating radiator is the source of power. Sound pressure level (SPL) is measured in *newtons per square metre* ($N \, m^{-2}$). A convenient reference level is set for sound pressure and intensity measurements, this being referred to as 0 dB.

FACT FILE **1.2**   The decibel

Basic decibels
The decibel is based on the logarithm of the ratio between two numbers. It describes how much larger or smaller one value is than the other. It can also be used as an absolute unit of measurement if the reference value is fixed and known. Some standardised references have been established for decibel scales in different fields of sound engineering (see below).

The decibel is strictly ten times the logarithm to the base ten of the ratio between the *powers* of two signals:

$$dB = 10 \log_{10}(P_1/P_2)$$

For example, the difference in decibels between a signal with a power of 1 watt and one of 2 watts is 10 log (2/1) = 3 dB.

If the decibel is used to compare values other than signal powers, the relationship to signal power must be taken into account. Voltage has a square relationship to power (from Ohm's law: $W = V^2/R$); thus to compare two voltages:

$$dB = 10 \log(V_1^2/V_2^2), \text{ or } 10 \log (V_1/V_2)^2,$$
$$\text{or } 20 \log(V_1/V_2)$$

For example, the difference in decibels between a signal with a voltage of 1 volt and one of 2 volts is 20 log (2/1) = 6 dB. So a doubling in voltage gives rise to an increase of 6 dB, and a doubling in power gives rise to an increase of 3 dB. A similar relationship applies to acoustical sound pressure (analogous to electrical voltage) and sound power (analogous to electrical power).

Decibels with a reference
If a signal level is quoted in decibels, then a reference must normally be given, otherwise the figure means nothing; e.g.: 'Signal level = 47 dB' cannot have a meaning unless one knows that the signal is 47 dB above a known point. '+8 dB ref. 1 volt' has a meaning since one now knows that the level is 8 dB higher than 1 volt, and thus one could calculate the voltage of the signal.

There are exceptions in practice, since in some fields a reference level is accepted as implicit. Sound pressure levels (SPLs) are an example, since the reference level is defined worldwide as 2×10^{-6} N m^{-2} (20 μPa). Thus to state 'SPL = 77 dB' is probably acceptable, although confusion can still arise due to misunderstandings over such things as weighting curves (see Fact File 1.4). In sound recording, 0 dB or 'zero level' is a nominal reference level used for aligning equipment and setting recording levels, often corresponding to 0.775 volts (0 dBu) although this is subject to variations in studio centres in different locations. (Some studios use +4 dBu as their reference level, for example.) '0 dB' does not mean 'no signal', it means that the signal concerned is at the same level as the reference.

Often a letter is placed after 'dB' to denote the reference standard in use (e.g.: 'dBm'), and a number of standard abbreviations are in use, some examples of which are given below. Sometimes the suffix denotes a particular frequency weighting characteristic used in the measurement of noise (e.g.: 'dBA').

Abbrev.	Ref. Level
dBV	1 volt
dBu	0.775 volt (Europe)
dBv	0.775 volt (USA)
dBm	1 milliwatt (see section 12.9)
dBA	dB SPL, A-weighted response

A full listing of suffixes is given in CCIR Recommendation 5741,1982.

Useful decibel ratios to remember (voltages or SPLs)
It is more common to deal in terms of voltage or SPL ratios than power ratios in audio systems. Here are some useful dB equivalents

of different voltage or SPL relationships and multiplication factors:

dB	Multiplication factor
0	1
+3	$\sqrt{2}$
+6	2
+20	10
+60	1000

This level of 0 dB is approximately equivalent to the threshold of hearing (the quietest sound perceivable by an average person) at a frequency of 1 kHz, and corresponds to an SPL of 2×10^{-5} N m^{-2}, which in turn is equivalent to an intensity of approximately 10^{-12} W m^{-2} in the free field (see below).

Sound pressure levels are often quoted in dB (e.g.: SPL = 63 dB means that the SPL is 63 dB above 2×10^{-5} N m^{-2}). The SPL in dB may not accurately represent the *loudness* of a sound, and thus a subjective unit of loudness has been derived

---

FACT FILE 1.3 The inverse-square law

The law of decreasing power per unit area (intensity) of a wavefront with increasing distance from the source is known as the *inverse-square law*, because intensity drops in proportion to the inverse square of the distance

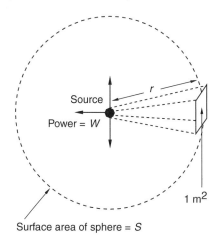

Source
Power = W

1 m^2

Surface area of sphere = S

from the source. Why is this? It is because the sound power from a point source is spread over the surface area of a sphere (S), which from elementary maths is given by:

$$S = 4\pi r^2$$

where *r* is the distance from the source or the radius of the sphere, as shown in the diagram.

If the original power of the source is *W* watts, then the intensity, or power per unit area (*I*) at distance *r* is:

$$I = W/(4\pi r^2)$$

For example, if the power of a source was 0.1 watt, the intensity at 4 m distance would be:

$$I = 0.1 \div$$
$$(4 \times 3.14 \times 16) \ 0.0005 \ \text{W m}^{-2}$$

The sound intensity level (SIL) of this signal in decibels can be calculated by comparing it with the accepted reference level of 10^{-12} W m^{-2} :

$$SIL(\text{dB}) = 10\log((5 \times 10^{-4}) \div (10^{-12}))$$
$$= 87 \text{ dB}$$

Figure 1.14 The sound power which had passed through 1 m² of space at distance *r* from the source will pass through 4 m² at distance 2*r*, and thus will have one quarter of the intensity

from research data, called the *phon*. This is discussed further in Chapter 2. Some methods of measuring sound pressure levels are discussed in Fact File 1.4.

1.12 Free and reverberant fields

The free field in acoustic terms is an acoustical area in which there are no reflections. Truly free fields are rarely encountered in reality, because there are nearly always reflections of some kind, even if at a very low level. If the reader can imagine the sensation of being suspended out-of-doors, way above the ground, away from any buildings or other surfaces, then he or she will have an idea of the experience of a free-field condition. The result is an acoustically 'dead' environment. Acoustic experiments are sometimes performed in *anechoic chambers*, which are rooms specially treated so as to produce almost no reflections at any frequency – the surfaces are totally absorptive – and these attempt to create near free-field conditions.

In the free field all the sound energy from a source is radiated away from the source and none is reflected; thus the inverse-square law (Fact File 1.3) entirely dictates the level of sound at any distance from the source. Of course the source may be directional, in which case its *directivity factor* must be taken into account. A source with a directivity factor of 2 on its axis of maximum radiation radiates twice as much power in this direction as it would have if it had been radiating omnidirectionally. The directivity *index* of a source is measured in dB, giving the above example a directivity index of 3 dB. If calculating the intensity at a given distance from a directional source (as shown in Fact File 1.3), one must take into account its directivity factor on the axis concerned by multiplying the power of the source by the directivity factor before dividing by $4\pi r^2$.

In a room there is both direct and reflected sound. At a certain distance from a source contained within a room the acoustic field is said to be *diffuse* or *reverberant*, since reflected sound energy predominates over direct sound. A short time after the source has begun to generate sound a diffuse pattern of reflections will have built up throughout the room, and the reflected sound energy will become roughly constant at any point in the room. Close to the source the direct sound energy is still at quite a high level, and thus the reflected sound makes a smaller contribution to the total. This region is called the *near field*. (It is popular in sound recording to make use of

Measuring SPLs

Typically a sound pressure level (SPL) meter is used to measure the level of sound at a particular point. It is a device that houses a high quality omnidirectional (pressure) microphone (see section 4.4.1) connected to amplifiers, filters and a meter (see diagram).

Weighting filters
The microphone's output voltage is proportional to the SPL incident upon it, and the weighting filters may be used to attenuate low and high frequencies according to a standard curve such as the 'A'-weighting curve, which corresponds closely to the sensitivity of human hearing at low levels (see Chapter 2). SPLs quoted simply in dB are usually unweighted – in other words all frequencies are treated equally – but SPLs quoted in dBA will have been A-weighted and will correspond more closely to the *perceived* loudness of the signal. A-weighting was originally designed to be valid up to a loudness of 55 phons, since the ear's frequency response becomes flatter at higher levels; between 55 and 85 phons the 'B' curve was intended to be used; above 85 phons the 'C' curve was used. The 'D' curve was devised particularly for measuring aircraft engine noise at very high level.

Now most standards suggest that the 'A' curve may be used for measuring noise at any SPL, principally for ease of comparability of measurements, but there is still disagreement in the industry about the relative merits of different curves. The 'A' curve attenuates low and high frequencies and will therefore under-read quite substantially for signals at these frequencies. This is an advantage in some circumstances and a disadvantage in others. The 'C' curve is recommended in the USA and Japan for aligning sound levels using noise signals in movie theatres, for example. This only rolls off the very extremes of the audio spectrum and is therefore quite close to an unweighted reading. Some researchers have found that the 'B' curve produces results that more closely relate measured sound signal levels to subjective loudness of those signals.

Noise criterion or rating (NC or NR)
Noise levels are often measured in rooms by comparing the level of the noise across the audible range with a standard set of curves called the *noise criteria (NC)* or *noise rating (NR)* curves. These curves set out how much noise is acceptable in each of a number of narrow frequency bands for the noise to meet a certain criterion. The noise criterion is then that of the nearest curve above which none of the measured results rises. NC curves are used principally in the USA, whereas NR curves are used principally in Europe. They allow considerably higher levels in low-frequency bands than in middle- and high-frequency bands, since the ear is less sensitive at low frequencies.

In order to measure the NC or NR of a location it is necessary to connect the measuring microphone to a set of filters or a spectrum analyser which is capable of displaying the SPL in one octave or one-third octave bands.

Further reading
British Standard 5969. *Specification for sound level meters.*
British Standard 6402. *Sound exposure meters.*

so-called 'near-field monitors', which are loudspeakers mounted quite close to the listener, such that the direct sound predominates over the effects of the room.)

The exact distance from a source at which a sound field becomes dominated by reverberant energy depends on the reverberation time of the room, and this in turn depends on the amount of absorption in the room, and the room's volume (see Fact File 1.5). Figure 1.15 shows how the SPL changes as distance increases from a source in three different rooms. Clearly, in the acoustically 'dead' room, the conditions approach that of the free field (with sound intensity dropping at close to the expected 6 dB per doubling in distance), since the amount of reverberant energy is very small. The *critical distance* at which the contribution from direct sound equals that from reflected sound is further from the source than when the room is very reverberant. In the reverberant room the sound pressure level does not change much with distance from the source because reflected sound energy predominates after only a short distance. This is important in room design, since although a short reverberation time may be desirable in a recording control room, for example, it has the disadvantage that the change in SPL with distance from the speakers will be quite severe, requiring very highly powered amplifiers and heavy-duty speakers to provide the necessary level. A slightly longer reverberation time makes the room less disconcerting to work in, and relieves the requirement on loudspeaker power.

1.13 Standing waves

The wavelength of sound varies considerably over the audible frequency range, as indicated in Fact File 1.5. At high frequencies, where the wavelength is small, it is appropriate to consider a sound wavefront rather like light – as a ray. Similar rules apply, such as the angle of incidence of a sound wave to a wall is the same as the angle of reflection. At low frequencies where the wavelength is comparable with the dimensions of the room it is necessary to consider other factors, since the room behaves more as a complex resonator, having certain frequencies at which strong pressure peaks and dips are set up in various locations.

Standing waves or *eigentones* (sometimes also called *room modes*) may be set up when half the wavelength of the sound or a multiple is equal to one of the dimensions of the room (length, width or height). In such a case (see Figure 1.16) the reflected wave from the two surfaces involved is in phase with the incident wave and a pattern of summations and cancellations is set up, giving rise to points in the room at which the sound pressure is very high, and other points where it is very low. For the first mode (pictured), there is a peak at the two walls and a trough in the centre of the room. It is easy to experience such modes by generating a low-frequency sine tone into a room from an oscillator connected to an amplifier and loudspeaker placed in a corner. At selected low frequencies the room will resonate strongly and the pressure peaks may be experienced by walking around the room. There are always peaks towards the boundaries of the room, with troughs distributed at regular intervals between them. The

FACT FILE 1.5 — Absorption, reflection and RT

Absorption

When a sound wave encounters a surface some of its energy is absorbed and some reflected. The *absorption coefficient* of a substance describes, on a scale from 0 to 1, how much energy is absorbed. An absorption coefficient of 1 indicates total absorption, whereas 0 represents total reflection. The absorption coefficient of substances varies with frequency.

The total amount of absorption present in a room can be calculated by multiplying the absorption coefficient of each surface by its area and then adding the products together. All of the room's surfaces must be taken into account, as must people, chairs and other furnishings. Tables of the performance of different substances are available in acoustics references (see *Recommended further reading*). Porous materials tend to absorb high frequencies more effectively than low frequencies, whereas resonant membrane- or panel-type absorbers tend to be better at low frequencies. Highly tuned artificial absorbers (Helmholtz absorbers) can be used to remove energy in a room at specific frequencies. The trends in absorption coefficient are shown in the diagram below.

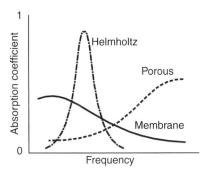

Reflection

The size of an object in relation to the wavelength of a sound is important in determining whether the sound wave will bend round it or be reflected by it. When an object is large in relation to the wavelength the object will act as a partial barrier to the sound, whereas when it is small the sound will bend or diffract around it. Since sound wavelengths in air range from approximately 18 metres at low frequencies to just over 1 cm at high frequencies, most commonly encountered objects will tend to act as barriers to sound at high frequencies but will have little effect at low frequencies.

Reverberation time

W. C. Sabine developed a simple and fairly reliable formula for calculating the reverberation time (RT_{60}) of a room, assuming that absorptive material is distributed evenly around the surfaces. It relates the volume of the room (V) and its total absorption (A) to the time taken for the sound pressure level to decay by 60 dB after a sound source is turned off.

$$RT_{60} = (0.16V)/A \text{ seconds}$$

In a large room where a considerable volume of air is present, and where the distance between surfaces is large, the absorption of the air becomes more important, in which case an additional component must be added to the above formula:

$$RT_{60} = (0.16V)/(A + xV) \text{ seconds}$$

where x is the absorption factor of air, given at various temperatures and humidities in acoustics references.

The Sabine formula has been subject to modifications by such people as Eyring, in an attempt to make it more reliable in extreme cases of high absorption, and it should be realised that it can only be a guide.

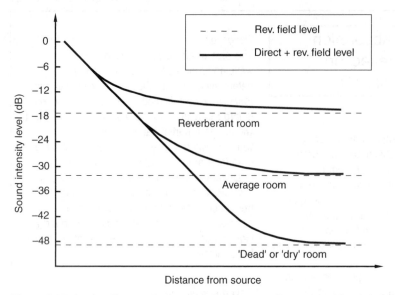

Figure 1.15 As the distance from a source increases direct sound level drops but reverberant sound level remains roughly constant. The resultant sound level experienced at different distances from the source depends on the reverberation time of the room, since in a reverberant room the level of reflected sound is higher than in a 'dead' room

positions of these depend on whether the mode has been created between the walls or between the floor and ceiling. The frequencies (f) at which the strongest modes will occur is given by:

$$f = (c/2) \times (n/d)$$

where c is the speed of sound, d is the dimension involved (distance between walls or floor and ceiling), and n is the number of the mode.

A more complex formula can be used to predict the frequencies of all the modes in a room, including those secondary modes formed by reflections between four and six surfaces (*oblique* and *tangential* modes). The secondary modes typically have lower amplitudes than the primary modes (the *axial* modes) since they experience greater absorption. The formula is:

$$f = (c/2)\sqrt{((p/L)^2 + (q/W)^2 + (r/H)^2)}$$

where p, q and r are the mode numbers for each dimension (1, 2, 3 ...) and L, W and H are the length, width and height of the room. For example, to calculate the first axial mode involving only the length, make $p = 1$, $q = 0$ and $r = 0$. To calculate the first oblique mode involving all four walls, make $p = 1$, $q = 1$, $r = 0$, and so on.

Some quick sums will show, for a given room, that the modes are widely spaced at low frequencies and become more closely spaced at high frequencies. Above a certain frequency, there arise so many modes per octave that it is hard to identify them separately. As a rule-of-thumb, modes tend only to be particularly problema-

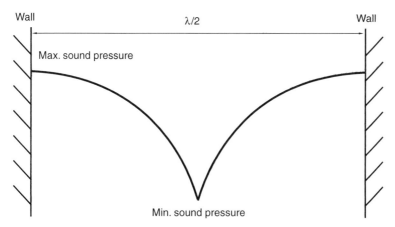

Figure 1.16 When a standing wave is set up between two walls of a room there arise points of maximum and minimum pressure. The first simple mode or *eigentone* occurs when half the wavelength of the sound equals the distance between the boundaries, as illustrated, with pressure maxima at the boundaries and a minimum in the centre

tical up to about 200 Hz The larger the room the more closely spaced the modes. Rooms with more than one dimension equal will experience so-called *degenerate modes* in which modes between two dimensions occur at the same frequency, resulting in an even stronger resonance at a particular frequency than otherwise. This is to be avoided.

Since low-frequency room modes cannot be avoided, except by introducing total absorption, the aim in room design is to reduce their effect by adjusting the ratios between dimensions to achieve an even spacing. A number of 'ideal' mode-spacing criteria have been developed by acousticians, but there is not the space to go into these in detail here. Larger rooms are generally more pleasing than small rooms, since the mode spacing is closer at low frequencies, and individual modes tend not to stick out so prominently, but room size has to be traded off against the target reverberation time. Making walls non-parallel does not prevent modes from forming (since oblique and tangential modes are still possible); it simply makes their frequencies more difficult to predict.

The practical difficulty with room modes results from the unevenness in sound pressure throughout the room at mode frequencies. Thus a person sitting in one position might experience a very high level at a particular frequency whilst other listeners might hear very little. A room with prominent LF modes will 'boom' at certain frequencies, and this is unpleasant and undesirable for critical listening. The response of the room modifies the perceived frequency response of a loudspeaker, for example, such that even if the loudspeaker's own frequency response may be acceptable it may become unacceptable when modified by the resonant characteristics of the room.

Room modes are not the only results of reflections in enclosed spaces, and some other examples are given in Fact File 1.6.

Echoes and reflections

Early reflections
Early reflections are those echoes from nearby surfaces in a room which arise within the first few milliseconds (up to about 50 ms) of the direct sound arriving at a listener from a source (see the diagram). It is these reflections which give the listener the greatest clue as to the size of a room, since the delay between the direct sound and the first few reflections is related to the distance of the major surfaces in the room from the listener. Artificial reverberation devices allow for the simulation of a number of early reflections before the main body of reverberant sound decay, and this gives different reverberation programs the characteristic of different room sizes.

Echoes
Echoes may be considered as discrete reflections of sound arriving at the listener after about 50 ms from the direct sound. These are perceived as separate arrivals, whereas those up to around 50 ms are normally integrated by the brain with the first arrival, not being perceived consciously as echoes. Such echoes are normally caused by more distant surfaces which are strongly reflective, such as a high ceiling or distant rear wall. Strong echoes are usually annoying in critical listening situations and should be suppressed by dispersion and absorption.

Flutter echoes
A flutter echo is sometimes set up when two parallel reflective surfaces face each other in a room, whilst the other surfaces are absorbent. It is possible for a wavefront to become 'trapped' into bouncing back and forth between these two surfaces until it decays, and this can result in a 'buzzing' or 'ringing' effect on transients (at the starts and ends of impulsive sounds such as hand claps).

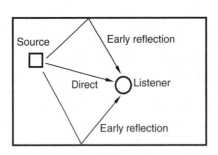

Recommended further reading

General

Eargle, J. (1995) *Music, Sound, Technology.* 2nd edition. Van Nostrand Rheinhold
Howard, D. and Angus, J. (2000) *Acoustics and Psychoacoustics.* 2nd edition. Focal Press
Rossing, T. D. (1989) *The Science of Sound,* 2nd Edition. Addison-Wesley

Architectural acoustics

Egan, M. D. (1988) *Architectural Acoustics.* McGraw-Hill
Rettinger, M. (1988) *Handbook of Architectural Acoustics and Noise Control.* TAB Books
Templeton, D. and Saunders, D. (1987) *Acoustic Design.* Butterworth Architecture

Musical acoustics

Alton Everest, F. (1994) *The Master Handbook of Acoustics.* McGraw-Hill
Benade, A. H. (1991) *Fundamentals of Musical Acoustics.* Oxford University Press
Campbell, M. and Greated, C. (1987) *The Musician's Guide to Acoustics.* Dent
Hall, D. E. (1991) *Musical Acoustics,* 2nd Edition. Brooks/Cole Publishing Co.

Chapter 2

Auditory perception

In this chapter the mechanisms by which sound is perceived will be introduced. The human ear often modifies the sounds presented to it before they are presented to the brain, and the brain's interpretation of what it receives from the ears will vary depending on the information contained in the nervous signals. An understanding of loudness perception is important when considering such factors as the perceived frequency balance of a reproduced signal, and an understanding of directional perception is relevant to the study of stereo recording techniques. Below, a number of aspects of the hearing process will be related to the practical world of sound recording and reproduction.

2.1 The hearing mechanism

Although this is not intended to be a lesson in physiology, it is necessary to invest-igate the basic components of the ear, and to look at how information about sound signals is communicated to the brain. Figure 2.1 shows a diagram of the ear mechanism, not anatomically accurate but showing the key mechanical compo-nents. The outer ear consists of the pinna (the visible skin and bone structure) and the auditory canal, and is terminated by the *tympanic membrane* or 'ear drum'. The middle ear consists of a three-bone lever structure which connects the tympanic membrane to the inner ear via the oval window (another membrane). The inner ear is a fluid-filled bony spiral device known as the *cochlea*, down the centre of which runs a flexible membrane known as the basilar membrane. The cochlea is shown here as if 'unwound' into a straight chamber for the purposes of description. At the end of the basilar membrane, furthest from the middle ear, there is a small gap called the *helicotrema* which allows fluid to pass from the upper to the lower chamber. There are other components in the inner ear, but those noted above are the most significant.

The ear drum is caused to vibrate in sympathy with the air in the auditory canal when excited by a sound wave, and these vibrations are transferred via the bones of the middle ear to the inner ear, being subject to a multiplication of force of the order of 15:1 by the lever arrangement of the bones. The lever arrangement, coupled with

Figure 2.1 A simplified mechanical diagram of the ear

the difference in area between the tympanic membrane and the oval window, helps to match the impedances of the outer and inner ears so as to ensure optimum transfer of energy. Vibrations are thus transferred to the fluid in the inner ear in which pressure waves are set up. The basilar membrane is not uniformly stiff along its length (it is narrow and stiff at the oval window end and wider and more flexible at the far end), and the fluid is relatively incompressible; thus a high-speed pressure wave travels through the fluid and a pressure difference is created across the basilar membrane.

2.2 Frequency perception

The motion of the basilar membrane depends considerably on the frequency of the sound wave, there being a peak of motion which moves closer towards the oval window the higher the frequency (see Figure 2.2).

At low frequencies the membrane has been observed to move as a whole, with the maximum amplitude of motion at the far end, whilst at higher frequencies there arises a more well-defined peak. It is interesting to note that for every octave (i.e.: for every doubling in the frequency) the position of this peak of maximum vibration moves a similar length up the membrane, and this may explain the human preference for displaying frequency-related information on a logarithmic frequency scale, which represents increase in frequency by showing octaves as equal increments along a frequency axis.

Figure 2.2 The position of maximum vibration on the basilar membrane moves towards the oval window as frequency increases

Frequency information is transmitted to the brain in two principal ways. At low frequencies hair cells in the inner ear are stimulated by the vibrations of the basilar membrane, causing them to discharge small electrical impulses along the auditory nerve fibres to the brain. These impulses are found to be synchronous with the sound waveform, and thus the period of the signal can be measured by the brain. Not all nerve fibres are capable of discharging once per cycle of the sound waveform (in fact most have spontaneous firing rates of a maximum of 150 Hz with many being much lower than this). Thus at all but the lowest frequencies the period information is carried in a *combination* of nerve fibre outputs, with at least a few firing on every cycle (see Figure 2.3). There is evidence to suggest that nerve fibres may re-trigger faster if they are 'kicked' harder – that is, the louder the sound the more regularly they may be made to fire. Also, whilst some fibres will trigger with only a low level of stimulation, others will only fire at high sound levels.

The upper frequency limit at which nerve fibres appear to cease firing synchronously with the signal is around 4 kHz, and above this frequency the brain relies increasingly on an assessment of the position of maximum excitation of the membrane to decide on the pitch of the signal. There is clearly an overlap region in the middle-frequency range, from about 200 Hz upwards, over which the brain has both synchronous discharge information and 'position' information on which to base its measurement of frequency. It is interesting to note that one is much less able to determine the precise musical pitch of a note when its frequency is above the synchronous discharge limit of 4 kHz.

The frequency selectivity of the ear has been likened to a set of filters, and this concept is described in more detail in Fact File 2.1. It should be noted that there is an unusual effect whereby the perceived pitch of a note is related to the loudness of the sound, such that the pitch shifts slightly with increasing sound level. This is sometimes noticed as loud sounds decay, or when removing headphones, for example. The effect of 'beats' may also be noticed when two pure tones of very similar

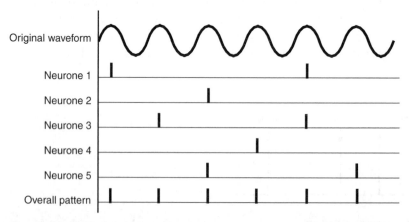

Figure 2.3 Although each neurone does not normally fire on every cycle of the causatory sound wave, the outputs of a combination of neurones firing on different cycles represent the period of the wave

---

FACT FILE

2.1

Critical bandwidth

The basilar membrane appears to act as a rough mechanical spectrum analyser, providing a spectral analysis of the incoming sound to an accuracy of between one-fifth and one-third of an octave in the middle frequency range (depending on which research data is accepted). It acts rather like a bank of overlapping filters of a fixed bandwidth. This analysis accuracy is known as the *critical bandwidth*, that is the range of frequencies passed by each notional filter.

The critical band concept is important in understanding hearing because it helps to explain why some signals are 'masked' in the presence of others (see Fact File 2.3). Fletcher, working in the 1940s, suggested that only signals lying within the same critical band as the wanted signal would be capable of masking it, although other work on masking patterns seems to suggest that a signal may have a masking effect on frequencies well above its own.

With complex signals, such as noise or speech for example, the total loudness of the signal depends to some extent on the number of critical bands covered by a signal. It can be demonstrated by a simple experiment that the loudness of a constant power signal does not begin to increase until its bandwidth extends over more than the relevant critical bandwidth, which appears to support the previous claim. (A useful demonstration of this phenomenon is to be found on the Compact Disc entitled *Auditory Demonstrations* described at the end of this chapter.)

Although the critical band concept helps to explain the first level of frequency analysis in the hearing mechanism, it does not account for the fine frequency selectivity of the ear which is much more precise than one-third of an octave. One can detect *changes* in pitch of only a few hertz, and in order to understand this it is necessary to look at the ways in which the brain 'sharpens' the aural tuning curves. For this the reader is referred to Moore (1989), as detailed at the end of this chapter.

---

frequency are sounded together, resulting in a pattern of addition and cancellation as they come in and out of phase with each other. The so-called 'beat frequency' is the difference frequency between the two signals, such that signals at 200 Hz and 201 Hz would result in a cyclic modulation of the overall level, or beat, at 1 Hz. Combined signals slightly further apart in frequency result in a 'roughness' which disappears once the frequencies of the two signals are further than a critical band apart.

2.3 Loudness perception

The subjective quantity of 'loudness' is not directly related to the SPL of a sound signal (see section 1.11). The ear is not uniformly sensitive at all frequencies, and a set of curves has been devised which represents the so-called equal-loudness contours of hearing (see Fact File 2.2). This is partially due to the resonances of the

FACT FILE
2.2

Equal-loudness contours

Fletcher and Munson devised a set of curves to show the sensitivity of the ear at different frequencies across the audible range. They derived their results from tests on a large number of subjects who were asked to adjust the level of test tones until they appeared equally as loud as a reference tone with a frequency of 1 kHz. The test tones were spread across the audible spectrum. From these results could be drawn curves of average 'equal loudness', indicating the SPL required at each frequency for a sound to be perceived at a particular loudness level (see diagram).

Loudness is measured in *phons,* the zero phon curve being that curve which passes through 0 dB SPL at 1 kHz – in other words, the threshold of hearing curve. All points along the 0 phon curve will sound equally loud, although clearly a higher SPL is required at extremes of the spectrum than in the middle. The so-called Fletcher–Munson curves are not the only equal-loudness curves in existence – Robinson and Dadson, amongst others, have published revised curves based upon different test data. The shape of the curves depends considerably on the type of sound used in the test, since filtered noise produces slightly different results to sine tones.

It will be seen that the higher-level curves are flatter than the low-level curves, indicating that the ear's frequency response changes with signal level. This is important when considering monitoring levels in sound recording (see text).

outer ear which have a peak in the middle-frequency region, thus increasing the effective SPL at the ear drum over this range.

The unit of loudness is the *phon.* If a sound is at the threshold of hearing (just perceivable) it is said to have a loudness of 0 phons, whereas if a sound is at the threshold of pain it will probably have a loudness of around 140 phons. Thus the ear has a dynamic range of approximately 140 phons, representing a range of sound pressures with a ratio of around 10 million to one between the loudest and quietest sounds perceivable. As indicated in Fact File 1.4, the 'A'-weighting curve is often used when measuring sound levels because it shapes the signal spectrum to represent more closely the subjective loudness of low-level signals. A noise level quoted in dBA is very similar to a loudness level in phons.

To give an idea of the loudnesses of some common sounds, the background noise of a recording studio might be expected to measure at around 20 phons, a low-level conversation perhaps at around 50 phons, a busy office at around 70 phons, shouted speech at around 90 phons, and a full symphony orchestra playing loudly at around 120 phons. These figures of course depend on the distance from the sound source, but are given as a guide.

The loudness of a sound depends to a great extent on its nature. Broad-band sounds tend to appear louder than narrow-band sounds, because they cover more critical bands (see Fact File 2.1), and distorted sounds appear psychologically to be louder than undistorted sounds, perhaps because one associates distortion with system overload. If two music signals are played at identical levels to a listener, one with severe distortion and the other without, the listener will judge the distorted signal to be louder.

A further factor of importance is that the threshold of hearing is raised at a particular frequency in the presence of another sound at a similar frequency. In other words, one sound may 'mask' another – a principle described in more detail in Fact File 2.3.

In order to give the impression of a doubling in perceived loudness, an increase of some 9–10 dB is required. Although 6 dB represents a doubling of the actual sound pressure, the hearing mechanism appears to require a greater increase than this for the signal to appear to be twice as loud. Another subjective unit, rarely used in practice, is that of the *sone*: 1 sone is arbitrarily aligned with 40 phons, and 2 sones is twice as loud as 1 sone, representing approximately 49 phons; 3 sones is three times as loud, and so on. Thus the sone is a true indication of the relative loudness of signals on a linear scale, and sone values may be added together to arrive at the total loudness of a signal in sones.

The ear is by no means a perfect transducer; in fact it introduces considerable distortions into sound signals due to its non-linearity. At high signal levels, especially for low-frequency sounds, the amount of harmonic and intermodulation distortion (see Appendix 1) produced by the ear can be high.

2.4 Practical implications of equal-loudness contours

The non-linear frequency response of the ear presents the sound engineer with a number of problems. Firstly, the perceived frequency balance of a recording will depend on how loudly it is replayed, and thus a balance made in the studio at one level may sound different when replayed in the home at another. In practice, if a recording is replayed at a much lower level than that at which it was balanced it will sound lacking in bass and extreme treble – it will sound thin and lacking warmth.

FACT FILE

2.3

Masking

Most people have experienced the phenomenon of masking, although it is often considered to be so obvious that it does not need to be stated. As an example: it is necessary to raise your voice in order for someone to hear you if you are in noisy surroundings. The background noise has effectively raised the perception threshold so that a sound must be louder before it can be heard. If one looks at the masking effect of a pure tone, it will be seen that it raises the hearing threshold considerably for frequencies which are the same as or higher than its own (see diagram). Frequencies below the masking tone are less affected. The range of

frequencies masked by a tone depends mostly on the area of the basilar membrane set into motion by the tone, and the pattern of motion of this membrane is more extended towards the HF end than towards the LF end. If the required signal produces more motion on the membrane than the masking tone produces at that point then it will be perceived.

The phenomenon of masking has many practical uses in audio engineering. It is used widely in noise reduction systems, since it allows the designer to assume that low-level noise which exists in the same frequency band as a high-level music signal will be effectively masked by the music signal. It is also used in digital audio data compression systems, since it allows the designer to use lower resolution in some frequency bands where the increased noise will be effectively masked by the wanted signal.

Conversely, if a signal is replayed at a higher level than that at which it was balanced it will have an increased bass and treble response, sounding boomy and overbright.

A 'loudness' control is often provided on hi-fi amplifiers to boost low and high frequencies for low-level listening, but this should be switched out at higher levels. Rock-and-roll and heavy-metal music often sounds lacking in bass when replayed at moderate sound levels because it is usually balanced at extremely high levels in the studio.

Some types of noise will sound louder than others, and hiss is usually found to be most prominent due to its considerable energy content at middle–high frequencies. Rumble and hum may be less noticeable because the ear is less sensitive at low frequencies, and a low-frequency noise which causes large deviations of the meters in a recording may not sound particularly loud in reality. This does not mean, of course, that rumble and hum are acceptable.

Recordings equalised to give a strong mid-frequency content often sound rather 'harsh', and listeners may complain of listening fatigue, since the ear is particularly sensitive in the range between about 1 and 5 kHz.

2.5 Spatial perception

Spatial perception principles are important when considering stereo sound reproduction (see Chapters 17 and 18) and when designing PA rigs for large auditoria, since an objective in both these cases is to give the illusion of directionality and spaciousness.

2.5.1 Sound source localisation

Most research into the mechanisms underlying directional sound perception conclude that there are two primary mechanisms at work, the importance of each depending on the nature of the sound signal and the conflicting environmental cues that may accompany discrete sources. These broad mechanisms involve the detection of timing or phase differences between the ears, and of amplitude or spectral differences between the ears. The majority of spatial perception is dependent on the listener having two ears, although certain monoaural cues have been shown to exist – in other words it is mainly the *differences* in signals received by the two ears that matter.

2.5.2 Time-based cues

A sound source located off the 0° (centre front) axis will give rise to a time difference between the signals arriving at the ears of the listener that is related to its angle of incidence, as shown in Figure 2.4. This rises to a maximum for sources at the side of the head, and enables the brain to localise sources in the direction of the earlier ear. The maximum time delay between the ears is of the order of 650 μs or 0.65 ms and is called the binaural delay. It is apparent that humans are capable of resolving direction down to a resolution of a few degrees by this method. There is no obvious way of distinguishing between front and rear sources or of detecting elevation by this method, but one way of resolving this confusion is by taking into account the effect of head movements. Front and rear sources at the same angle of offset from centre to one side, for example, will result in opposite changes in time of arrival for a given direction of head turning.

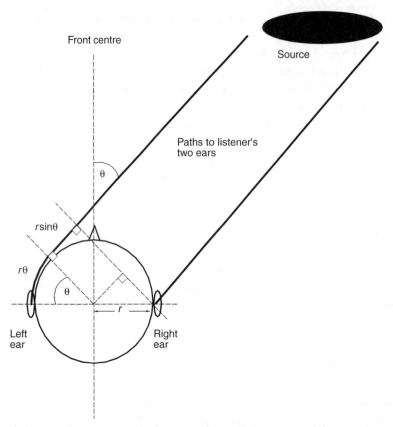

Figure 2.4 The interaural time difference (ITD) for a listener depends on the angle of incidence of the source, as this affects the additional distance that the sound wave has to travel to the more distant ear. In this model the ITD is given by $r(\theta + \sin\theta)/c$ (where $c = 340$ m/s, the speed of sound, and θ is in radians)

Time difference cues are particularly registered at the starts and ends of sounds (onsets and offsets) and seem to be primarily based on the low-frequency content of the sound signal. They are useful for monitoring the differences in onset and offset of the overall envelope of sound signals at higher frequencies.

Timing differences can be expressed as phase differences when considering sinusoidal signals. The ear is sensitive to interaural phase differences only at low frequencies and the sensitivity to phase begins to deteriorate above about 1 kHz. At low frequencies the hair cells in the inner ear fire regularly at specific points in the phase of the sound cycle, but at high frequencies this pattern becomes more random and not locked to any repeatable point in the cycle. Sound sources in the lateral plane give rise to phase differences between the ears that depend on their angle of offset from the $0°$ axis (centre front). Because the distance between the ears is constant, the phase difference will depend on the frequency and location of the source. (Some sources also show a small difference in the time delay between the ears at LF and HF.) Such a phase difference model of directional perception is only really relevant for continuous sine waves auditioned in anechoic environments, which are rarely heard except in laboratories. It also gives ambiguous information above about 700 Hz where the distance between the ears is equal to half a wavelength of the sound, because it is impossible to tell which ear is lagging and which is leading. Also there arise frequencies where the phase difference is zero. Phase differences can also be confusing in reflective environments where room modes and other effects of reflections may modify the phase cues present at the ears.

When two or more physically separated sources emit similar sounds the precedence effect is important in determining the apparent source direction, as explained in Fact File 2.4.

FACT FILE

2.4

The precedence effect

The precedence effect is important for understanding sound localisation when two or more sources are emitting essentially the same sound (e.g. a person speaking and a loudspeaker in a different place emitting an amplified version of their voice). It is primarily a feature of transient sounds rather than continuous sounds. In such an example both ears hear both the person and the loudspeaker. The brain tends to localise based on the interaural delay arising from the earliest arriving wavefront, the source appearing to come from a direction towards that of the earliest arriving signal (within limits).

This effect operates over delays between the sources that are somewhat greater than the interaural delay, of the order of a few milliseconds. Similar sound arriving within up to 50 ms of each other tend to be perceptually fused together, such that one is not perceived as an echo of the other. The time delay over which this fusing effect obtains depends on the source, with clicks tending to separate before complex sounds like music or speech. The timbre and spatial qualities of this 'fused sound', though, may be affected.

One form of precedence effect is sometimes referred to as the Haas Effect after the Dutch scientist who conducted some of the original experiments. It was originally identified in experiments designed to determine what would happen to the perception of speech in the presence of a single echo. Haas determined that the delayed 'echo' could be made substantially louder than the earlier sound before it was perceived to be equally loud, as shown in the approximation below. The effect depends considerably on the spatial separation of the two or more sources involved. This has important implications for recording techniques where time and intensity differences between channels are used either separately or combined to create spatial cues.

2.5.3 Amplitude and spectral cues

The head's size makes it an appreciable barrier to sound at high frequencies but not at low frequencies. Furthermore, the unusual shape of the pinna (the visible part of the outer ear) gives rise to reflections and resonances that change the spectrum of the sound at the eardrum depending on the angle of incidence of a sound wave. Reflections off the shoulders and body also modify the spectrum to some extent. A final amplitude cue that may be relevant for spherical wave sources close to the head is the level difference due to the extra distance travelled between the ears by off-centre sources. For sources at most normal distances from the head this level difference is minimal, because the extra distance travelled is negligible compared with that already travelled.

The sum of all of these effects is a unique *head-related transfer function* or HRTF for every source position and angle of incidence, including different elevations and front–back positions. Some examples of HRTFs at different angles are shown in Figure 2.5. It will be seen that there are numerous spectral peaks and dips, particularly at high frequencies, and common features have been found that characterise certain source positions. This, therefore, is a unique form of directional encoding that the brain can learn. Typically, sources to the rear give rise to a reduced high-frequency response in both ears compared to those at the front, owing to the slightly forward-facing shape of the pinna. Sources to one side result in an increased high-frequency difference between the ears, owing to the shadowing effect of the head.

These HRTFs are superimposed on the natural spectra of the source themselves. It is therefore hard to understand how the brain might use the monoaural spectral characteristics of sounds to determine their positions as it would be difficult to separate the timbral characteristics of sources from those added by the HRTF. Monaural cues are likely to be more detectable with moving sources, because moving sources allow the brain to track changes in the spectral characteristics that should be independent of a source's own spectrum. For lateralisation it is most likely to be *differences* in HRTFs between the ears that help the brain to localise sources, in conjunction with the associated interaural time delay. Monaural cues may be more relevant for localisation in the median plane where there are minimal differences between the ears.

There are remarkable differences in HRTFs between individuals, although common features can be found. Figure 2.6 shows just two HRTF curves measured by Begault for different subjects, illustrating the problem of generalisation in this respect.

The so-called *concha* resonance (that created by the main cavity in the centre of the pinna) is believed to be responsible for creating a sense of externalisation – in other words a sense that the sound emanates from outside the head rather than within. Sound-reproducing systems that disturb or distort this resonance, such as certain headphone types, tend to create in-the-head localisation as a result.

Figure 2.5 Monaural transfer functions of the left ear for several directions in the horizontal plane, relative to sound incident from the front; anechoic chamber, 2 m loudspeaker distance, impulse technique, 25 subjects, complex averaging (Blauert, 1997). (a) Level difference; (b) time difference. (Courtesy of MIT Press)

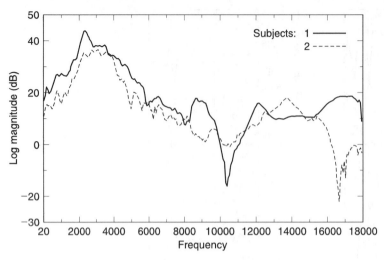

Figure 2.6 HRTFs of two subjects for a source at 0° azimuth and elevation. Note considerable HF differences. (Begault, 1991)

2.5.4 Effects of reflections

Reflections arising from sources in listening spaces affect spatial perception significantly, as discussed in Fact File 2.5. Reflections in the early time period after direct sound (up to 50–80 ms) typically have the effect of broadening or deepening the spatial attributes of a source. They are unlikely to be individually localisable. In the period up to about 20 ms they can cause severe timbral coloration if they are at high levels. After 80 ms they tend to contribute more to the sense of envelopment or spaciousness of the environment.

2.5.5 Interaction between hearing and other senses

Some spatial cues are context dependent and may be strongly influenced by the information presented by other senses, particularly vision. Learned experience leads the brain to expect certain cues to imply certain spatial conditions, and if this is contradicted then confusion may arise. For example, it is unusual to experience the sound of a plane flying along beneath one, but the situation can occasionally arise when climbing mountains. Generally one expects planes to fly above, and most people will look up or duck when played loud binaural recordings of planes flying over, even if the spectral cues do not imply this direction.

It is normal to rely quite heavily on the visual sense for information about events within the visible field, and it is interesting to note that most people, when played binaural recordings (see Chapter 16) of sound scenes without accompanying visual information or any form of head tracking, localise the scene primarily behind them rather than in front. In fact obtaining front images from any binaural system using headphones is surprisingly difficult. This may be because one is used to using the hearing sense to localise things where they cannot be seen, and that if something

FACT FILE 2.5 — Reflections affect spaciousness

The subjective phenomenon of apparent or auditory source width (ASW) has been studied for a number of years, particularly by psychoacousticians interested in the acoustics of concert halls. ASW relates to the issue of how large a space a source appears to occupy from a sonic point of view (ignoring vision for the moment), as shown below. Individual source width should be distinguished from overall 'sound stage width' (in other words, the distance perceived between the left and right limits of a stereophonic scene).

Early reflected energy in a space (up to about 80 ms) appears to modify the ASW of a source by broadening it somewhat, depending on the magnitude and time delay of early reflections. Concert hall experiments seem to show that subjects prefer larger amounts of ASW, but it is not clear what is the *optimum*

degree of ASW (presumably sources that appeared excessively large would be difficult to localise and unnatural).

Envelopment, spaciousness and sometimes 'room impression' are typically spatial features of a reverberant environment rather than individual sources, and are largely the result of late reflected sound (particularly lateral reflections after about 80 ms). Spaciousness is used most often to describe the sense of open space or 'room' in which the subject is located, usually as a result of some sound sources such as musical instruments playing in that space. It is also related to the sense of 'externalisation' perceived – in other words whether the sound appears to be outside the head rather than constrained to a region close to or inside it. Envelopment is a similar term and is used to describe the sense of immersivity and involvement in a (reverberant) soundfield, with that sound appearing to come from all around. It is regarded as a positive quality that is experienced in good concert halls.

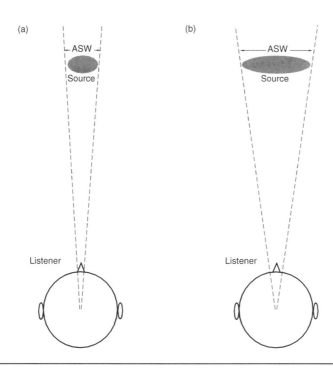

cannot be seen it is likely to be behind. In the absence of the ability to move the head to resolve front–back conflicts the brain tends to assume a rear sound image. So-called 'reversals' in binaural audio systems are consequently very common.

2.5.6 Resolving conflicting cues

In environments where different cues conflict in respect of the implied location of sound sources, the hearing process appears to operate on a sort of majority decision logic basis. In other words it evaluates the available information and votes on the most likely situation, based on what it can determine. Auditory perception has been likened to a hypothesis generation and testing process, whereby likely scenarios are constructed from the available information and tested against subsequent experience (often over a very short time interval). Context-dependent cues and those from other senses are quite important here. Since there is a strong precedence effect favouring the first-arriving wavefront, the direct sound in a reflective environment (which arrives at the listener first) will tend to affect localisation most, while subsequent reflections may be considered less important. Head movements will also help to resolve some conflicts, as will visual cues. Reflections from the nearest surfaces, though, particularly the floor, can aid the localising process in a subtle way. Moving sources also tend to provide more information than stationary ones, allowing the brain to measure changes in the received information that may resolve some uncertainties.

2.5.7 Distance and depth perception

Apart from lateralisation of sound sources, the ability to perceive distance and depth of sound images is crucial to our subjective appreciation of sound quality. Distance is a term specifically related to how far away an individual source appears to be, whereas depth can describe the overall front–back distance of a scene and the sense of perspective created. Individual sources may also appear to have depth.

A number of factors appear to contribute to distance perception, depending on whether one is working in reflective or 'dead' environments. Considering for a moment the simple differences between a sound source close to a listener and the same source further away, the one further away will have the following differences:

- Quieter (extra distance travelled)
- Less high-frequency content (air absorbtion)
- More reverberant (in reflective environment)
- Less difference between time of direct sound and first-floor reflection
- Attenuated ground reflection

Numerous studies have shown that absolute distance perception, using the auditory sense alone, is very unreliable in non-reflective environments, although it is possible for listeners to be reasonably accurate in judging relative distances (since there is then a reference point with known distance against which other sources can be compared). In reflective environments, on the other hand, there is substantial additional information available to the brain. The ratio of direct to reverberant sound is directly related to source distance. The reverberation time and the early reflection timing tells the brain a lot about the size of the space and the distance to the surfaces, thereby giving it boundaries beyond which sources could not reasonably be expected to lie.

2.5.8 Naturalness in spatial hearing

The majority of spatial cues received in reproduced sound environments are similar to those received in natural environments, although their magnitudes and natures may be modified somewhat. There are, nonetheless, occasional phenomena that might be considered as specifically associated with reproduced sound, being rarely or never encountered in natural environments. The one that springs most readily to mind is the 'out-of-phase' phenomenon, in which two sound sources such as loudspeakers or headphones are oscillating exactly 180° out of phase with each other – usually the result of a polarity inversion somewhere in the signal chain. This creates an uncomfortable sensation with a strong but rather unnatural sense of spaciousness, and makes phantom sources hard to localise. The out-of-phase sensation never arises in natural listening and many people find it quite disorientating and uncomfortable. Its unfamiliarity makes it hard to identify for naïve listeners, whereas for expert audio engineers its sound is unmistakeable. Naïve listeners may even quite like the effect, and extreme phase effects have sometimes been used in low-end audio products to create a sense of extra stereo width.

Audio engineers also often refer to problems with spatial reproduction as being 'phasy' in quality. Usually this is a negative term that can imply abnormal phase differences between the channels, or an unnatural degree of phase difference that may be changing with time. Anomalies in signal processing or microphone technique can create such effects and they are unique to reproduced sound, so there is in effect no natural anchor or reference point against which to compare these experiences.

Recommended further reading

Blauert, J. (1997) *Spatial Hearing*. 2nd Edition. Translated by J. S. Allen. MIT Press
Howard, D. and Angus, J. (1996) *Acoustics and Psychoacoustics*. Focal Press
Moore, B. C. J. (1997) *An Introduction to the Psychology of Hearing*. 4th Edition. Academic Press

Recommended listening

Auditory Demonstrations (Compact Disc). Philips Cat. No. 1126-061. Available from the Acoustical Society of America

Chapter 3

A guide to the audio signal chain

Sound signals start life as vibrations carried through the air, and this is also how they end up after they have passed through the recording and reproduction process with all its stages. In broadcasting the signal is not only recorded and reproduced, it is also transmitted by means of radio waves. In the following chapter an overview will be given of the various stages involved in the recording and broadcasting chain, and in subsequent chapters the technical and operational aspects involved in audio systems will be examined in greater detail. An introduction to the differences in level which may be encountered at stages in the signal chain is also included here. More detailed coverage will be given to the matter of line-up, metering and signal levels in appropriate chapters.

3.1 A short history

3.1.1 Early recording machines

When Edison and Berliner first developed recording machines in the last years of the nineteenth century they involved little or no electrical apparatus. Certainly the recording and reproduction process itself was completely mechanical or 'acoustic', the system making use of a small horn terminated in a stretched, flexible diaphragm attached to a stylus which cut a groove of varying depth into the malleable tin foil on Edison's 'phonograph' cylinder or of varying lateral deviation in the wax on Berliner's 'gramophone' disk (see Figure 3.1). On replay, the undulations of the groove caused the stylus and diaphragm to vibrate, thus causing the air in the horn to move in sympathy, thus reproducing the sound – albeit with a very limited frequency range and very distorted.

Cylinders for the phonograph could be recorded by the user, but they were difficult to duplicate for mass production, whereas disks for the gramophone were normally replay only, but they could be duplicated readily for mass production. For this reason disks fairly quickly won the day as the mass-market pre-recorded music medium. There was no such thing as magnetic recording tape at the time, and thus recordings were made directly on to a master disk, lasting for the

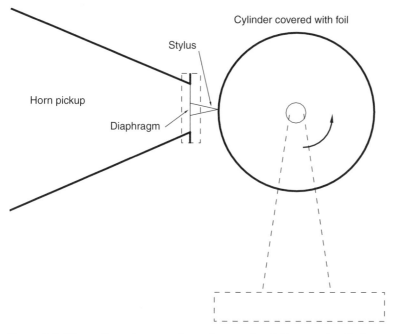

Cylinder covered with foil

Stylus

Horn pickup

Diaphragm

Figure 3.1 The earliest phonograph used a rotating foil-covered cylinder and a stylus attached to a flexible diaphragm. The recordist spoke or sang into the horn causing the stylus to vibrate, thus inscribing a modulated groove into the surface of the soft foil. On replay the modulated groove would cause the stylus and diaphragm to vibrate, resulting in a sound wave being emitted from the horn

duration of the side of the disk – a maximum of around 4 minutes – with no possibility for editing. Recordings containing errors were either remade or they were passed with mistakes intact. A long item of music would be recorded in short sections with gaps to change the disk, and possibilities arose for discontinuities between the sections, as well as variations in pitch and tempo.

Owing to the deficiencies of the acoustic recording process, instruments had to be grouped quite tightly around the pick-up horn in order for them to be heard on the recording, and often louder instruments were substituted for quieter ones (the double bass was replaced by the tuba, for example) in order to correct for the poor frequency balance. It is perhaps partly because of this that much of the recorded music of the time consisted of vocal soloists and small ensembles, since these were easier to record than large orchestras.

3.1.2 Electrical recording

During the 1920s, when broadcasting was in its infancy, electrical recording became more widely used, based on the principles of electromagnetic transduction (see Fact File 3.1). The possibility for a microphone to be connected remotely to a recording machine meant that microphones could be positioned in more suitable places, connected by wires to a complementary transducer at the other end of the wire which

FACT FILE

3.1

Electro-
magnetic
transducers

Electromagnetic transducers facilitate the conversion of acoustic signals into electrical signals, and they also act to convert electrical signals back into acoustic sound waves. The principle is very simple: if a wire can be made to move in a magnetic field, perpendicular to the lines of flux linking the poles of the magnet, then an electric current is induced in the wire (see diagram). The direction of motion governs the direction of current flow in the wire, and thus if the wire can be made to move back and forth then an alternating current can be induced in the wire, related in frequency and amplitude to the motion of the wire. Conversely, if a current is made to flow through a wire which cuts the lines of a magnetic field then the wire will move.

It is a short step from here to see how acoustic sound signals may be converted into electrical signals and vice versa. A simple moving-coil microphone, as illustrated in Fact File 4.1, involves a wire moving in a magnetic field, by means of a coil attached to a flexible diaphragm which vibrates in sympathy with the sound wave. The output of the microphone is an alternating electrical current, whose frequency is the same as that of the sound wave which caused the diaphragm to vibrate. The amplitude of the electrical signal generated depends on the mechanical characteristics of the transducer, but is proportional to the *velocity* of the coil.

Vibrating systems, such as transducer diaphragms, with springiness (compliance) and mass, have a *resonant frequency* (a natural frequency of free vibration). If the driving force's frequency is below this resonant frequency then the motion of the system depends principally on its stiffness; at resonance the motion is dependent principally on its damping (resistance); and above resonance it is mass controlled. Damping is used in transducer diaphragms to control the amplitude of the resonant response peak, and to ensure a more even response around resonance. Stiffness and mass control are used to ensure as flat a frequency response as possible in the relevant frequency ranges.

Exactly the reverse process occurs in a loudspeaker, where an alternating current is fed *into* a coil attached to a diaphragm, there being a similar magnet around the coil. This time the diaphragm moves in sympathy with the frequency and magnitude of the incoming electrical audio signal, causing compression and rarefaction of the air.

Current in wire

Magnet   S   N   Magnet

Motion of wire

drove the stylus to cut the disk. Even more usefully, the outputs of microphones could be mixed together before being fed to the disk cutter, allowing greater flexibility in the balance. Basic variable resistors could be inserted into the signal chain in order to control the levels from each microphone, and valve amplifiers would be used to increase the electrical level so that it would be suitable to drive the cutting stylus (see Figure 3.2).

The sound quality of electrical recordings shows a marked improvement over acoustic recordings, with a wider frequency range and a greater dynamic range (see

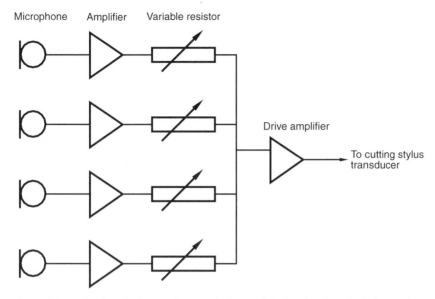

Figure 3.2 Early electrical recording made it possible for the electrical signals from microphones to be fed through variable resistors to alter their gains, and then be combined to feed an electrical cutting stylus

Appendix 1). Experimental work took place both in Europe and the USA on stereo recording and reproduction, but it was not to be until much later that stereo took its place as a common consumer format, nearly all records and broadcasts being in mono at that time.

3.1.3 Later developments

During the 1930s work progressed on the development of magnetic recording equipment, and examples of experimental wire recorders and tape recorders began to appear, based on the principle of using a current flowing through a coil to create a magnetic field which would in turn magnetise a moving metal wire or tape coated with magnetic material. The 1940s, during wartime, saw the introduction of the first AC-biased tape recorders (see section 8.2) which brought with them good sound quality and the possibility for editing. Tape itself, though, was first made of paper coated with metal oxide which tended to deteriorate rather quickly, and only later of plastics which proved longer lasting and easier to handle.

In the 1950s the microgroove LP record appeared, with markedly lower surface noise and improved frequency response, having a playing time of around 25 minutes per side. This was an ideal medium for distribution of commercial stereo recordings, which began to appear in the late 1950s, although it was not until the 1960s that stereo really took hold. In the early 1960s the first multitrack tape recorders appeared, the Beatles making use of an early four-track recorder for their *Sergeant Pepper's Lonely Hearts Club Band* album. The machine offered the unprecedented flexibility of allowing sources to be recorded separately, and the results in

CHESTER COLLEGE LIBRARY

the stereo mix are panned very crudely to left and right in somewhat 'gimmicky' stereo.

Mixing equipment in the 1950s and the 1960s was often quite basic, compared with today's sophisticated consoles, and rotary faders were the norm. There simply was not the quantity of tracks involved as exists today.

Recent recording history has seen the birth of high-quality digital recording (see Chapter 10), with Compact Disc and digital tape systems for consumer use, allowing full frequency-range recordings with minimal distortions of all kinds to be distributed commercially. Broadcast sound is now capable of high quality, either using FM or digital transmissions with radio and television. Studio recording has the benefit of an almost unlimited number of tracks and outboard effects if required.

3.2 The modern recording chain

In a modern music recording, sound signals begin life either as real acoustic sources (e.g.: vocals, a piano, acoustic guitars) or as sources of electrical sound signals (e.g.: synthesisers, electric guitars, drum machines). The latter may be amplified and reproduced via a loudspeaker, but can also be fed directly into a mixer (see Chapter 6). Acoustic sources will be picked up by microphones and fed into the mic inputs of a mixer (which incorporates amplifiers to raise the low-voltage output from microphones), whilst other sources usually produce so-called 'line level' outputs (see section 3.4) which can be connected to the mixer without extra amplification.

In the mixer, sources are combined in proportions controlled by the engineer and recorded. It is at this point that different techniques diverge, since much 'classical' music recording makes use of very straightforward recording techniques, whereas 'pop' sessions usually involve multitrack recording, separate mixing down of the multitrack tape, the addition of effects and the compilation of the album or single master.

3.2.1 'Straight-to-stereo'

Figure 3.3 shows the typical signal chain through the stages of a production which is being recorded 'straight-to-stereo', such as a classical music recording for CD or broadcast. Microphone sources are mixed 'live' without recording to multitrack tape, creating a stereo *session master*, either analogue or digital, which is the collection of original recordings, often consisting of a number of *takes* of the musical material. The balance between the sources must be correct at this stage, and often only a small number of carefully positioned microphones is used. The session master tapes will then proceed to the editing stage where takes are assembled in an artistically satisfactory manner, under the control of the producer, to create a final master which will be transmitted or made into a commercial release. This final master could be made into a number of production masters which will be used to make LPs, cassettes and CDs. An LP master requires special equalisation to prepare it for cutting on to disk, whereas a cassette master is usually made on to a half-inch

Mics   Mix to stereo   Stereo recording

Broadcast
CD
LP
Cassette

Editing   Mastering

Figure 3.3 A diagrammatic representation of 'straight-to-stereo' music production

'loop-bin master' for high-speed duplication. Digital masters require the addition of 'PQ subcodes' to mark the starts and ends of tracks, as well as other cueing information. This stage is known, not surprisingly, as 'mastering'.

In the case of 'straight-to-stereo' productions the mixing console used may be a simpler affair than that used for multitrack recording, since the mixer's job is to take multiple inputs and combine them to a single stereo output, with intervening processing such as equalisation. This method of production is clearly cheaper and less time consuming than multitrack recording, but requires skill to achieve a usable balance quickly. It also limits flexibility in post-production. Occasionally, classical music is recorded on to multitrack tape, especially in the case of complex operas or large-force orchestral music with a choir and soloists, where to get a correct balance at the time of the session could be costly and time-consuming. In such a case, the production process becomes more similar to the pop recording situation described below.

3.2.2 Multitrack recording

'Pop' music is rarely recorded live, except at live events such as concerts, but is created in the recording studio. Acoustic and electrical sources are fed into a mixer and recorded on to multitrack tape, often a few tracks at a time, gradually building up a *montage* of sounds (see Figure 3.4). The resulting tape thus contains a collection of individual sources on multiple tracks which must subsequently be mixed into the final format (usually stereo). Individual songs or titles are recorded in separate places on the tape, to be compiled later.

It is not so common these days to record multitrack pop titles in 'takes' for later editing, as with classical music, since mixer automation allows the engineer to work on a song in sections for automatic execution in sequence by a computer (see Chapter 14). In any case, multitrack tape machines have comprehensive 'drop-in' facilities for recording short inserted sections on individual tracks without introducing clicks, and a pop-music master is usually built up by laying down backing

Mics   Track-laying via mixer   Multitrack recording

Other electronic sources

Mixdown   Stereo recording   Mastering

Figure 3.4 A diagrammatic representation of multitrack music production

tracks for a complete song (drums, keyboards, rhythm guitars, etc.) after which lead lines are over-dubbed using drop-in facilities. Occasionally multitrack tapes are edited early on during a recording session to compile an acceptable backing track from a number of takes, after which further layers are added, but this is considered by some to be a potentially risky business.

Considerable use is made these days of computer-sequenced electronic instruments, under MIDI control (see Chapter 14), often in conjunction with multitrack tape or disk recording. The computer controlling the electronic instruments is synchronised to the tape machine using time-code (see Chapter 15) and the outputs of the instruments are fed to the mixer to be combined with the non-sequenced sources. Increasingly, MIDI-controlled sources are not recorded on to multitrack tape at all, the information about 'which-notes-occurred-when' being stored in the computer to be synchronised with the multitrack recorder only during the mixdown. This has the obvious disadvantage that all the same electronic instruments are required in the control room on mixdown as were used in the session, but may have the advantage of reducing the number of tracks required on tape and allows first-generation sound from sequenced sources to be fed into the mix. This technique is more common in 'home studio' and budget sessions than in 'no-expense-spared' music production, since in the latter the producer may wish to take a tape to the other side of the world for mixdown and will tend to feel happier if he or she knows that everything he or she needs is stored on one tape.

Once the session is completed, the multitrack master tape is mixed down. This is often done somewhere different from the original session, and involves feeding the outputs of each track into individual line inputs of the mixer, treating each track as if it were an original source. The balance between the tracks, and the positioning of the tracks in the stereo image, can then be carried out at leisure (within the budget constraints of the project!), often without all the musicians present, under control of the producer. During the mixdown, further post-production takes place such as the addition of effects from outboard equipment (see Chapter 13) to enhance the mix. An automation system is often used to memorise fader and mute movements on the console, since the large number of channels involved in modern recording makes it difficult if not impossible for the engineer to mix a whole song correctly in one go.

Following mixdown, the stereo master which results will be edited very basically, in order to compile titles in the correct order for the production master. The compiled tape will then be mastered for the various distribution media, as described in section 3.2.1.

3.2.3 Film sound

In commercial film production, sound is often recorded separately to picture, using a quarter-inch portable tape recorder such as the Nagra IV-S. A synchronising signal is also recorded on the sound tape, related to the speed of the film camera, to allow for the subsequent locking of sound to picture in post-production. Sound tracks from field recordings are copied on to sprocket-holed magnetic tape in the studio in order that they may be edited in the same way as pictures (see Figure 3.5). Individual sprocketed sound 'tracks' can be locked together very simply using a device known as a 'Steenbeck' which involves a mechanical linkage between multiple sprockets which can be declutched and slipped physically in relation to each other – thus the sound tracks can be moved relatively in time (see Figure 3.6).

In film sound editing the original sound and various added tracks are cut to match the edited picture, with blank stock added to make all the tracks the same length, and to make the timing of entries correct. Overdubbed dialogue may be added on a separate track to improve the clarity of scenes where the original sound

Figure 3.5 A diagrammatic representation of film sound production

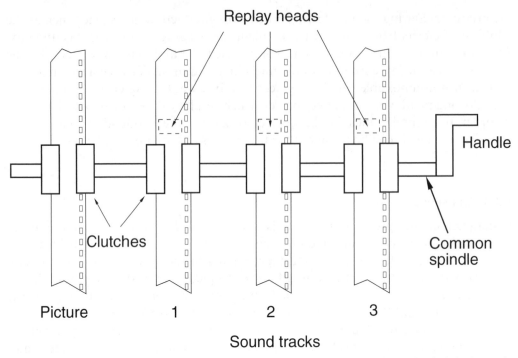

Figure 3.6 Film sound tracks are often synchronised during editing using a mechanical device such as that shown here

was poor, and music and sound effects (FX) are added on further tracks. A dubbing sheet is built up showing which tracks are active at which times, and what is contained in each section. During the 'dub', the tracks are all synchronised on a number of sprocketed film transports, electrically locked together, and the resulting sound is mixed against the dubbing sheet so as to be suitable for the anticipated release format (see Figure 3.7). Often sound is produced in a number of formats depending on the destination, there being a number of different cinema sound systems, such as Academy Mono, 35 mm Dolby Stereo, and 70 mm Dolby Stereo.

More recently, digital hard-disk editing systems (see section 10.9) have been adopted for film sound dubbing, and the number of tracks available on such systems is increasing to a point where fairly complex productions are possible. The independence of tracks in such systems has close parallels with traditional film sound techniques, and thus the technology is fairly easily assimilated by film sound editors.

3.2.4 Sound for video

Professional video recorders (VTRs) have a limited number of sound tracks, and until recently the sound quality has been only average. The recent development of digital VTRs has brought with it an improvement in the number and quality of

Time	Picture	Dialogue 1	Dialogue 2	FX1	Music
33'15"	Long shot restuarant	Margaret: 'Over here wtr.!'		Bkgnd. restrnt.	Quiet Palm Court Orch. 1920s dance music
33'24"	Close-up – Nigel		Nigel: 'Mine's a G&T, etc...'		
33'30"			Waiter: Just a moment sir.'		
33'33"	Nigel + Margaret		M. 'Now, about our little problem...'		

Figure 3.7 A fictitious example of a simple dubbing chart, showing picture action, two dialogue tracks, an effects track and a music track

sound tracks, but it is still normal for additional tracks to be required for all but the simplest programme productions.

Original sound in a video production may be recorded directly on to the audio tracks of the VTR concerned, since there is not the tradition of separate sound recording in this field, as there is with film sound. Original sound is then edited alongside pictures, using an electronic editing system, often using a separate time-code-locked quarter-inch tape recorder or solid-state store to 'lay off' sound around edit points so that sound overlaps may be made between outgoing and incoming takes (see Figure 3.8). The edited master may then proceed to post-production for

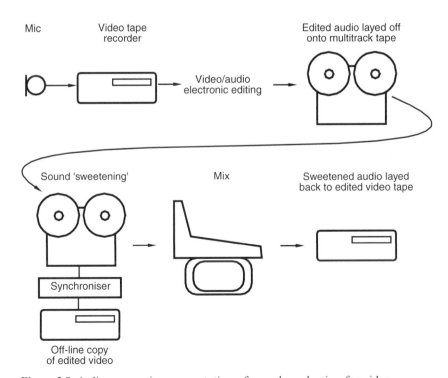

Figure 3.8 A diagrammatic representation of sound production for video

'sweetening' or 'dubbing', during which ambience tracks, over-dubbed dialogue, effects and music may be added. Traditionally a multitrack tape machine has been used for this purpose, but increasingly hard-disk editors are used (see section 10.9) since they offer greater flexibility in track positioning and allow a library of effects to be accessed quickly and easily.

Once the dub is completed, the result is mixed and 'layed back' on to the edited video tape in synchronism with the picture, the original edited sound now being replaced by the dubbed sound which includes parts of the original.

3.3 Broadcast distribution

Radio and television sound may go through any of the production processes introduced above, but will experience the additional problems of distribution around a potentially large network of studio centres and transmitter sites, requiring that it is carried over a variety of links, both analogue and digital.

Over its life, a typical television sound signal from an outside broadcast location may have travelled over a large number of miles (see Figure 3.9). A radio microphone (see section 4.10) may transmit locally to the outside broadcast vehicle, which may in turn use a microwave radio link to send the programme sound back to the studio centre. The signal may then travel through a length of internal cabling at the studio centre, finally to be connected to the network which distributes signals around the country – to other studio centres and to transmitter sites. This network may be in the form of so-called 'land lines', which are hard-wired connections, equalised to compensate for the losses encountered over the distances involved,

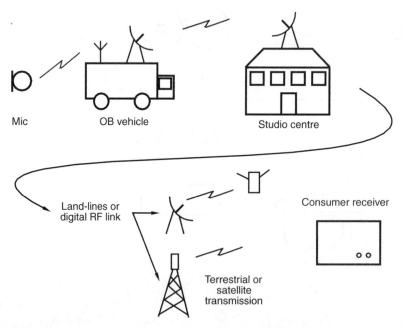

Figure 3.9 A diagrammatic representation of the typical path followed by a television sound signal

or it may be in the form of radio-frequency links, perhaps digitally encoded. From the transmitter site the programme is distributed to the consumer via a further radio-frequency link, with the additional possibility that the signal may be beamed to a satellite for national or international transmission.

3.4 An introduction to signal levels

Looking at the diagram in Figure 3.10 it will be seen that an audio signal passing from a microphone through a mixer to a tape recorder and also to a loudspeaker system encounters a number of changes in overall level (signal voltage). Other potential sources with their nominal output levels are also shown to illustrate the differences involved.

The nominal output voltage of a microphone is usually quite low, depending on the microphone type (see Chapter 4), being of the order of a millivolt or so (1 mV = 0.001 V). Clearly the actual level depends on the audio signal concerned. This is often known as a 'mic level' signal, to distinguish it from a 'line level' signal which has a nominal level of nearer to 1 volt. Line level signals are normally quoted with reference to a level of 0.775 volts – given the designation 0 dBu (see Fact File 1.2). Tape recorders, effects devices, mixers and other studio electrical devices normally have line level outputs. Thus mic level signals are some 60 dB lower in voltage than line level and such signals require amplification to bring them up to line level. Nearly all mixers have microphone amplifiers built into the channels, and a variable gain control is provided to offer amplification of between around 30 and 80 dB, depending on the microphone's output level and the output level of the acoustic source it is facing.

The output of a mixer is at line level, and this is fed to a power amplifier which in turns feeds the loudspeakers. Loudspeakers require a relatively high voltage to drive them sufficiently hard to produce the level of sound required, the precise voltage depending on the efficiency of the loudspeaker and the loudness required. The level of the signal driving the loudspeaker might be, say, around 10 volts nominal.

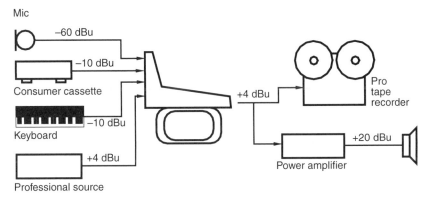

Figure 3.10 Typical signal levels encountered at stages in the recording chain

An analogue tape recorder is driven with a line level electrical signal but converts this to a magnetic signal which it records on the tape (see section 8.2). There are a number of accepted magnetic reference levels, as discussed in Fact File 8.5, and it is important to realise that there is a variable relationship between the magnetic level on tape and the electrical input level to a tape machine which depends on how the machine is aligned.

Consumer sources such as FM tuners, cassette machines, and some cheaper effects devices will have lower output levels than 0 dBu, these being nominally at around 100 mV or sometimes −10 dBV. Line level amplifiers can be obtained to bring such signals up to line level, often at the same time as balancing the unbalanced output of the consumer device (see section 12.2), or it may be possible to connect such signals directly to line level balanced inputs with the possibility for interference and slightly increased noise. Consumer hi-fi amplifiers usually operate with the lower-level input signals, and all of the inputs on such an amplifier except that set aside for the LP turntable (sometimes labelled 'RIAA' or 'Phono') will normally be at this level. Sometimes a special CD/DAT input is provided with a slightly higher nominal input level, since consumer digital equipment tends to produce a higher maximum output voltage than other consumer equipment. If connecting professional equipment outputs to consumer hi-fi equipment it is important not to overload the inputs, and again amplifiers or transformers (see section 12.1) are available to modify the levels accordingly.

LP turntables are somewhat unique, since a typical moving-magnet pickup cartridge produces a similarly low signal level as a microphone, but the signal is designed to be equalised according to the RIAA curve (see Appendix A2.2). A filter is installed in the RIAA pre-amplifier of the system on which the record is replayed. For this reason the LP input on an amplifier should not be used for any other purpose, neither should the output of a turntable be connected to an ordinary mic level input since its frequency balance will be uncorrected.

The meter indications corresponding to different electrical levels in a professional system depend on the type of meter, the operating level of the studio, and any other conventions adhered to, and since this is a large subject it will be covered in greater detail in section 7.2.

Recommended further reading

Historical

Gelatt, R. (1977) *The Fabulous Phonograph*. Cassell and Co., London
Read, O. (1976) *From Tinfoil to Stereo*

Audio systems and signals (general)

BS 6840. *Audio Equipment*. British Standards Office
CCIR Rec. 574–1 (1982) (re signal levels and dB suffixes). Vol. 13. Green Book

See also *General further reading* at the end of this book

Chapter 4

Microphones

A microphone is a transducer which converts acoustical sound energy into electrical energy. It performs the opposite function to a loudspeaker which converts electrical energy into acoustical energy. The three most common principles of operation are the moving coil or 'dynamic', the ribbon, and the capacitor or condenser. The principles are described in Fact Files 4.1, 4.2 and 4.3 respectively.

4.1 The moving-coil or dynamic microphone

The moving-coil microphone is widely used in the sound reinforcement industry, its robustness making it particularly suitable for hand-held vocal use. Wire-mesh bulbous wind shields are usually fitted to such models, and contain foam material which attenuates wind noise and 'p-blasting' from the vocalist's mouth. Built-in bass attenuation is also often provided to compensate for the effect known as bass tip-up, a phenomenon whereby sound sources at a distance of less than 50 cm or so are reproduced with accentuated bass if the microphone has a directional response (see Fact File 4.4). The frequency response of the moving-coil mic tends to show a resonant peak of several decibels in the upper-mid frequency or 'presence' range, at around 5 kHz or so, accompanied by a fairly rapid fall-off in response above 8 or 10 kHz. This is due to the fact that the moving mass of the coil–diaphragm structure is sufficient to impede the diaphragm's rapid movement necessary at high frequencies. The shortcomings have actually made the moving coil a good choice for vocalists since the presence peak helps to lift the voice and improve intelligibility. Its robustness has also meant that it is almost exclusively used as a bass drum mic in the rock industry. Its sound quality is restricted by its slightly uneven and limited frequency response, but it is extremely useful in applications such as vocals, drums, and the micing-up of guitar amplifiers.

One or two high-quality moving-coil mics have appeared with an extended and somewhat smoother frequency response, and one way of achieving this has been to use what are effectively two mic capsules in one housing, one covering mid and high frequencies, one covering the bass.

FACT FILE 4.1 — Dynamic microphone – principles

The moving-coil microphone functions like a moving-coil speaker in reverse. As shown in the diagram, it consists of a rigid diaphragm, typically 20–30 mm in diameter, which is suspended in front of a magnet. A cylindrical former is attached to the diaphragm on to which is wound a coil of very fine-gauge wire. This sits in the gap of a strong permanent magnet. When the diaphragm is made to vibrate by sound waves the coil in turn moves to and fro in the magnet's gap, and an

alternating current flows in the coil, producing the electrical output (see Fact File 3.1). Some models have sufficient windings on the coil to produce a high enough output to be fed directly to the output terminals, whereas other models use fewer windings, the lower output then being fed to a step-up transformer in the microphone casing and then to the output. The resonant frequency of dynamic microphone diaphragms tends to be in the middle frequency region.

The standard output impedance of professional microphones is 200 ohms. This value was chosen because it is high enough to allow useful step-up ratios to be employed in the output transformers, but low enough to allow a microphone to drive long lines of 100 metres or so. It is possible, though, to encounter dynamic microphones with output impedances between 50 and 600 ohms. Some moving-coil models have a transformer that can be wired to give a high-level, high-impedance output suitable for feeding into the lower-sensitivity inputs found on guitar amplifiers and some PA amplifiers. High-impedance outputs can, however, only be used to drive cables of a few metres in length, otherwise severe high-frequency loss results. (This is dealt with fully in Chapter 12.)

Output leads / Magnet N / S / Diaphragm / S / N / Suspension / Coil

4.2 The ribbon microphone

The ribbon microphone at its best is capable of very high-quality results. The comparatively 'floppy' suspension of the ribbon gives it a low-frequency resonance at around 40 Hz, below which its frequency response fairly quickly falls away. At the high-frequency end the frequency response remains smooth. However, the moving mass of ribbon itself means that it has difficulty in responding to very high frequencies, and there is generally a roll-off above 14 kHz or so. Reducing the size (therefore the mass) of ribbon reduces the area for the sound waves to work upon and it's electrical output becomes unacceptably low. One manufacturer has adopted a 'double-ribbon' principle which goes some way towards removing this dilemma. Two ribbons, each half the length of a conventional ribbon are mounted one above the other and are connected in series. They are thus analogous to a conventional

FACT FILE

4.2

Ribbon microphone – principles

The ribbon microphone consists of a long thin strip of conductive metal foil, pleated to give it rigidity and 'spring', lightly tensioned between two end clamps, as shown in the diagram. The opposing magnetic poles create a magnetic field across the ribbon such that when it is excited by sound waves a current is induced into it (see Fact File 3.1). The electrical output of the ribbon is very small, and a transformer is built into the microphone which steps up the output. The step-up ratio of a particular ribbon design is chosen so that the resulting output impedance is the standard 200 ohms, this also

giving an electrical output level comparable with that of moving-coil microphones. The resonant frequency of ribbon microphones is normally at the bottom of the audio spectrum.

ribbon that has been 'clamped' in the centre. Each ribbon now has half the moving mass and thus a better top-end response. Both of them working together still maintain the necessary output.

The ribbon mic is rather more delicate than the moving coil, and it is better suited to applications where its smooth frequency response comes into its own, such as the micing of acoustic instruments and classical ensembles. There are, however, some robust models which look like moving-coil vocal mics and can be interchanged with them. Micing a rock bass drum with one is still probably not a good idea, due to the very high transient sound pressure levels involved.

4.3 The capacitor or condenser microphone

4.3.1 Basic capacitor microphone

The great advantage of the capacitor mic's diaphragm over moving-coil and ribbon types is that it is not attached to a coil and former, and it does not need to be of a shape and size which makes it suitable for positioning along the length of a magnetic field. It therefore consists of an extremely light disk, typically 12–25 mm in diameter, frequently made from polyester coated with an extremely thin vapour-deposited metal layer so as to render it conductive. Sometimes the diaphragm itself is made of a metal such as titanium. The resonant frequency of the diaphragm is typically in the 12–20 kHz range, but the increased output here is rather less prominent than with moving coils due to the diaphragm's very light weight.

FACT FILE 4.3   Capacitor microphone – principles

The capacitor (or condenser) microphone operates on the principle that if one plate of a capacitor is free to move with respect to the other, then the capacitance (the ability to hold electrical charge) will vary. As shown in the diagram, the capacitor consists of a flexible diaphragm and a rigid back plate, separated by an insulator, the diaphragm being free to move in sympathy with sound waves incident upon it. The 48 volts DC phantom power (see section 4.9) charges the capacitor via a very high resistance. A DC blocking capacitor simply prevents the phantom power from entering the head amplifier, allowing only audio signals to pass.

When sound waves move the diaphragm the capacitance varies, and thus the voltage across the capacitor varies proportionally, since the high resistance only allows very slow leakage of charge from the diaphragm (much slower than the rate of change caused by audio frequencies). This voltage modulation is fed to the head amplifier (via the blocking capacitor) which converts the very high impedance output of the capacitor capsule to a much lower impedance. The output transformer balances this signal (see section 12.4) and conveys it to the microphone's output terminals. The resonant frequency of a capacitor mic diaphragm is normally at the upper end of the audio spectrum.

The head amplifier consists of a field-effect transistor (FET) which has an almost infinitely high input impedance. Other electronic components are also usually present which perform tasks such as voltage regulation and output stage duties. Earlier capacitor microphones had valves built into the housing, and were somewhat more bulky affairs than their modern counterparts. Additionally, extra wiring had to be incorporated in the mic leads to supply the valves with HT (high-tension) and valve-heater voltages. They were thus not particularly convenient to use, but such is the quality of sound available from capacitor mics that they quickly established themselves. Today, the capacitor microphone is the standard top-quality type, other types being used for relatively specialised applications. The electrical current requirement of capacitor microphones varies from model to model, but generally lies between 0.5 mA and 8 mA, drawn from the phantom power supply.

| FACT FILE 4.4 | Bass tip-up |

Pressure-gradient microphones are susceptible to a phenomenon known as bass tip-up, meaning that if a sound source is close to the mic (less than about a metre) the low frequencies become unnaturally exaggerated. In normal operation, the driving force on a pressure-gradient microphone is related almost totally to the phase difference of the sound wave between front and rear of the diaphragm (caused by the extra distance travelled by the wave). For a fixed path-length difference between front and rear, therefore, the phase difference increases with frequency. At LF the phase difference is small and at MF to HF it is larger.

Close to a small source, where the microphone is in a field of roughly spherical waves, sound pressure drops as distance from the source increases (see Fact File 1.3). Thus, in addition to the phase difference between front and rear of the mic's diaphragm, there is a pressure difference due to the natural level-drop with distance from the source. Since the driving force on the diaphragm due to phase difference is small at LF, this pressure drop makes a significant additional contribution, increasing the overall output level at LF. At HF the phase difference is larger, and thus the contribution made by pressure difference is smaller as a proportion of the total driving force.

At greater distances from the source, the sound field approximates more closely to one of plane waves, and the pressure drop over the front–back distance may be considered insignificant as a driving force on the diaphragm, making the mic's output related only to front–back phase difference.

Occasionally capacitor microphones are capable of being switched to give a line level output, this being simple to arrange since an amplifier is built into the mic anyway. The high-level output gives the signal rather more immunity to interference when very long cables are employed, and it also removes the need for microphone amplifiers at the mixer or tape recorder. Phantom power does, however, still need to be provided (see section 4.9.1).

4.3.2 Electret designs

A much later development was the so-called 'electret' or 'electret condenser' principle. The need to polarise the diaphragm with 48 volts is dispensed with by introducing a permanent electrostatic charge into it during manufacture. In order to achieve this the diaphragm has to be of a more substantial mass, and its audio performance is therefore closer to a moving-coil than to a true capacitor type. The power for the head amplifier is supplied either by a small dry-cell battery in the stem of the mic or by phantom power. The electret principle is particularly suited to applications where compact size and light weight are important, such as in small portable cassette machines (all built-in mics are now electrets) and tie-clip microphones which are ubiquitous in television work. They are also made in vast quantities very cheaply.

Later on, the so-called 'back electret' technique was developed. Here, the diaphragm is the same as that of a true capacitor type, the electrostatic charge being induced into the rigid back plate instead. Top-quality examples of back electrets are therefore just as good as conventional capacitor mics with their 48 volts of polarising voltage.

4.3.3 RF capacitor microphone

Still another variation on the theme is the RF (Radio Frequency) capacitor mic, in which the capacitor formed by the diaphragm and back plate forms part of a tuned circuit to generate a steady carrier frequency which is much higher than the highest audio frequency. The sound waves move the diaphragm as before, and this now causes modulation of the tuned frequency. This is then demodulated by a process similar to the process of FM radio reception, and the resulting output is the required audio signal. (It must be understood that the complete process is carried out within the housing of the microphone and it does not in itself have anything to do with radio microphone systems, as discussed in section 4.10.)

4.4 Directional responses and polar diagrams

Microphones are designed to have a specific directional response pattern, described by a so-called 'polar diagram'. The polar diagram is a form of two-dimensional contour map, showing the magnitude of the microphone's output at different angles of incidence of a sound wave. The distance of the polar plot from the centre of the graph (considered as the position of the microphone diaphragm) is usually calibrated in decibels, with a nominal 0 dB being marked for the response at zero degrees at 1 kHz. The further the plot is from the centre, the greater the output of the microphone at that angle.

4.4.1 Omnidirectional pattern

Ideally, an omnidirectional or 'omni' microphone picks up sound equally from all directions. The omni polar response is shown in Figure 4.1, and is achieved by leaving the microphone diaphragm open at the front, but completely enclosing it at the rear, so that it becomes a simple pressure transducer, responding only to the change of air pressure caused by the sound waves. This works extremely well at low and mid frequencies, but at high frequencies the dimensions of the microphone capsule itself begin to be comparable with the wavelength of the sound waves, and a shadowing effect causes high frequencies to be picked up rather less well to the rear and sides of the mic. A pressure increase also results for high-frequency sounds from the front. Coupled with this is the possibility for cancellations to arise when a high-frequency wave, whose wavelength is comparable with the diaphragm diameter, is incident from the side of the diaphragm. In such a case positive and negative peaks of the wave may result in opposing forces on the diaphragm.

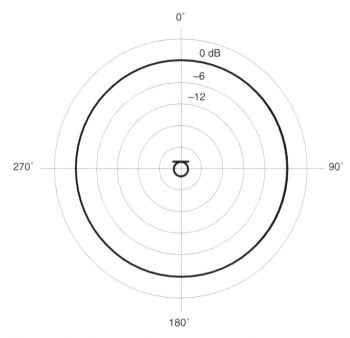

Figure 4.1 Idealised polar diagram of an omnidirectional microphone

Figure 4.2 shows the polar response plot which can be expected from a real omnidirectional microphone with a capsule half an inch (13 mm) in diameter. It is perfectly omnidirectional up to around 2 kHz, but then it begins to lose sensitivity at the rear; at 3 kHz its sensitivity at 180° will typically be 6 dB down compared with lower frequencies. Above 8 kHz, the 180° response could be as much as 15 dB down, and the response at 90° and 270° could show perhaps a 10 dB loss. As a consequence, sounds which are being picked up significantly off axis from the microphone will be reproduced with considerable treble loss, and will sound dull. It is at its best on axis and up to 45° either side of the front of the microphone.

High-quality omnidirectional microphones are characterised by their wide, smooth frequency response extending both to the lowest bass frequencies and the high treble with minimum resonances or coloration. This is due to the fact that they are basically very simple in design, being just a capsule which is open at the front and completely enclosed at the rear. (In fact a very small opening is provided to the rear of the diaphragm in order to compensate for overall changes in atmospheric pressure which would otherwise distort the diaphragm.) The small tie-clip microphones which one sees in television work are usually omnidirectional electret types which are capable of very good performance. The smaller the dimensions of the mic, the better the polar response at high frequencies, and mics such as these have quarter-inch diaphragms which maintain a very good omnidirectional response right up to 10 kHz.

Omni microphones are usually the most immune to handling and wind noise of all the polar patterns, since they are only sensitive to absolute sound pressure.

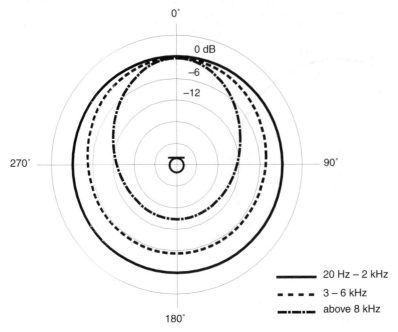

Figure 4.2 Typical polar diagram of an omnidirectional microphone at a number of frequencies

Patterns such as figure-eight (especially ribbons) and cardioid, described below, are much more susceptible to handling and wind noise than omnis because they are sensitive to the large pressure *difference* created across the capsule by low-frequency movements such as those caused by wind or unwanted diaphragm motion. A pressure-gradient microphone's mechanical impedance (the diaphragm's resistance to motion) is always lower at LF than that of a pressure (omni) microphone, and thus it is more susceptible to unwanted LF disturbances.

4.4.2 Figure-eight or bidirectional pattern

The figure-eight or bidirectional polar response is shown in Figure 4.3. Such a microphone has an output proportional to the mathematical cosine of the angle of incidence. One can quickly draw a figure-eight plot on a piece of graph paper, using a protractor and a set of cosine tables or pocket calculator. Cos $0° = 1$, showing a maximum response on the forward axis (this will be termed the 0 dB reference point). Cos $90° = 0$, so at 90° off axis no sound is picked up. Cos 180° is -1, so the output produced by a sound which is picked up by the rear lobe of the microphone will be 180° out of phase compared with an identical sound picked up by the front lobe. The phase is indicated by the $+$ and $-$ signs on the polar diagram. At 45° off axis, the output of the microphone is 3 dB down (cos 45° represents 0.707 or $1/\sqrt{2}$ times the maximum output) compared with the on-axis output.

Traditionally the ribbon microphone has sported a figure-eight polar response, and the ribbon has been left completely open both to the front and to the rear. Such

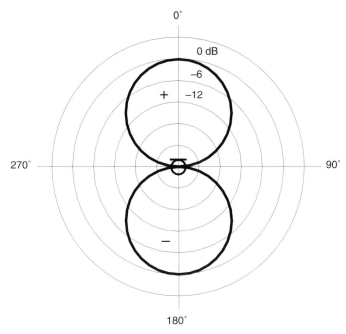

Figure 4.3 Idealised polar diagram of a figure-eight microphone

a diaphragm operates on the pressure-gradient principle, responding to the *difference* in pressure between the front and the rear of the microphone. Consider a sound reaching the mic from a direction 90° off axis to it. The sound pressure will be of equal magnitude on both sides of the diaphragm and so no movement will take place, giving no output. When a sound arrives from the 0° direction a phase difference arises between the front and rear of the ribbon, due to the small additional distance travelled by the wave. The resulting difference in pressure produces movement of the diaphragm and an output results.

At very low frequencies, wavelengths are very long and therefore the phase difference between front and rear of the mic is very small, causing a gradual reduction in output as the frequency gets lower. In ribbon microphones this is compensated for by putting the low-frequency resonance of the ribbon to good use, using it to prop up the bass response. Single-diaphragm capacitor mic designs which have a figure-eight polar response do not have this option, since the diaphragm resonance is at a very high frequency, and a gradual roll-off in the bass can be expected unless other means such as electronic frequency correction in the microphone design have been employed. Double-diaphragm switchable types which have a figure-eight capability achieve this by combining a pair of back-to-back cardioids (see section 4.4.3) that are mutually out of phase.

Like the omni, the figure-eight can give very clear uncoloured reproduction. The polar response tends to be very uniform at all frequencies, except for a slight narrowing above 10 kHz or so, but it is worth noting that a ribbon mic has a rather better polar response at high frequencies in the horizontal plane than in the vertical plane, due to the fact that the ribbon is long and thin. A high-frequency sound coming from a direction somewhat above the plane of the microphone will suffer partial cancellation, since at frequencies where the wavelength begins to be comparable with the length of the ribbon the wave arrives partially out of phase at the lower portion compared with the upper portion, therefore reducing the effective acoustical drive of the ribbon compared with mid frequencies. Ribbon figure-eight microphones should therefore be orientated either upright or upside-down with their stems vertical so as to obtain the best polar response in the horizontal plane, vertical polar response usually being less important.

Although the figure-eight picks up sound equally to the front and to the rear, it must be remembered that the rear pickup is out of phase with the front, and so correct orientation of the mic is required.

4.4.3 Cardioid or unidirectional pattern

The cardioid pattern is described mathematically as $1 + \cos\vartheta$, where ϑ is the angle of incidence of the sound. Since the omni has a response of 1 (equal all round) and the figure-eight has a response represented by $\cos\vartheta$, the cardioid may be considered theoretically as a product of these two responses. Figure 4.4(a) illustrates its shape. Figure 4.4(b) shows an omni and a figure-eight superimposed, and one can see that adding the two produces the cardioid shape: at $0°$, both polar responses are of equal amplitude and phase, and so they reinforce each other, giving a total output which is actually twice that of either separately. At $180°$, however, the two are of equal amplitude but opposite phase, and so complete cancellation occurs and there is no output. At $90°$ there is no output from the figure-eight, but just the contribution from the omni, so the cardioid response is 6 dB down at $90°$. It is 3 dB down at $65°$ off axis.

One or two early microphone designs actually housed a figure-eight and an omni together in the same casing, electrically combining their outputs to give a resulting cardioid response. This gave a rather bulky mic, and also the two diaphragms could not be placed close enough together to produce a good cardioid response at higher frequencies due to the fact that at these frequencies the wavelength of sound became comparable with the distance between the diaphragms. The designs did, however, obtain a cardioid from first principles.

The cardioid response is now obtained by leaving the diaphragm open at the front, but introducing various acoustic labyrinths at the rear which cause sound to reach the back of the diaphragm in various combinations of phase and amplitude to produce a resultant cardioid response. This is difficult to achieve at all frequencies simultaneously, and Figure 4.5 illustrates the polar pattern of a typical cardioid mic with a three-quarter-inch diaphragm. As can be seen, at mid frequencies the polar response is very good. At low frequencies it tends to degenerate towards omni, and at very high frequencies it becomes rather more directional than is desirable. Sound arriving from, say, $45°$ off axis will be reproduced with treble loss, and sounds arriving from the rear will not be completely attenuated, the low frequencies being picked up quite uniformly.

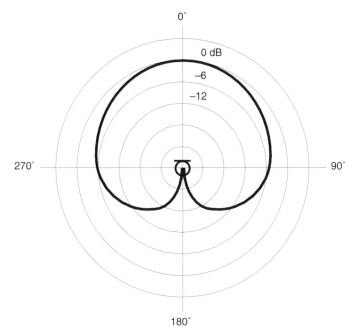

Figure 4.4 (a) Idealised polar diagram of a cardioid microphone

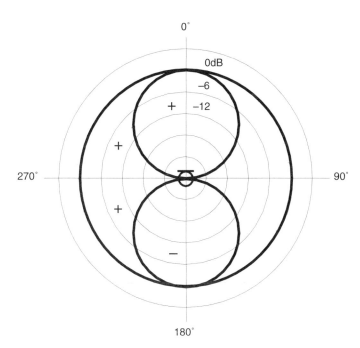

Figure 4.4 (b) A cardioid microphone can be seen to be the mathematical equivalent of an omni and a figure-eight response added together

Figure 4.5 Typical polar diagram of a cardioid microphone at low, middle and high frequencies

The above example is very typical of moving-coil cardioids, and they are in fact very useful for vocalists due to the narrow pickup at high frequencies helping to exclude off-axis sounds, and also the relative lack of pressure-gradient component at the bass end helping to combat bass tip-up. High-quality capacitor cardioids with half-inch diaphragms achieve a rather more ideal cardioid response. Owing to the presence of acoustic labyrinths, coloration of the sound is rather more likely, and it is not unusual to find that a relatively cheap electret omni will sound better than a fairly expensive cardioid.

4.4.4 Hypercardioid pattern

The hypercardioid, sometimes called 'cottage loaf' because of its shape, is shown in Figure 4.6. It is described mathematically by the formula $0.5 + \cos\vartheta$, i.e.: it is a combination of an omni attenuated by 6 dB, and a figure-eight. Its response is in between the cardioid and figure-eight patterns, having a relatively small rear lobe which is out of phase with the front lobe. Its sensitivity is 3 dB down at 55° off axis. Like the cardioid, the polar response is obtained by introducing acoustic labyrinths to the rear of the diaphragm. Because of the large pressure-gradient component it too is fairly susceptible to bass tip-up. Practical examples of hypercardioid microphones tend to have polar responses which are tolerably close to the ideal. The hypercardioid has the highest direct-to-reverberant ratio of the patterns described, which means that the ratio between the level of on-axis sound and the level of reflected sounds picked up from other angles is very high, and so it is good for excluding unwanted sounds such as excessive room ambience or unwanted noise.

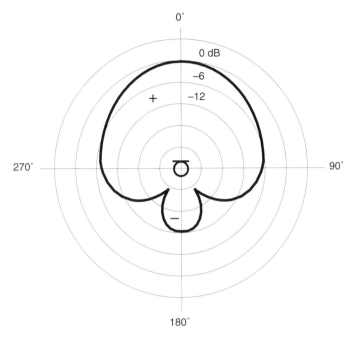

Figure 4.6 Idealised polar diagram of a hypercardioid microphone

4.5 Specialised microphone types

4.5.1 Rifle microphone

The rifle microphone is so called because it consists of a long tube of around three-quarters of an inch (1.9 cm) in diameter and perhaps 2 feet (61 cm) in length, and looks rather like a rifle barrel. The design is effectively an ordinary cardioid microphone to which has been attached a long barrel along which slots are cut in such a way that a sound arriving off axis enters the slots along the length of the tube and thus various versions of the sound arrive at the diaphragm at the bottom of the tube in relative phases which tend to result in cancellation. In this way, sounds arriving off axis are greatly attenuated compared with sounds arriving on axis. Figure 4.7 illustrates the characteristic club-shaped polar response. It is an extremely directional device, and is much used by news sound crews where it can be pointed directly at a speaking subject, excluding crowd noise. It is also used for wildlife recording, sports broadcasts, along the front of theatre stages in multiples, and in audience participation discussions where a particular speaker can be picked out. For outside use it is normally completely enclosed in a long, fat wind shield, looking like a very big cigar. Half-length versions are also available which have a polar response midway between a club shape and a hypercardioid. All versions, however, tend to have a rather wider pickup at low frequencies.

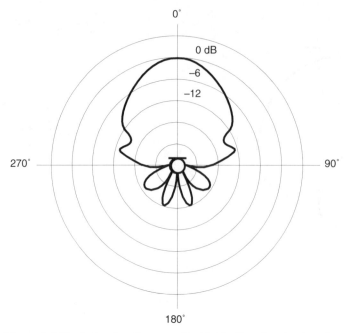

Figure 4.7 Typical polar diagram of a highly-directional microphone

4.5.2 Parabolic microphone

An alternative method of achieving high directionality is to use a parabolic dish, as shown in Figure 4.8. The dish has a diameter usually of between 0.5 and 1 metre, and a directional microphone is positioned at its focal point. A large 'catchment area' is therefore created in which the sound is concentrated at the head of the mic. An overall gain of around 15 dB is typical, but at the lower frequencies where the wavelength of sound becomes comparable with the diameter of the dish the response falls away. Because this device actually concentrates the sound rather than merely rejecting off-axis sounds, comparatively high outputs are achieved from distant sound sources. They are very useful for capturing bird song, and they are also sometimes employed around the boundaries of cricket pitches. They are, however, rather cumbersome in a crowd, and can also produce a rather coloured sound.

4.5.3 Boundary or 'pressure-zone' microphone

The so-called boundary or pressure-zone microphone (PZM) consists basically of an omnidirectional microphone capsule mounted on a plate usually of around 6 inches (15 cm) square or 6 inches in diameter such that the capsule points directly at the plate and is around 2 or 3 millimetres away from it. The plate is intended to be placed on a large flat surface such as a wall or floor, and it can also be placed on the underside of a piano lid for instance. Its polar response is hemispherical. Because the mic capsule is a simple omni, quite good-sounding versions are available with

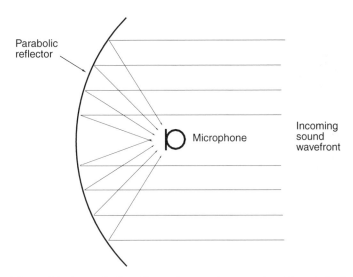

Parabolic reflector

Microphone

Incoming sound wavefront

Figure 4.8 A parabolic reflector is sometimes used to 'focus' the incoming sound wavefront at the microphone position, thus making it highly directional

electret capsules fairly cheaply, and so if one wishes to experiment with this unusual type of microphone one can do so without parting with a great deal of money. It is important to remember though that despite its looks it is not a contact mic – the plate itself does not transduce surface vibrations – and it should be used with the awareness that it is equivalent to an ordinary omnidirectional microphone pointing at a flat surface, very close to it. The frequency response of such a microphone is rarely as flat as that of an ordinary omni, but it can be unobtrusive in use.

4.6 Switchable polar patterns

The double-diaphragm capacitor microphone, such as the commercial example shown in Figure 4.9, is a microphone in which two identical diaphragms are employed, placed each side of a central rigid plate in the manner of a sandwich. Perforations in the central plate give both diaphragms an essentially cardioid response. When the polarising voltage on both diaphragms is the same, the electrically combined output gives an omnidirectional response due to the combination of the back-to-back cardioids in phase. When the polarising voltage of one diaphragm is opposite to that of the other, and the potential of the rigid central plate is midway between the two, the combined output gives a figure-eight response (back-to-back cardioids mutually out of phase). Intermediate combinations give cardioid and hypercardioid polar responses. In this way the microphone is given a switchable polar response which can be adjusted either by switches on the microphone itself or via a remote control box. Some microphones with switchable polar patterns achieve this by employing a conventional single diaphragm around which is placed appropriate mechanical labyrinths which can be switched to give the various patterns.

Figure 4.9 A typical double-diaphragm condenser microphone with switchable polar pattern: the AKG C4141B-ULS. (Courtesy of AKG Acoustics GmbH)

Another method manufacturers have used is to make the capsule housing on the end of the microphone detachable, so that a cardioid capsule, say, can be unscrewed and removed to be replaced with, say, an omni. This also facilitates the use of extension tubes whereby a long thin pipe of around a metre or so in length with suitably threaded terminations is inserted between the main microphone body and the capsule. The body of the microphone is mounted on a short floor stand and the thin tube now brings the capsule up to the required height, giving a visually unobtrusive form of microphone stand.

4.7 Stereo microphones

Stereo microphones, such as the example shown in Figure 4.10, are available in which two microphones are built into a single casing, one capsule being rotatable with respect to the other so that the angle between the two can be adjusted. Also, each capsule can be switched to give any desired polar response. One can therefore adjust the mic to give a pair of figure-eight microphones angled at, say, 90°, or a pair of cardioids at 120°, and so on. Some stereo mics, such as that pictured in Figure 4.11, are configured in a sum-and-difference arrangement, instead of as a left–right pair, with a 'sum' capsule pointing forwards and a figure-eight 'difference' capsule facing sideways. The sum-and-difference or 'middle and side' (M and S) signals are combined in a matrix box to produce a left–right stereo signal by adding M and S to give the left channel and subtracting M and S to give the right channel. This is discussed in more detail in Fact File 4.5.

Figure 4.10 A typical stereo microphone: the Neumann SM69. (Courtesy of FWO Bauch Ltd)

Figure 4.11 A typical 'sum-and-difference' stereo microphone: the Shure VP88. (Courtesy of HW International)

FACT FILE

4.5

Sum and difference processing

MS signals may be converted to conventional stereo very easily, either using three channels on a mixer, or using an electrical matrix. M is the mono sum of two conventional stereo channels, and S is the difference between them. Thus:

$$M = (L + R) \div 2$$
$$S = (L - R) \div 2$$

and

$$L = (M + S) \div 2$$
$$R = (M - S) \div 2$$

A pair of transformers may be used wired as shown in the diagram to obtain either MS from LR, or vice versa. Alternatively, a pair of summing amplifiers may be used, with the M and S (or L and R) inputs to one being wired in phase (so that they add) and to the other out of phase (so that they subtract).

The mixer configuration shown in the diagram may also be used. Here the M signal is panned centrally (feeding L and R outputs), whilst the S signal is panned left (M + S = L). A post-fader insertion feed is taken from the S channel to a third channel which is phase reversed to give −S. The gain of this channel is set at 0 dB and is panned right (M − S = R). If the S fader is varied in level, the width of the stereo image and the amount of rear pickup can be varied.

A sophisticated stereo microphone is the Soundfield Research microphone. In this design, four 'subcardioid' capsules (i.e.: between omni and cardioid) are arranged in a tetrahedral array such that their outputs can be combined in various ways to give four outputs, termed 'B format'. The raw output from the four capsules is termed 'A format'. The four B-format signals consist of a forward-facing figure-eight ('X'), a sideways-facing figure-eight ('Y'), an up-and-down-facing figure-eight ('Z'), and an omnidirectional output ('W'). These are then appropriately combined to produce any configuration of stereo microphone output, each channel being fully adjustable from omni through cardioid to figure-eight, the angles between the capsules also being fully adjustable. The tilt angle of the microphone, and also the 'dominance' (the front-to-back pickup ratio) can also be controlled. All of this is achieved electronically by a remotely sited control unit. Additionally, the raw B-format signals can be recorded on a four-channel tape recorder, later to be replayed through the control unit where all of the above parameters can be chosen after the recording session.

The ST250 is a second generation stereo microphone based on soundfield principles, designed to be smaller and to be usable either 'end-fire' or 'side-fire' (see Figure 4.12). It can be electronically inverted and polar patterns and capsule angles are variable remotely.

4.8 Microphone performance

Professional microphones have a balanced low-impedance output usually via a three-pin XLR-type plug in their base. The impedance, which is usually around 200 ohms but sometimes rather lower, enables long microphone leads to be used. Also, the balanced configuration, discussed in section 12.4, gives considerable immunity from interference. Other parameters which must be considered are sensitivity (see Fact File 4.6) and noise (see Fact File 4.7).

Figure 4.12 The Soundfield ST250 microphone is based on soundfield principles, and can be operated either end- or side-fire, or upside-down, using electrical matrixing of the capsules within the control unit. (Courtesy of SoundField Ltd)

FACT FILE 4.6 Microphone sensitivity

The sensitivity of a microphone is an indication of the electrical output which will be obtained for a given acoustical sound pressure level (SPL). The standard SPL is either 74 dB ($= 1\ \mu B$) or 94 dB ($= 1$ Pascal or 10 μB) (μB = microbar). One level is simply ten times greater than the other, so it is easy to make comparisons between differently specified models. 74 dB is roughly the level of moderately loud speech at a distance of 1 metre. 94 dB is 20 dB or ten times higher than this, so a microphone yielding 1 mV μB^{-1} will yield 10 mV in a sound field of 94 dB. Other ways of specifying sensitivity include expressing the output as being so many decibels below a certain voltage for a specified SPL. For example, a capacitor mic may have a sensitivity figure of -60 dBV Pa^{-1} meaning that its output level is 60 dB below 1 volt for a 94 dB SPL, which is 1 mV (60 dB = times 1000).

Capacitor microphones are the most sensitive types, giving values in the region of 5–15 mV Pa^{-1}, i.e.: a sound pressure level of 94 dB will give between 5 and 15 millivolts of electrical output. The least sensitive microphones are ribbons, having typical sensitivities of 1–2 mV Pa^{-1}, i.e.: around 15 or 20 dB lower than capacitor types. Moving coils are generally a little more sensitive than ribbons, values being typically 1.5–3 mV Pa^{-1}.

FACT FILE 4.7 Microphone noise specifications

All microphones inherently generate some noise. The common way of expressing capacitor microphone noise is the 'A'-weighted equivalent self-noise. A typical value of 'A'-weighted self-noise of a high-quality capacitor microphone is around 18 dBA. This means that its output noise voltage is equivalent to the microphone being placed in a sound field with a loudness of 18 dBA. A self-noise in the region of 25 dBA from a microphone is rather poor, and if it were to be used to record speech from a distance of a couple of metres or so the hiss would be noticeable on the recording. The very best capacitor microphones achieve self-noise values of around 12 dBA.

When comparing specifications one must make sure that the noise specification is being given in the same units. Some manufacturers give a variety of figures, all taken using different weighting systems and test meter characteristics, but the 'A'-weighted self-noise discussed will normally be present among them. Also, a signal-to-noise ratio is frequently quoted for a 94 dB reference SPL, being 94 minus the self-noise, so a mic with a self-noise of 18 dBA will have a signal-to-noise ratio of 76 dBA for a 94 dB SPL, which is also a very common way of specifying noise.

4.8.1 Microphone sensitivity in practice

The consequence of mics having different sensitivity values is that rather more amplification is needed to bring ribbons and moving coils up to line level than is the case with capacitors. For example, speech may yield, say, 0.15 mV from a ribbon. To amplify this up to line level (775 mV) requires a gain of around $\times 5160$ or 74 dB. This is a lot, and it taxes the noise performance of the equipment and will also cause considerable amplification of any interference that manages to get into the microphone cables.

Consider now the same speech recording, made using a capacitor microphone of 1 mV μB^{-1} sensitivity. Now only $\times 775$ or 57 dB of gain is needed to bring this up to line level, which means that any interference will have a rather better chance of being unnoticed, and also the noise performance of the mixer will not be so severely taxed. This does not mean that high-output capacitor microphones should always be used, but it illustrates that high-quality mixers and microphone cabling are required to get the best out of low-output mics.

4.8.2 Microphone noise in practice

The noise coming from a capacitor microphone is mainly caused by the head amplifier. Since ribbons and moving coils are purely passive devices one might think that they would therefore be noiseless. This is not the case, since a 200 ohm passive resistance at room temperature generates a noise output between 20 Hz and 20 kHz of 0.26 μV (μV = microvolts). Noise in passive microphones is thus due to thermal excitation of the charge carriers in the microphone ribbon or voice coil, and the output transformer windings. To see what this means in equivalent self-noise terms so that ribbons and moving coils can be compared with capacitors, one must relate this to sensitivity.

Take a moving coil with a sensitivity of 0.2 mV μB^{-1}, which is 2 mV for 94 dB SPL. The noise is 0.26 μV or 0.000 26 mV. The signal-to-noise ratio is given by dividing the sensitivity by the noise:

$$2 \div 0.000\ 26 \approx 7600$$

and then expressing this in decibels:

$$dB = 20 \log 7600 = 77\,dB$$

This is an *unweighted* figure, and 'A' weighting will usually improve it by a couple of decibels. However, the microphone amplifier into which the mic needs to be plugged will add a bit of noise, so it is a good idea to leave this figure as it is to give a fairly good comparison with the capacitor example. (Because the output level of capacitor mics is so much higher than that of moving coils, the noise of a mixer's microphone amplifier does not figure in the noise discussion as far as these are concerned. The noise generated by a capacitor mic is far higher than noise generated by good microphone amplifiers and other types of microphone.)

A 200 ohm moving-coil mic with a sensitivity of 0.2 mV μB^{-1} thus has a signal-to-noise ratio of about 77 dB, and therefore an equivalent self-noise of $94 - 77 =$

17 dB which is comparable with high-quality capacitor types, providing that high-quality microphone amplifiers are also used. A low-output 200 ohm ribbon microphone could have a sensitivity of 0.1 mV μB^{-1}, i.e.: 6 dB less than the above moving-coil example. Because its 200 ohm thermal noise is roughly the same, its equivalent self noise is therefore 6 dB worse, i.e.: 23 dB. This would probably be just acceptable for recording speech and classical music if an ultra-low-noise microphone amplifier were to be used which did not add significantly to this figure.

The discussion of a few decibels here and there may seem a bit pedantic, but in fact self-noises in the low twenties are just on the borderline of being acceptable if one wishes to record speech or the quieter types of classical music. Loud music, and mic positions close to the sound sources such as is the practice with rock music, generate rather higher outputs from the microphones and here noise is rarely a problem. But the high output levels generated by close micing of drums, guitar amps and the like can lead to overload in the microphone amplifiers. For example, if a high-output capacitor microphone is used to pick up a guitarist's amplifier, outputs as high as 150 mV or more can be generated. This would overload some fixed-gain microphone input stages, and an in-line attenuator which reduces the level by an appropriate amount such as 10–20 dB would have to be inserted at the mixer or tape recorder end of the microphone line. Attenuators are available built into a short cylindrical tube which carries an XLR-type plug at one end and a socket at the other end. It is simply inserted between the mixer or tape recorder input and the mic lead connector. It should not be connected at the microphone end because it is best to leave the level of signal along the length of the mic lead high to give it greater immunity from interference.

4.9 Microphone powering options

4.9.1 Phantom power

Consideration of capacitor microphones reveals the need for supplying power to the electronics which are built into the casing, and also the need for a polarising voltage across the diaphragm of many capacitor types. It would obviously be inconvenient and potentially troublesome to incorporate extra wires in the microphone cable to supply this power, and so an ingenious method was devised whereby the existing wires in the cable which carry the audio signal could also be used to carry the DC voltage necessary for the operation of capacitor mics – hence the term 'phantom power', since it is invisibly carried over the audio wires. Furthermore, this system does not preclude the connection of a microphone not requiring power to a powered circuit. The principle is outlined in Fact File 4.8.

It will be appreciated that if, for instance, a ribbon microphone is connected to the line in place of a capacitor mic, no current will flow into the microphone because there will be no centre tap provided on the microphone's output transformer. Therefore, it is perfectly safe to connect other types of balanced microphone to

FACT FILE 4.8 Phantom powering

The diagram below illustrates the principle of phantom powering. Arrows indicate the path of the phantom power current. (Refer to section 12.4 for details of the balanced line system.) Here 48 volts DC is supplied to the capacitor microphone as follows: the voltage is applied to each of the audio lines in the microphone cable via two equal value resistors, 6800 (6k8) ohms being the standard value. The current then travels along both audio lines and into the microphone. The microphone's output transformer secondary has either a 'centre tap' – that is, a wire connected half-way along the transformer winding, as shown in the diagram – or two resistors as in the arrangement shown at the other end of the line. The current thus travels towards the centre of the winding from each end, and then via the centre tap to the electronic circuit and diaphragm of the microphone. To complete the circuit, the return path for the current is provided by the screening braid of the microphone cable.

this line. The two 6k8 resistors are necessary for the system because if they were replaced simply by two wires directly connected to the audio lines, these wires would short-circuit the lines together and so no audio signal would be able to pass. The phantom power could be applied to a centre tap of the input transformer, but if a short circuit were to develop along the cabling between one of the audio wires and the screen, potentially large currents could be drawn through the transformer windings and the phantom power supply, blowing fuses or burning out components. Two 6k8 resistors limit the current to around 14 mA, which should not cause serious problems. The 6k8 value was chosen so as to be high enough not to load the microphone unduly, but low enough for there to be only a small DC voltage drop across them so that the microphone still receives nearly the full 48 volts. Two real-life examples will be chosen to investigate exactly how much voltage drop occurs due to the resistors.

Firstly, the current flows through both resistors equally and so the resistors are effectively 'in parallel'. Two equal-value resistors in parallel behave like a single resistor of half the value, so the two 6k8 resistors can be regarded as a single 3k4 resistor as far as the 48 V phantom power is concerned. Ohm's law (see Fact File 1.1) states that the voltage drop across a resistor is equal to its resistance multiplied by the current passing through it. Now a Calrec 1050C microphone draws 0.5 milliamps ($= 0.0005$ amps) through the resistors, so the voltage drop is $3400 \times 0.0005 = 1.7$ volts. Therefore the microphone receives $48 - 1.7$ volts, i.e.: 46.3 volts. The Schoeps CMC-5 microphone draws 4 mA so the voltage drop is $3400 \times 0.004 = 13.6$ volts. Therefore the microphone receives $48 - 13.6$ volts, i.e.: 34.4 volts. The manufacturer normally takes this voltage drop into account in the design of the microphone, although examples exist of mics which draw so much current that they load down the phantom voltage of a mixer to a point where it is no longer adequate to power the mics. In such a case some mics become very noisy, some will not work at all, and yet others may produce unusual noises or oscillation. A stand-alone dedicated power supply or internal battery supply may be the solution in difficult cases.

The universal standard is 48 volts, but some capacitor microphones are designed to operate on a range of voltages down to 9 volts, and this can be advantageous for instance when using battery-powered equipment on location, or out of doors away from a convenient source of mains power.

Figure 4.13 illustrates the situation with phantom powering when electronically balanced circuits are used, as opposed to transformers. Capacitors are used to block the DC voltage from the power supply, but they present a very low impedance to the audio signal.

4.9.2 A–B powering

Another form of powering for capacitor microphones which is sometimes encountered is A–B powering. Figure 4.14 illustrates this system schematically. Here, the power is applied to one of the audio lines via a resistor and is taken to the microphone electronics via another resistor at the microphone end. The return path is provided by the other audio line as the arrows show. The screen is not used for carrying any current. There is a capacitor at the centre of the winding of each transformer. A capacitor does not allow DC to pass, and so these capacitors prevent

Figure 4.13 A typical 48 volt phantom powering arrangement in an electronically balanced circuit

Figure 4.14 A typical 12 volt A–B powering arrangement

the current from short-circuiting via the transformer windings. The capacitors have a very low impedance at audio frequencies, so as far as the audio signal is concerned they are not there. The usual voltage used in this system is 12 volts.

Although, like phantom power, the existing microphone lines are used to carry the current, it is dangerous to connect another type of microphone in place of the one illustrated. If, say, a ribbon microphone were to be connected, its output transformer would short-circuit the applied current. Therefore 12 volt A–B powering should be switched off before connecting any other type of microphone, and this is clearly a disadvantage compared with the phantom powering approach. It is encountered most commonly in location film sound recording equipment.

4.10 Radio microphones

Radio microphones are widely used in film, broadcasting, theatre and other industries, and it is not difficult to think of circumstances in which freedom from trailing microphone cables can be a considerable advantage in all of the above.

4.10.1 Principles

The radio microphone system consists of a microphone front end (which is no different from an ordinary microphone); an FM (Frequency Modulation) transmitter, either built into the housing of the mic or housed in a separate case into which the mic plugs; a short aerial via which the signal is transmitted; and a receiver which is designed to receive the signal from a particular transmitter. Only one specified transmission frequency is picked up by a given receiver. The audio output of the receiver then feeds a mixer or tape machine in the same manner as any orthodox microphone or line level source would. The principle is illustrated in Figure 4.15.

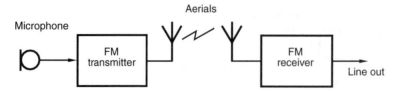

Figure 4.15 A radio microphone incorporates an FM transmitter, resulting in no fixed link between microphone and mixer

The transmitter can be built into the stem of the microphone, or it can be housed in a separate case, typically the size of a packet of cigarettes, into which the microphone or other signal source is plugged. A small battery which fits inside the casing of the transmitter provides the power, and this can also supply power to those capacitor mics which are designed to operate at the typical 9 volts of the transmitter battery. The transmitter is of the FM type (see Fact File 4.9), as this offers high-quality audio performance.

Frequently, two or more radio microphones need to be used. Each transmitter must transmit at a different frequency, and the spacing between each adjacent frequency must not be too close otherwise they will interfere with each other. In practice, channels with a minimum spacing of 0.2 MHz are used. Although only one transmitter can be used at a given frequency, any number of receivers can of course be used, as is the case with ordinary radio reception.

4.10.2 Facilities

Transmitters are often fitted with facilities which enable the operator to set the equipment up for optimum performance. A 1 kHz line-up tone is sometimes encountered which sends a continuous tone to the receiver to check continuity.

FACT FILE 4.9 Frequency modulation

In FM systems the transmitter radiates a high-frequency radio wave (the carrier) whose frequency is modulated by the amplitude of the audio signal. The positive-going part of the audio waveform causes the carrier frequency to deviate upwards, and the negative-going part causes it to deviate downwards. At the receiver, the modulated carrier is demodulated, converting variations in carrier frequency back into variations in the amplitude of an audio signal.

Audio signals typically have a wide dynamic range, and this affects the degree to which the carrier frequency is modulated. The carrier deviation must be kept within certain limits, and manufacturers specify the maximum deviation permitted. The standard figure for a transmitter with a carrier frequency of around 175 MHz is ±75 kHz, meaning that the highest-level audio signal modulates the carrier frequency between 175.075 MHz and 174.925 MHz. The transmitter incorporates a limiter to ensure that these limits are not exceeded.

Input gain controls are useful, with an indication of peak input level, so that the transmitter can be used with mics and line level sources of widely different output levels. It is important that the optimum setting is found, as too great an input level may cause a limiter (see section 13.2) to come into action much of the time, which can cause compression and 'pumping' noises as the limiter operates. Too weak a signal gives insufficient drive, and poor signal-to-noise ratios can result.

The receiver will have a signal strength indicator. This can be very useful for locating 'dead spots'; transmitter positions which cause unacceptably low meter readings should be avoided, or the receiving aerial should be moved to a position which gives better results. Another useful facility is an indicator which tells the condition of the battery in the transmitter. When the battery voltage falls below a certain level, the transmitter sends out an inaudible warning signal to the receiver which will then indicate this condition. The operator then has a warning that the battery will soon fail, which is often within 15 minutes of the indication.

4.10.3 Licences

Transmitting equipment usually requires a licence for its operation, and governments normally rigidly control the frequency bands over which a given user can operate. This ensures that local and network radio transmitters do not interfere with police, ambulance and fire brigade equipment, etc. In the UK the frequency band for which radio mics do *not* have to be licensed is between 173.8 MHz and 175 MHz. Each radio mic transmitter needs to be spaced at least 0.2 MHz apart, and commonly used frequencies are 173.8, 174.1, 174.5, 174.8, and 175.0 MHz. An additional requirement is that the frequencies must be crystal controlled, which ensures that they cannot drift outside tightly specified limits. Maximum transmitter power is limited to 10 milliwatts, which gives an effective radiated power (ERP) at the aerial of 2 milliwatts which is very low, but adequate for the short ranges over which radio mics are operated.

In recent years radio mics in the UHF band have become available operating across the 800 MHz band of frequencies, for which licences are required. Aerials are correspondingly smaller than those for the VHF models giving greater convenience for the wearer. The UHF band has proved very reliable, and notably free from interference problems.

4.10.4 Aerials

The dimensions of the transmitting aerial are related to the wavelength of the transmitted frequency. The wavelength (λ) in an electrical conductor at a frequency of 174.5 MHz is approximately 64 inches (160 cm). To translate this into a suitable aerial length, it is necessary to discuss the way in which a signal resonates in a conductor. It is convenient to consider a simple dipole aerial, as shown in Figure 4.16. This consists of two conducting rods, each a quarter of a wavelength long, fed by the transmitting signal as shown. The centre of the pair is the nodal point and

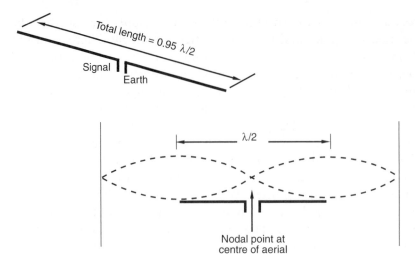

Figure 4.16 A simple dipole aerial configuration

exhibits a characteristic impedance of about 70 ohms. For a radio mic, we need a total length of $\lambda/2$, i.e.: $64/2 = 32$ inches (80 cm).

A 32 inch dipole will therefore allow the standard range of radio mic frequencies to resonate along its length to give efficient radiation, the precise length not being too critical. Consideration also has to be given to the radiated polar response (this is not the same as the microphone's polar response). Figure 4.17 shows the polar response for a dipole. As can be seen, it is a figure-eight with no radiation in the directions the two halves are pointing in. Another factor is polarisation of the signal. Electromagnetic waves consist of an electric wave plus a magnetic wave radiating at right angles to each other, and so if a transmitting aerial is orientated

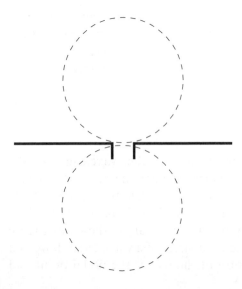

Figure 4.17 The dipole has a figure-eight radiation pattern

vertically, the receiving aerial should also be orientated vertically. This is termed vertical polarisation.

The radio mic transmitter therefore has a transmitting aerial of about 16 inches long: half of a dipole. The other half is provided by the earth screen of the audio input lead, and will be in practice rather longer than 16 inches. The first-mentioned half is therefore looked upon as being the aerial proper, and it typically hangs vertically downwards. The screened signal input cable will generally be led upwards, but other practical requirements tend to override its function as part of the aerial system.

Another type which is often used for hand-held radio mics is the helical aerial. This is typically rather less than half the length of the 16 inch aerial, and has a diameter of a centimetre or so. It protrudes from the base of the microphone. It consists of a tight coil of springy wire housed in a plastic insulator, and has the advantage of being both smaller and reasonably tolerant of physical abuse. Its radiating efficiency is, however, less good than the 16 inch length of wire. At the receiver, a similar aerial is required. The helical aerial is very common here, and its short stubby form is very convenient for outside broadcast and film crews. A 16 inch length of metal tubing, rather like a short car aerial, can be a bit unwieldy although it is a more efficient receiver.

Other aerial configurations exist, offering higher gain and directionality. In the two-element aerial shown in Figure 4.18 the reflector is slightly larger than the dipole, and is spaced behind it at a distance which causes reflection of signal back on to it. It increases the gain, or strength of signal output, by 3 dB. It also attenuates signals approaching from the rear and sides. The three-element 'Yagi', named after its Japanese inventor and shown in Figure 4.19, uses the presence of a director and reflector to increase the gain of a conventional dipole, and a greatly elongated rectangle called a folded dipole is used, which itself has a characteristic impedance of about 300 ohms. The other elements are positioned such that the final impedance is reduced to the standard 50 ohms. The three-element Yagi is even more directional than the dipole, and has increased gain. It can be useful in very difficult reception conditions, or where longer distances are involved such as receiving the signal from a transmitter carried by a rock climber for running commentary! The

Figure 4.18 A simple two-element aerial incorporates a dipole and a reflector for greater directionality than a dipole

Figure 4.19 The three-element 'Yagi' configuration

multi-element, high-gain, highly directional UHF television aerial is of course a familiar sight on our roof-tops.

These aerials can also be used for transmitting, the principles being exactly the same. Their increased directionality also helps to combat multipath problems. The elements should be vertically orientated, because the transmitting aerial will normally be vertical, and the 'direction of maximum sensitivity' arrows on the figures show the direction the aerials should be pointed in.

Another technique for improving the signal-to-noise ratio under difficult reception conditions is noise reduction, which operates as follows. Inside the transmitter there is an additional circuit which compresses the incoming audio signal, thus reducing its overall dynamic range. At the receiver, a reciprocal circuit expands the audio signal, after reception and demodulation, and as it pushes the lower-level audio signals back down to their correct level it also therefore pushes the residual noise level down. Previously unacceptable reception conditions will often yield usable results when such transmitters and receivers are employed. It should be noted though that the system does not increase signal strength, and all the problems of transmission and reception still apply. (Noise reduction systems are covered further in Chapter 9.)

4.10.5 Aerial siting and connection

It is frequently desirable to place the receiving aerial itself closer to the transmitter than the receiver, in order to pick up a strong signal. To do this an aerial is rigged at a convenient position close to the transmitter, for example in the wings of a theatre stage, or on the front of a balcony, and then an aerial lead is run back to the receiver. A helical dipole aerial is frequently employed. In such a situation, characteristic impedance must be considered. As discussed in section 12.9.1, when the wavelength of the electrical signal in a conductor is similar to the length of the conductor, reflections can be set up at the receiving end unless the cable is properly terminated. Therefore, impedance matching must be employed between the aerial and the transmitter or receiver, and additionally the connecting lead needs to have the correct characteristic impedance.

The standard value for radio microphone equipment is 50 ohms, and so the aerial, the transmitter, the receiver, the aerial lead and the connectors must all be rated at this value. This cannot be measured using a simple test meter, but an aerial and cable can be tuned using an SWR (Standing Wave Ratio) meter to detect the level of the reflected signal. The aerial lead should be a good-quality, low-loss type, otherwise the advantage of siting the aerial closer to the transmitter will be wasted by signal loss along the cable. Poor signal reception causes noisy performance, because the receiver has a built-in automatic gain control (AGC), which sets the amplification of the carrier frequency to an appropriate value. Weak signals simply require higher amplification and therefore higher noise levels result.

The use of several radio microphones calls for a complementary number of receivers which all need an aerial feed. It is common practice to use just one aerial which is plugged into the input of an aerial distribution amplifier. This distribution unit has several outputs which can be fed into each receiver. It is not possible simply to connect an aerial to all the inputs in parallel due to the impedance mismatch that this would cause.

Apart from obvious difficulties such as metallic structures between transmitter and receiver, there are two phenomena which cause the reception of the radio signal to be less than perfect. The first phenomenon is known as multi-path (see Figure 4.20). When the aerial transmits, the signal reaches the receiving aerial by a number of routes. Firstly, there is the direct path from aerial to aerial. Additionally, signals bounce off the walls of the building and reach the receiving aerial via a longer route. So the receiving aerial is faced with a number of signals

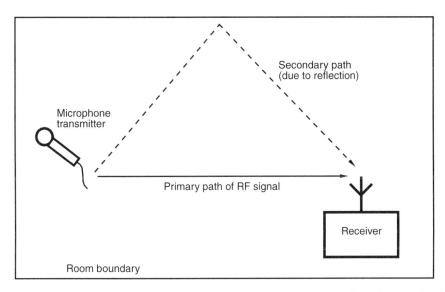

Figure 4.20 Multipath distortion can arise between source and receiver due to reflections

of more or less random phase and strength, and these will sometimes combine to cause severe signal cancellation and consequently very poor reception. The movement of the transmitter along with the person wearing it will alter the relationship between these multipath signals, and so 'dead spots' are sometimes encountered where particular combinations of multipath signals cause signal 'drop-out'. The solution is to find out where these dead spots are by trial and error, re-siting the receiving aerial until they are minimised or eliminated. It is generally good practice to site the aerial close to the transmitter so that the direct signal will be correspondingly stronger than many of the signals arriving from the walls. Metal structures should be kept clear of wherever possible due to their ability to reflect and screen RF signals. Aerials can be rigged on metal bars, but at right angles to them, not parallel.

The other phenomenon is signal cancellation from other transmitters when a number of channels are in use simultaneously. Because the transmitting frequencies of the radio mics will be quite close together, partial cancellation of all the signals takes place. The received signals are therefore weaker than for a single transmitter on its own. Again, siting the receiving aerial close to the transmitters is a good idea. The 'sharpness' or 'Q' of the frequency tuning of the receivers plays a considerable part in obtaining good reception in the presence of a number of signals. A receiver may give a good performance when only one transmitter is in use, but a poor Q will vastly reduce the reception quality when several are used. This should be checked for when systems are being evaluated, and the testing of one channel on its own will not of course show up these kinds of problems.

4.10.6 Diversity reception

A technique known as 'spaced diversity' goes a good way towards combatting the above problems. In this system, two aerials feed two identical receivers for each radio channel. A circuit continuously monitors the signal strength being received by each receiver and automatically selects the one which is receiving the best signal (see Figure 4.21). When they are both receiving a good signal, the outputs of the two are mixed together. A crossfade is performed between the two as one RF signal fades and the other becomes strong.

The two aerials are placed some distance apart, in practice several metres gives good results, so that the multipath relationships between a given transmitter position and each aerial will be somewhat different. A dead spot for one aerial is therefore unlikely to coincide with a dead spot for the other one. A good diversity system overcomes many reception problems, and the considerable increase in reliability of performance is well worth the extra cost. The point at which diversity becomes desirable is when more than two radio microphones are to be used, although good performance from four channels in a non-diversity installation is by no means out of the question. Good radio microphones are very expensive, a single channel of a quality example costing over a thousand pounds today. Cheaper ones exist, but experience suggests that no radio microphone at all is vastly preferable to a cheap one.

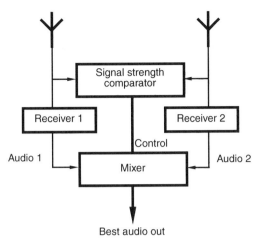

Figure 4.21 A diversity receiver incorporates two aerials spaced apart and two receivers. The signal strength from each aerial is used to determine which output will have the higher quality

Recommended further reading

AES (1979) *Microphones: An Anthology*. Audio Engineering Society
Bartlett, B. (1991) *Stereo Microphone Techniques*. Focal Press
Eargle, J. (2001) *The Microphone Book*. Focal Press
Gayford, M. (1994) ed., *Microphone Engineering Handbook*. Focal Press
Huber, D. (1988) *Microphone Manual – Design and Application*. Focal Press
Nisbett, A. (1989) *Use of Microphones*. Focal Press
Robertson, A. E. (1963) *Microphones*. Hayden

See also *General further reading* at the end of this book.

Chapter 5

Loudspeakers

A loudspeaker is a transducer which converts electrical energy into acoustical energy. A loudspeaker must therefore have a diaphragm of some sort which is capable of being energised in such a way that it vibrates to produce sound waves which are recognisably similar to the original sound from which the energising signal was derived. To ask a vibrating plastic loudspeaker cone to reproduce the sound of, say, a violin is to ask a great deal, and it is easy to take for granted how successful the best examples have become. Continuing development and refinement of the loudspeaker has brought about a more or less steady improvement in its general performance, but it is a sobering thought that one very rarely mistakes a sound coming from a speaker for the real sound itself, and that one nevertheless has to use these relatively imperfect devices to assess the results of one's work. Additionally, it is easy to hear significant differences between one model and another. Which is right? It is important not to tailor a sound to suit a particular favourite model. There are several principles by which loudspeakers can function, and the commonly employed ones will be briefly discussed.

A word or two must be said about the loudspeaker enclosure. The box can have as big an influence on the final sound of a speaker system as can the drivers themselves. At first sight surprising, this fact can be more readily appreciated when one remembers that a speaker cone radiates virtually the same amount of sound into the cabinet as out into the room. The same amount of acoustical energy that is radiated is therefore also being concentrated in the cabinet, and the sound escaping through the walls and also back out through the speaker cone has a considerable influence upon the final sound of the system.

5.1 The moving-coil loudspeaker

The moving-coil principle is by far the most widely used, as it can be implemented in very cheap transistor radio speakers, PA (Public Address) systems, and also top-quality studio monitors, plus all performance levels and applications in between. Figure 5.1 illustrates a cutaway view of a typical moving-coil loudspeaker. Such a device is also known as a drive unit or driver, as it is the

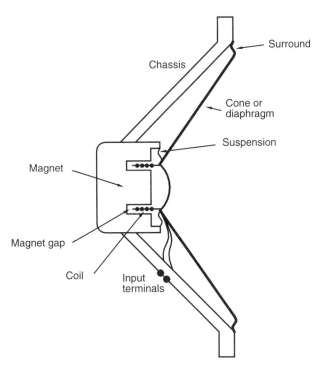

Figure 5.1 Cross-section through a typical moving-coil loudspeaker

component of a complete speaker system which actually produces the sound or 'drives' the air. Basically, the speaker consists of a powerful permanent magnet which has an annular gap to accommodate a coil of wire wound around a cylindrical former. This former is attached to the cone or diaphragm which is held in its rest position by a suspension system which usually consists of a compliant, corrugated, doped (impregnated) cloth material and a compliant surround around the edge of the cone which can be made of a type of rubber, doped fabric, or it can even be an extension of the cone itself, suitably treated to allow the required amount of movement of the cone.

The chassis usually consists either of pressed steel or a casting, the latter being particularly desirable where large heavy magnets are employed, since the very small clearance between the coil and the magnet gap demands a rigid structure to maintain the alignment, and a pressed steel chassis can sometimes be distorted if the loudspeaker is subject to rough handling as is inevitably the case with portable PA systems and the like. (A properly designed pressed steel chassis should not be overlooked though.) The cone itself can in principle be made of almost any material, common choices being paper pulp (as used in many PA speaker cones for its light weight, giving good efficiency), plastics of various types (as used in many hi-fi speaker cones due to the greater consistency achievable than with paper pulp, and the potentially lower coloration of the sound, usually at the expense of increased weight and therefore lower efficiency which is not crucially important in a domestic loudspeaker), and sometimes metal foil.

The principle of operation is based on the principle of electromagnetic transducers described in Fact File 3.1, and is the exact reverse of the process involved in the moving-coil microphone (see Fact File 4.1). The cone vibration sets up sound waves in the air which are an acoustic analogue of the electrical input signal. Thus in principle the moving-coil speaker is a very crude and simple device, but the results obtained today are incomparably superior to the original 1920s Kellog and Rice design. It is, however, a great tribute to those pioneers that the principle of operation of what is still today's most widely used type of speaker is still theirs.

5.2 Other loudspeaker types

The electrostatic loudspeaker first became commercially viable in the 1950s, and is described in Fact File 5.1. The electrostatic principle is far less commonly employed than is the moving coil, since it is difficult and expensive to manufacture and will not produce the sound levels available from moving-coil speakers. The sound quality of the best examples, such as the Quad ESL 63 pictured in Figure 5.2, is, however, rarely equalled by other types of speaker.

FACT FILE 5.1   Electrostatic loudspeaker – principles

The electrostatic loudspeaker's drive unit consists of a large, flat diaphragm of extremely light weight, placed between two rigid plates. The diagram shows a side view. There are

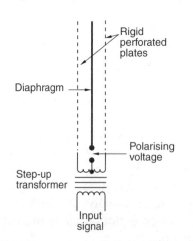

parallels between this loudspeaker and the capacitor microphone described in Chapter 4. The diaphragm has a very high resistance, and a DC polarising voltage in the kilovolt (kV) range is applied to the centre tap of the secondary of the input transformer, and charges the capacitor formed by the narrow gap between the diaphragm and the plates. The input signal appears (via the transformer) across the two rigid plates and thus modulates the electrostatic field. The diaphragm, being the other plate of the capacitor, thus experiences a force which alters according to the input signal. Being free to move within certain limits with respect to the two rigid plates, it thus vibrates to produce the sound.

There is no cabinet as such to house the speaker, and sound radiates through the holes of both plates. Sound therefore emerges equally from the rear and the front of the speaker, but not from the sides. Its polar response is therefore a figure-eight, similar to a figure-eight microphone with the rear lobe being out of phase with the front lobe.

Figure 5.2 The Quad ESL63 electrostatic loudspeaker. (Courtesy of Quad Electroacoustics Ltd)

Another technique in producing a panel-type speaker membrane has been to employ a light film on which is attached a series of conductive strips which serve as the equivalent of the coil of a moving-coil cone speaker. The panel is housed within a system of strong permanent magnets, and the drive signal is applied to the conductive strips. Gaps in the magnets allow the sound to radiate. Such systems tend to be large and expensive like the electrostatic models, but again very high-quality results are possible. In order to get adequate bass response and output level from such panel speakers the diaphragm needs to be of considerable area.

The ribbon loudspeaker principle has sometimes been employed in high-frequency applications ('tweeters') and has recently also been employed in large full-range models. Figure 5.3 illustrates the principle. A light corrugated aluminium ribbon, clamped at each end, is placed between two magnetic poles, one north, one south. The input signal is applied, via a step-down transformer, to each end of the ribbon. The alternating nature of the signal causes an alternating magnetic field around the ribbon, which behaves like a single turn of a coil in a moving-coil speaker. The magnets each side thus cause the ribbon to vibrate, producing sound waves. The impedance of the ribbon is often extremely low, and an amplifier cannot drive it directly. A transformer is therefore used which steps up the impedance of the ribbon. The ribbon itself produces a very low acoustic output and often has a horn in front of it to improve its acoustical matching with the air, giving a higher output for a given electrical input. Some ribbons are, however, very long – half a metre or more – and drive the air directly.

A recent panel-type of speaker is the so-called 'distributed mode loudspeaker' (DML), developed by the NXT company following the UK's Defence Evaluation

Figure 5.3 A ribbon loudspeaker mechanism

and Research Agency's discovery that certain lightweight composite panels used in military aircraft could act as efficient sound radiators (Figure 5.4). Its operating principle is the antithesis of conventional wisdom: whereas it is normal practice to strive for 'pistonic' motion of a cone driver or panel, the complete area of the radiating surface moving backwards and forwards as a whole with progressively smaller areas of the surface moving as frequency increases, the DML panel is deliberately made very flexible so that a multiplicity of bending modes or resonances, equally distributed in frequency, are set up across its surface. This creates a large number of small radiating areas which are virtually independent of each

Figure 5.4 DML loudspeaker. (Courtesy of New Transducers Ltd)

other, giving an uncorrelated set of signals but summing to give a resultant output. The panel is driven not across its whole area but usually at a strategically placed point by a moving-coil transducer. Because of the essentially random-phase nature of the radiating areas, the panel is claimed not to suffer from the higher-frequency beaming effects of conventional panels, and also there is not the global 180° out-of-phase radiation from the rear.

Further research into DML materials has brought the promise of integrated audio-visual panels, a single screen radiating both sound and vision simultaneously.

There are a few other types of speaker in use, but these are sufficiently uncommon for descriptions not to be merited in this brief outline of basic principles.

5.3 Mounting and loading drive units

5.3.1 'Infinite baffle' systems

The moving-coil speaker radiates sound equally in front of and to the rear of the diaphragm or cone. As the cone moves forwards it produces a compression of the air in front of it but a rarefaction behind it, and vice versa. The acoustical wave-forms are therefore 180° out of phase with each other and when they meet in the surrounding air they tend to cancel out, particularly at lower frequencies where diffraction around the cone occurs. A cabinet is therefore employed in which the drive unit sits, which has the job of preventing the sound radiated from the rear of the cone from reaching the open air. The simplest form of cabinet is the sealed box (commonly, but wrongly, known as the 'infinite baffle') which will usually have some sound-absorbing material inside it such as plastic foam or fibre wadding. A true 'infinite baffle' would be a very large flat piece of sheet material with a circular hole cut in the middle into which the drive unit would be mounted. Diffraction around the baffle would then only occur at frequencies below that where the wave-length approached the size of the baffle, and thus cancellation of the two mutually out-of-phase signals would not occur over most of the range, but for this to be effective at the lowest frequencies the baffle would have to measure at least 3 or 4 metres square. The only practical means of employing this type of loading is to mount the speaker in the dividing wall between two rooms, but this is rarely encountered for obvious reasons.

5.3.2 Bass reflex systems

Another form of loading is the bass reflex system, as shown in Figure 5.5. A tunnel, or port, is mounted in one of the walls of the cabinet, and the various parameters of cabinet internal volume, speaker cone weight, speaker cone suspension compliance, port dimensions, and thus mass of air inside the port are chosen so that at a specified low frequency the air inside the port will resonate, which reduces the movement of the speaker cone at that frequency. The port thus produces low-frequency output of its own, acting in combination with the driver. In this manner increased low-frequency output, increased efficiency, or a combination of the two

Figure 5.5 A ported bass reflex cabinet construction

can be achieved. However, it is worth remembering that at frequencies lower than the resonant frequency the driver is acoustically unloaded because the port now behaves simply as an open window. If extremely low frequencies from, say, mis-handled microphones or record player arms reach the speaker they will cause considerable excursion of the speaker cone which can cause damage. The air inside a closed box system, however, provides a mechanical supporting 'spring' right down to the lowest frequencies.

A device known as an auxiliary bass radiator (ABR) is occasionally used as an alternative to a reflex port, and takes the form of a further bass unit without its own magnet and coil. It is thus undriven electrically. Its cone mass acts in the same manner as the air plug in a reflex port, but has the advantage that mid-range frequencies are not emitted, resulting in lower coloration.

A further form of bass loading is described in Fact File 5.2.

FACT FILE 5.2 Transmission line system

large cabinet is required for its proper functioning.

A form of bass loading is the acoustic labyrinth or 'transmission line', as shown in the diagram. A large cabinet houses a folded tunnel the length of which is chosen so that resonance occurs at a specified low frequency. Above that frequency, the tunnel, which is filled or partially filled with acoustically absorbent material, gradually absorbs the rear-radiated sound energy along its length. At resonance, the opening, together with the air inside the tunnel, behaves like the port of a bass reflex design. An advantage of this type of loading is the very good bass extension achievable, but a

5.3.3 Horn loading

Horn loading is a technique commonly employed in large PA loudspeaker systems, as described in Fact File 5.3. Here, a horn is placed in front of the speaker diaphragm.

The so-called 'long-throw' horn tends to beam the sound over an included angle of perhaps 90° horizontally and 40° vertically. The acoustical energy is therefore concentrated principally in the forward direction, and this is one reason for the horn's high efficiency. The sound is beamed forwards towards the rear of the hall with relatively little sound reaching the side walls. The 'constant directivity' horn aims to achieve a consistent spread of sound throughout the whole of its working frequency range, and this is usually achieved at the expense of an uneven frequency response. Special equalisation is therefore often applied to compensate for this.

The long-throw horn does not do much for those members of an audience who are close to the stage between the speaker stacks, and an acoustic lens is often employed, which, as its name suggests, diffracts the sound, such that the higher frequencies are spread out over a wider angle to give good coverage at the front.

FACT FILE

5.3

Horn loudspeaker – principles

A horn is an acoustic transformer, that is it helps to match the air impedance at the throat of the horn (the throat is where the speaker drive unit is) with the air impedance at the mouth. Improved acoustic efficiency is therefore achieved, and for a given electrical input a horn can increase the acoustical output of a driver by 10 dB or more compared with that driver mounted in a conventional cabinet. A horn functions over a relatively limited frequency range, and therefore relatively small horns are used for the high frequencies, larger ones for upper mid frequencies, and so on. This is very worthwhile where high sound levels need to be generated in large halls, rock concerts and open-air events.

Each design of horn has a natural lower cut-off frequency which is the frequency below which it ceases to load the driver acoustically. Very large horns indeed are needed to reproduce low frequencies, and one technique has been to fold the horn up by building it into a more conventional-looking cabinet. The horn principle is rarely employed at bass frequencies due to the necessarily large size. It is, however, frequently employed at mid and high frequencies, but the higher coloration of the sound it produces tends to rule it out for hi-fi and studio monitoring use other than at high frequencies if high sound levels are required. Horns tend to be more directional than conventional speakers, and this has further advantages in PA applications.

Front view   Side view

Figure 5.6 An example of an acoustic lens

Figure 5.6 shows a typical acoustic lens. It consists of a number of metal plates which are shaped and positioned with respect to each other in such a manner as to cause outward diffraction of the high frequencies. The downward slope of the plates is incidental to the design requirements and it is not incorporated to project the sound downwards. Because the available acoustic output is spread out over a wider area than is the case with the long-throw horn, the on-axis sensitivity tends to be lower.

The high efficiency of the horn has also been much exploited in those PA applications which do not require high sound quality, and their use for outdoor events such as fêtes, football matches and the like, as well as on railway station platforms, will have been noticed. Often, a contrivance known as a re-entrant horn is used, as shown in Figure 5.7. It can be seen that the horn has been effectively cut in half, and the half which carries the driver is turned around and placed inside the bell of the other. Quite a long horn is therefore accommodated in a compact structure, and this method of construction is particularly applicable to hand-held loudhailers.

The high-frequency horn is driven not by a cone speaker but by a 'compression driver' which consists of a dome-shaped diaphragm usually with a diameter of 1 or 2 inches (2.5 or 5 cm). It resembles a hi-fi dome tweeter but with a flange or thread in front of the dome for fixing on to the horn. The compression driver can easily be damaged if it is driven by frequencies below the cut-off frequency of the horn it is looking into.

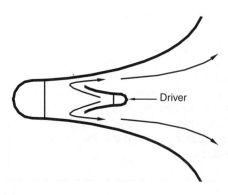

Driver

Figure 5.7 A re-entrant horn

5.4 Complete loudspeaker systems

5.4.1 Two-way systems

It is a fact of life that no single drive unit can adequately reproduce the complete frequency spectrum from, say, 30 Hz to 20 kHz. Bass frequencies require large drivers with relatively high cone excursions so that adequate areas of air can be set in motion. Conversely, the same cone could not be expected to vibrate at 15 kHz–15 000 times a second to reproduce very high frequencies. A double bass is much larger than a flute, and the strings of a piano which produce the low notes are much fatter and longer than those for the high notes.

The most widely used technique for reproducing virtually the whole frequency spectrum is the so-called two-way speaker system, which is employed at many quality levels from fairly cheap audio packages to very high-quality studio monitors. It consists of a bass/mid driver which handles frequencies up to around 3 kHz, and a high-frequency unit or 'tweeter' which reproduces frequencies from 3 kHz to 20 kHz or more. Figure 5.8 shows a cutaway view of a tweeter. Typically of around 1 inch (2.5 cm) in diameter, the dome is attached to a coil in the same way that a cone is in a bass/mid driver. The dome can be made of various materials, 'soft' or 'hard', and metal domes are also frequently employed. A bass/mid driver cannot adequately reproduce high frequencies as has been said. Similarly, such a small dome tweeter would actually be damaged if bass frequencies were fed to it; thus a crossover network is required to feed each drive unit with frequencies in the correct range, as described in Fact File 5.4.

In a basic system the woofer would typically be of around 8 inches (20 cm) in diameter for a medium-sized domestic speaker, mounted in a cabinet having several cubic feet internal volume. Tweeters are usually sealed at the rear, and therefore they are simply mounted in an appropriate hole cut in the front baffle of the enclosure. This type of speaker is commonly encountered at the cheaper end of the price range, but its simplicity makes it well worth study since it nevertheless incorporates the basic features of many much more costly designs. The latter differ in that they make use of more advanced and sophisticated drive units, higher-quality cabinet materials and constructional techniques, and a rather more sophis-

Figure 5.8 Cross-section through a typical dome tweeter

FACT FILE
5.4

A basic crossover network

A frequency-dividing network or 'crossover' is fitted into the speaker enclosure which divides the incoming signal into high frequencies (above about 3 kHz) and lower frequencies, sending the latter to the bass/mid unit or 'woofer' and the former to the tweeter. A simple example of the principle involved is illustrated in the diagram. In practical designs additional account should be taken of the fact that speaker drive units are not pure resistances.

The tweeter is fed by a capacitor. A capacitor has an impedance which is inversely proportional to frequency, that is at high frequencies its impedance is very low and at low frequencies its impedance is relatively high. The typical impedance of a tweeter is 8 ohms, and so for signals below the example of 3 kHz (the 'crossover frequency') a value of capacitor is chosen which exhibits an impedance of 8 ohms also at 3 kHz, and due to the nature of the voltage/current phase relationship of the

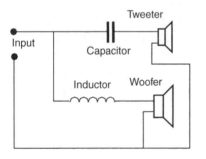

signal across a capacitor the power delivered to the tweeter is attenuated by 3 dB at that frequency. It then falls at a rate of 6 dB per octave thereafter (i.e.: the tweeter's output is 9 dB down at 1.5 kHz, 15 dB down at 750 Hz and so on) thus protecting the tweeter from lower frequencies. The formula which contains the value of the capacitor for the chosen 3 kHz frequency is:

$$f = 1/(2\pi RC)$$

where R is the resistance of the tweeter, and C is the value of the capacitor in farads.

The capacitor value will more conveniently be expressed in microfarads (millionths of a farad) and so the final formula becomes:

$$C = 159\,155 \div (8 \text{ ohms} \times 3000\,\text{Hz})$$
$$= 6.7\mu F$$

Turning now to the woofer, it will be seen that an inductor is placed in series with it. An inductor has an impedance which rises with frequency; therefore, a value is chosen that gives an impedance value similar to that of the woofer at the chosen crossover frequency. Again, the typical impedance of a woofer is 8 ohms. The formula which contains the value of the inductor is:

$$f = R/(2\pi L)$$

where L = inductance in henrys, R = speaker resistance, f = crossover frequency. The millihenry (one-thousandth of a henry, mH) is more appropriate, so this gives:

$$L = 8000 \div (2\pi \times 3000) = 0.42\,\text{mH}$$

ticated crossover which usually incorporates both inductors and capacitors in the treble and bass sections as well as resistors which together give much steeper filter slopes than our 6 dB/octave example. Also, the overall frequency response can be adjusted by the crossover to take account of, say, a woofer which gives more acoustic output in the mid range than in the bass: some attenuation of the mid range can give a flatter and better-balanced frequency response.

5.4.2 Three-way systems

Numerous three-way loudspeaker systems have also appeared where a separate mid-range driver is incorporated along with additional crossover components to restrict the frequencies feeding it to the mid range, for example between 400 Hz and 4 kHz. It is an attractive technique due to the fact that the important mid frequencies where much of the detail of music and speech resides are reproduced by a dedicated driver designed specially for that job. But the increased cost and complexity does not always bring about a proportional advance in sound quality.

5.5 Active loudspeakers

So far, only 'passive' loudspeakers have been discussed, so named because simple passive components – resistors, capacitors and inductors – are used to divide the frequency range between the various drivers. 'Active' loudspeakers are also encountered, in which the frequency range is divided by active electronic circuitry at line level, after which each frequency band is sent to a separate power amplifier and thence to the appropriate speaker drive unit. The expense and complexity of active systems has tended to restrict the active technique to high-powered professional PA applications where four-, five- and even six-way systems are employed, and to professional studio monitoring speakers, such as the Rogers LS5/8 system pictured in Figure 5.9. Active speakers are still comparatively rare in domestic audio.

Each driver has its own power amplifier, which of course immediately increases the cost and complexity of the speaker system, but the advantages include: lower distortion (due to the fact that the signal is now being split at line level, where only a volt or so at negligible current is involved, as compared with the tens of volts and several amps that passive crossovers have to deal with); greater system-design flexibility due to the fact that almost any combination of speaker components can be used because their differing sensitivities, impedances and power requirements can be compensated for by adjusting the gains of the separate power amplifiers or electronic crossover outputs; better control of final frequency response, since it is far easier to incorporate precise compensating circuitry into an electronic crossover design than is the case with a passive crossover; better clarity of sound and firmer bass simply due to the lack of passive components between power amplifiers and drivers; and an improvement in power amplifier performance due to the fact that each amplifier now handles a relatively restricted band of frequencies.

In active systems amplifiers can be better matched to loudspeakers, and the system can be designed as a whole, without the problems which arise when an unpredictable load is attached to a power amplifier. In passive systems, the designer has little or no control over which type of loudspeaker is connected to which type of amplifier, and thus the design of each is usually a compromise between adaptability and performance. Some active speakers have the electronics built into the speaker cabinet which simplifies installation.

Figure 5.9 Rogers LS5/8 high-quality active studio loudspeaker. (Courtesy of Swisstone Electronics Ltd)

5.6 Subwoofers

Good bass response from a loudspeaker requires a large internal cabinet volume so that the resonant frequency of the system can be correspondingly low, the response of a given speaker normally falling away below this resonant point. This implies the use of two large enclosures which are likely to be visually obtrusive in a living room for instance. A way around this problem is to incorporate a so-called 'subwoofer' system. A separate speaker cabinet is employed which handles only the deep bass frequencies, and it is usually driven by its own power amplifier. The signal to drive the power amp comes from an electronic crossover which subtracts the low bass frequencies from the feed to the main stereo amplifier and speakers, and sends the mono sum of the deep bass to the subwoofer system.

Freed from the need to reproduce deep bass, the main stereo speakers can now be small high-quality systems; the subwoofer can be positioned anywhere in the room according to the manufacturers of such systems since it only radiates frequencies below around 100 Hz or so, where sources tend to radiate only omnidirectionally anyway. Degradation of the stereo image has sometimes been noted when the subwoofer is a long way from the stereo pair, and a position close to one of these is probably a good idea.

Subwoofers are also employed in concert and theatre sound systems. It is difficult to achieve both high efficiency and a good bass response at the same time from a speaker intended for public address use, and quite large and loud examples often have little output below 70 Hz or so. Subwoofer systems, if properly integrated into the system as a whole, can make a large difference to the weight and scale of live sound.

5.7 Loudspeaker performance

5.7.1 Impedance

The great majority of loudspeaker drive units and systems are labelled 'Impedance = 8 ohms'. This is, however, a nominal figure, the impedance in practice varying widely with frequency (see section 1.8). A speaker system may indeed have an 8 ohm impedance at, say, 150 Hz, but at 50 Hz it may well be 30 ohms, and at 10 kHz it could be 4 ohms. Figure 5.10 shows the impedance plot of a typical two-way, sealed box, domestic hi-fi speaker.

The steep rise in impedance at a certain low-frequency is indicative of the low-frequency resonance of the system. Other undulations are indicative of the reactive nature of the speaker due to capacitive and inductive elements in the crossover components and the drive units themselves. Also, the driver/box interface has an effect, the most obvious place being at the already-mentioned LF resonant frequency.

Figure 5.11 shows an impedance plot of a bass reflex design. Here we see the characteristic 'double hump' at the bass end. The high peak at about 70 Hz is the bass driver/cabinet resonance point. The trough at about 40 Hz is the resonant frequency of the bass reflex port where maximum LF sound energy is radiated from the port itself and minimum energy is radiated from the bass driver. The low peak at about 20 Hz is virtually equal to the free-air resonance of the bass

Figure 5.10 Impedance plot of a typical two-way sealed-box domestic loudspeaker

Figure 5.11 Impedance plot of a typical bass reflex design

driver itself because at very low frequencies the driver is acoustically unloaded by the cabinet due to the presence of the port opening. A transmission-line design exhibits a similar impedance characteristic.

The DC resistance of an 8 ohm driver or speaker system tends to lie around 7 ohms, and this simple measurement is a good guide if the impedance of an un-labelled speaker is to be estimated. Other impedances encountered include 15 ohm and 4 ohm models. The 4 ohm speakers are harder to drive because for a given amplifier output voltage they draw twice as much current. The 15 ohm speaker is an easy load, but its higher impedance means that less current is drawn from the amplifier and so the power (volts × amps) driving the speaker will be correspondingly less. So a power amplifier may not be able to deliver its full rated power into this higher impedance. Thus 8 ohms has become virtually standard, and competently designed amplifiers can normally be expected to drive competently designed speakers. Higher-powered professional power amplifiers can also be expected to drive two 8 ohm speakers in parallel, giving a resultant nominal impedance of 4 ohms.

5.7.2 Sensitivity

A loudspeaker's sensitivity is a measure of how efficiently it converts electrical sound energy into acoustical sound energy. The principles are described in Fact File 5.5. Loudspeakers are very inefficient devices indeed. A typical high-quality domestic speaker system has an efficiency of less than 1 per cent, and therefore if 20 watts is fed into it the resulting acoustic output will be less than 0.2 acoustical watts. Almost all of the rest of the power is dissipated as heat in the voice coils of the drivers. Horn-loaded systems can achieve a much better efficiency, figures of around 10 per cent being typical. An efficiency figure is not in itself a very helpful thing to know, parameters such as sensitivity and power handling being much more useful. But it is as well to be aware that most of the power fed into a speaker has to be dissipated as heat, and prolonged high-level drive causes high voice-coil temperatures.

FACT FILE

5.5

Loudspeaker sensitivity

Sensitivity is defined as the acoustic sound output for a given voltage input. The standard conditions are an input of 2.83 volts (corresponding to 1 watt into 8 ohms) and an acoustic SPL measurement at a distance of 1 metre in front of the speaker. The input signal is pink noise which contains equal sound energy per octave (see section 1.6). A single frequency may correspond with a peak or dip in the speaker's response, leading to an inaccurate overall assessment. For example, a domestic speaker may have a quoted sensitivity of 86 dB W^{-1}, that is 1 watt of input will produce 86 dB output at 1 metre.

Sensitivities of various speakers differ quite widely and this is not an indication of the sound quality. A high-level professional monitor speaker may have a sensitivity of 98 dB W^{-1} suggesting that it will be very much louder than its domestic cousin, and this will indeed be the case. High-frequency PA horns sometimes achieve a value of 118 dB for just 1 watt input. Sensitivity is thus a useful guide when considering which types of speaker to choose for a given application. A small speaker having a quoted sensitivity of 84 dB W^{-1} and 40 watts power handling will not fill a large hall with sound. The high sound level capability of large professional models will be wasted in a living room.

It has been suggested that sensitivity is not an indication of quality. In fact, it is often found that lower-sensitivity models tend to produce a better sound. This is because refinements in sound quality usually come at the expense of reduced acoustical output for a given input, and PA speaker designers generally have to sacrifice absolute sound quality in order to achieve the high sensitivity and sound output levels necessary for the intended purpose.

5.7.3 Distortion

Distortion in loudspeaker systems is generally an order of magnitude or more higher than in other audio equipment. Much of it tends to be second-harmonic distortion (see section A1.3) whereby the loudspeaker will add frequencies an octave above the legitimate input signal. This is especially manifest at low frequencies where speaker diaphragms have to move comparatively large distances to reproduce them. When output levels of greater than 90 dB for domestic systems and 105 dB or so for high-sensitivity systems are being produced, low-frequency distortion of around 10 per cent is quite common, this consisting mainly of second-harmonic and partly of third-harmonic distortion.

At mid and high frequencies distortion is generally below 1 per cent, this being confined to relatively narrow bands of frequencies which correspond to areas such as crossover frequencies or driver resonances. Fortunately, distortion of this magnitude in a speaker does not indicate impending damage, and it is just that these transducers are inherently non-linear to this extent. Much of the distortion is at low frequencies where the ear is comparatively insensitive to it, and also the predominantly second-harmonic character is subjectively innocuous to the ear. Distortion levels of 10–15 per cent are fairly common in the throats of high-frequency horns.

5.7.4 Frequency response

The frequency response of a speaker also indicates how linear it is. Ideally, a speaker would respond equally well to all frequencies, producing a smooth 'flat' output response to an input signal sweeping from the lowest to the highest frequencies at a constant amplitude. In practice, only the largest speakers produce a significant output down to 20 Hz or so, but even the smallest speaker systems can respond to 20 kHz. The 'flatness' of the response, i.e.: how evenly a speaker responds to all frequencies, is a rather different matter. High-quality systems achieve a response that is within 6 dB of the 1 kHz level from 80 Hz to 20 kHz, and such a frequency response might look like Figure 5.12(a). Figure 5.12(b) is an example of a rather lower-quality speaker which has a considerably more ragged response and an earlier bass roll-off.

The frequency response can be measured using a variety of different methods, some manufacturers taking readings under the most favourable conditions to hide inadequacies. Others simply quote something like '±3 dB from 100 Hz to 15 kHz'. This does at least give a fairly good idea of the smoothness of the response. These specifications do not, however, tell you how a system will sound, and they must be used only as a guide. They tell nothing of coloration levels, or the ability to reproduce good stereo depth, or the smoothness of the treble, or the 'tightness' of the bass.

5.7.5 Power handling

Power handling is the number of watts a speaker can handle before unacceptable amounts of distortion ensue. It goes hand in hand with sensitivity in determining the maximum sound level a speaker can deliver. For example, a domestic speaker may be rated at 30 watts and have a sensitivity of 86 dB W^{-1}. The decibel increase of 30 watts over 1 watt is given by:

$$\text{dB increase} = 10 \log 30 \approx 15 \, \text{dB}$$

Therefore, the maximum output level of this speaker is $86 + 15 = 101$ dB at 1 metre for 30 watts input. This is loud, and quite adequate for domestic use. Consider now a PA speaker with a quoted sensitivity of 99 dB W^{-1}. 30 watts input now produces $99 + 15 = 114$ dB, some 13 dB more than with the previous example for the same power input. To get 114 dB out of the 86 dB W^{-1} speaker one would need to drive it with no less than 500 watts, which would of course be way beyond its capabilities. This dramatically demonstrates the need to be aware of the implications of sensitivity and power handling.

(a)

(b)

Figure 5.12 Typical loudspeaker frequency response plots. (a) A high-quality unit. (b) A lower-quality unit

A 30 watt speaker can, however, safely be driven even by a 500 watt amplifier providing that sensible precautions are taken with respect to how hard the amplifier is driven. Occasional peaks of more than 30 watts will be quite happily tolerated; it is sustained high-level drive which will damage a speaker. It is perfectly all right to drive a high power speaker with a low-power amplifier, but care must be taken that the latter is not overdriven otherwise the harsh distortion products can easily damage high-frequency horns and tweeters even though the speaker system may have quoted power handling well in excess of the amplifier. The golden rule is to listen carefully. If the sound is clean and unstressed, all will be well.

5.7.6 Directivity

Directivity, or dispersion, describes the angle of coverage of a loudspeaker's output. Very low frequencies radiated from a speaker are effectively omnidirectional, because the wavelength of the sound is large compared with the dimensions of the speaker and its enclosure, and efficient diffraction of sound around the latter is the result. As the frequency increases, wavelengths become comparable to the dimensions of the speaker's front surface, diffraction is curtailed, and the speaker's output is predominantly in the forwards direction. At still higher frequencies, an even narrower dispersion angle results as a further effect comes into play: off-axis phase cancellation. If one listens, say, 30° off-axis from the front of a speaker, a given upper frequency (with a short wavelength) arrives which has been radiated both from the closest side of the speaker cone to the listener and from the furthest side of the cone, and these two sound sources will not therefore be in phase with each other because of the different distances they are away from one another. Phase cancellation therefore occurs, perceived output level falls, and the effect becomes more severe as frequencies increase. The phenomenon is mitigated by designing for progressively smaller radiating areas of the speaker cone to be utilised as the frequency increases, finally crossing over to a tweeter of very small dimensions. By these means, fairly even dispersion of sound, at least in the mid and lower treble regions, can be maintained.

Various other methods have been used to control directivity (the acoustic lens has been covered) and one or two will be described. Low frequencies, which are normally omnidirectional, have been given a cardioid-like dispersion pattern by mounting large speaker drivers on essentially open baffles which by themselves give a figure-of-eight polar response, the output falling with falling frequency. To these was added a considerable amount of absorbent material to the rear, and together with appropriate bass boost to flatten the frequency response of the speakers, predominantly forwards-radiation of low frequencies was achieved. A more elegant technique has been to mount essentially open-baffle speakers (the rear radiation therefore being 180° out of phase with the front producing a figure-of-eight polar pattern, and with bass boost applied to flatten the frequency response) adjacent to closed-box omnidirectional speakers. Their combined acoustical outputs thereby produce a cardioid dispersion pattern, useful for throwing low frequencies forwards into an auditorium rather than across a stage where low-frequency feedback with microphones can be a problem.

Another fascinating technique, introduced by Philips in 1983, is the Bessel Array. It was developed to counteract the beaming effects of multiple-speaker systems. Essentially it makes use of Bessel coefficients to specify phase relationships and output level requirements from each of a horizontal row of speakers necessary to obtain an overall dispersion pattern from the row which is the same as one speaker on its own. Normally, path-length differences between off-axis listeners and the various speaker drivers result in phase cancellations and consequent loss of level, particularly in the upper frequency range. For a horizontal five-speaker row, labelled A, B, C, D and E, the Bessel function gives:

A:B:C:D:E = 1:2:2 − 2:1

In other words, speakers A and E are required to draw half the current of speakers B, C and D; and speaker D must be connected out of phase. A practical implementation would be to connect speakers A and E in series, with speakers B, C and D each connected straight across the system's input terminals but with D wired out of phase. The speaker drivers are mounted side by side very close together to give good results across the frequency range.

For a seven-speaker row, the Bessel function gives:

A:B:C:D:E:F:G = 1:2:2:0: − 2:2: − 1

Speaker D can therefore be omitted, but a space in the row must be left in its position so as to preserve the correct distance relationships between the others.

Both horizontal and vertical rows of speakers can be combined into a square arrangement so that an array of, for example, 25 speakers, together having potentially very high power handling and output level capability, can, however, give the same dispersion characteristics of one speaker on its own. The amplitude and phase relationships necessary in such an array are given by the numbers in the circles representing the speakers in Figure 5.13.

It is worth mentioning that the same technique can also be applied to microphones, offering potential for a high output, very low noise array whilst still maintaining a good polar response.

A highly directional speaker incorporating a parabolic reflector of about 1.3 metres in diameter has been developed by the Mayer loudspeaker company as their type SB-1. Designed to work between 500 Hz and 15 kHz, the system comprises an outrigger supporting a small horn and compression driver at the focus of the dish which fires into it, and a small hole at the dish's centre admits sound from a 12-inch cone driver. Claimed dispersion (−6 dB points) is 10° vertical and 10° horizontal, and maximum peak output at 100 metres distance is 110 dB.

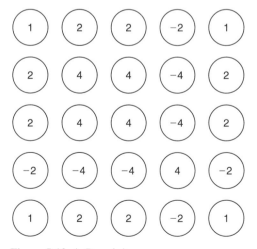

Figure 5.13 A Bessel Array

5.8 Setting up loudspeakers

5.8.1 Phase

Phase is a very important consideration when wiring up speakers. A positive-going voltage will cause a speaker cone to move in a certain direction, which is usually forwards, although at least two American and two British manufacturers have unfortunately adopted the opposite convention. It is essential that both speakers of a stereo pair, or all of the speakers of a particular type in a complete sound rig, are 'in phase', that is all the cones are moving in the same direction at any one time when an identical signal is applied. If two stereo speakers are wired up out of phase, this produces vague 'swimming' sound images in stereo, and cancellation of bass frequencies. This can easily be demonstrated by temporarily connecting *one* speaker in opposite phase and then listening to a mono signal source – speech from the radio is a good test. The voice will seem to come from nowhere in particular, and small movements of the head produce sudden large shifts in apparent sound source location. Now reconnect the speakers in phase and the voice will come from a definite position in between the speakers. It will also be quite stable when you move a few feet to the left or to the right.

Occasionally it is not possible to check the phase of an unknown speaker by listening. An alternative method is to connect a 1.5 V battery across the input terminals and watch which way the cone of the bass driver moves. If it moves forwards, then the positive terminal of the battery corresponds to the positive input terminal of the speaker. If it moves backwards as the battery is connected, then the positive terminal of the battery is touching the negative input terminal of the speaker. The terminals can then be labelled + and −.

5.8.2 Positioning

Loudspeaker positioning has a significant effect upon the performance. In smaller spaces such as control rooms and living rooms the speakers are likely to be positioned close to the walls, and 'room gain' comes into effect whereby the low frequencies are reinforced. This happens because at these frequencies the speaker is virtually omnidirectional, i.e.: it radiates sound equally in all directions. The rear- and side-radiated sound is therefore reflected off the walls and back into the room to add more bass power. As we move higher in frequency, a point is reached whereby the wavelength of lower mid frequencies starts to become comparable with the distance between the speaker and a nearby wall. At half wavelengths the reflected sound is out of phase with the original sound from the speaker and some cancellation of sound is caused. Additionally, high-frequency 'splash' is often caused by nearby hard surfaces, this often being the case in control rooms where large consoles, tape machines, outboard processing gear, etc. can be in close proximity to the speakers. Phantom stereo images can thus be generated which distort the perspective of the legitimate sound. A loudspeaker which has an encouragingly flat frequency response can therefore often sound far from neutral in a real listening environment. It is therefore essential to give consideration to loudspeaker place-ment, and a position such that the speakers are at head height when viewed from the listening position (high-frequency dispersion is much narrower than at lower frequencies, and therefore a speaker should be listened to on axis) and also away from room boundaries will give the most tonally accurate sound.

Some speakers, however, are designed to give of their best when mounted directly against a wall, the gain in bass response from such a position being allowed for in the design. A number of professional studio monitors are designed to be let into a wall such that their drivers are then level with the wall's surface. The manufacturer's instructions should be heeded, in conjunction with experimentation and listening tests. Speech is a good test signal. Male speech is good for revealing boominess in a speaker, and female speech reveals treble splash from hard-surfaced objects nearby. Electronic music is probably the least helpful since it has no real-life reference by which to assess the reproduced sound. It is worth emphasising that the speaker is the means by which the results of previous endeavour are judged, and that time spent in both choosing and siting is time well spent.

Speakers are of course used in audio–visual work, and one frequently finds that it is desirable to place a speaker next to a video monitor screen. But the magnetic field from the magnets can affect the picture quality by pulling the internal electron beams off course. Some speakers are specially magnetically screened so as to avoid this.

Loudspeaker positioning issues affecting two-channel stereo and surround sound reproduction are covered in greater detail in Chapters 16 and 17.

Recommended further reading

Borwick, J. (2001) ed. *Loudspeaker and Headphone Handbook*, 3rd Edition. Focal Press

Colloms, M. (1997) *High Performance Loudspeakers*, 5th Edition. Pentech Press/ Wiley

Earl, J. (1973) *Pickups and Loudspeakers*. Fountain Press

Gayford, M. (1970) *Loudspeakers*. Newnes–Butterworth

Sinclair, I. (1993) ed. *Audio and Hi-Fi Handbook*, 2nd Edition; Chapter 15: *Loudspeakers*, by Stan Kelly. Butterworth-Heinemann

Mixers 1

In its simplest form an audio mixer combines several incoming signals into a single output signal. This cannot be achieved simply by connecting all the incoming signals in parallel and then feeding them into a single input because they may influence each other. The signals need to be isolated from each other. Individual control of at least the level of each signal is also required.

In practice, mixers also do rather more things than simply mix. They can provide phantom power for capacitor microphones (see section 4.3); pan control (whereby each signal can be placed in any desired position in a stereo image); filtering and equalisation; routing facilities; and monitoring facilities, whereby one of a number of sources can be routed to a pair of loudspeakers for listening, often without affecting the mixer's main output.

6.1 A simple six-channel mixer

6.1.1 Overview

By way of example, a simple six-channel mixer will be considered, having six inputs and two outputs (for stereo). Figure 6.1 illustrates such a notional six-into-two mixer with basic facilities. It also illustrates the back panel. The inputs illustrated are via XLR-type three-pin latching connectors, and are of a balanced configuration. Separate inputs are provided for microphone and line level signals, although it is possible to encounter systems which simply use one socket switchable to be either mic or line. Many cheap mixers have unbalanced inputs via quarter-inch jack sockets, or even 'phono' sockets such as are found on hi-fi amplifiers. Some mixers employ balanced XLR inputs for microphones, but unbalanced jack or phono inputs for line level signals, since the higher-level line signal is less susceptible to noise and interference, and will probably have travelled a shorter distance.

On some larger mixers a relatively small number of multipin connectors are provided, and multicore cables link these to a large jackfield which consists of rows of jack sockets mounted in a rack, each being individually labelled. All inputs and outputs will appear on this jackfield, and patch cords of a metre or so in length

110

Figure 6.1 Front panel and rear connectors of a typical simple six-channel mixer

with GPO-type jack plugs at each end enable the inputs and outputs to be interfaced with other equipment and tie-lines in any appropriate combination. (The jackfield is more fully described in sections 6.4.9 and 12.12.)

The outputs are also on three-pin XLR-type connectors. The convention for these audio connections is that inputs have sockets or holes, outputs have pins. This means that the pins of the connectors 'point' in the direction of the signal, and therefore one should never be confused as to which connectors are inputs and which are outputs. The microphone inputs also have a switch each for supplying 48 V phantom power to the microphones if required. Sometimes this is found on the input module itself, or sometimes on the power supply, switching 48 V for all the inputs at once.

6.1.2 Input channels

All the input channels in this example are identical, and so only one will be described. The first control in the signal chain is input gain or sensitivity. This control adjusts the degree of amplification provided by the input amplifier, and is often labelled in decibels, either in detented steps or continuously variable. Inputs are normally switchable between mic and line. In 'mic' position, depending on the output level of the microphone connected to the channel (see section 4.8), the input gain is adjusted to raise the signal to a suitable line level, and up to 80 dB or so of gain is usually available here (see section 6.4.1, below). In 'line' position little amplification is used and the gain control normally provides adjustment either side of unity gain (0 dB), perhaps ±20 dB either way, allowing the connection of high-level signals from such devices as CD players, tape machines and musical keyboards.

The equalisation or EQ section which follows (see section 6.4.4) has only two bands in this example – treble and bass – and these provide boost and cut of around ±12 dB over broad low and high-frequency bands (e.g.: centred on 100 Hz and 10 kHz). This section can be used like the tone controls on a hi-fi amplifier to adjust the spectral balance of the signal. The fader controls the overall level of the channel, usually offering a small amount of gain (up to 12 dB) and infinite attenuation. The law of the fader is specially designed for audio purposes (see Fact File 6.1). The pan control divides the mono input signal between left and right mixer outputs, in order to position the signal in a virtual stereo sound stage (see Fact File 6.2).

6.1.3 Output section

The two main output faders (left and right) control the overall level of all the channel signals which have been summed on the left and right mix buses, as shown in the block diagram (Figure 6.2). The outputs of these faders (often called the group outputs) feed the main output connectors on the rear panel, and an internal feed is taken from the main outputs to the monitor selector. The monitor selector on this simple example can be switched to route either the main outputs or the PFL bus (see Fact File 6.3) to the loudspeakers. The monitor gain control adjusts the loudspeaker output level *without affecting* the main line output level, but of course any changes made to the main fader gain will affect the monitor output.

The slate facility on this example allows for a small microphone mounted in the mixer to be routed to the main outputs, so that comments from the engineer (such as take numbers) can be recorded on a tape machine connected to the main outputs. A rotary control adjusts the slate level.

FACT FILE

6.1

Fader
facts

Fader law

Channel and output faders, and also rotary level controls, can have one of two laws: linear or logarithmic (the latter sometimes also termed 'audio taper'). A linear law means that a control will alter the level of a signal (or the degree of cut and boost in a tone control circuit) in a linear fashion: that is, a control setting midway between maximum and minimum will attenuate a signal by half its voltage, i.e.: −6 dB. But this is not a very good law for an audio level control because a 6 dB drop in level does not produce a subjective halving of loudness. Additionally, the rest of the scaling (−10 dB, −20 dB, −30 dB and so on) has to be accommodated within the lower half of the control's travel, so the top half gives control over a mere 6 dB, the bottom half all the rest.

For level control, therefore, the logarithmic or 'log' law is used whereby a non-linear voltage relationship is employed in order to produce an approximately even spacing when the control is calibrated in decibels, since the decibel scale is logarithmic. A log fader will therefore attenuate a signal by 10 dB at a point approximately a quarter of the way down from the top of its travel. Equal dB increments will then be fairly evenly spaced below this point. A rotary log pot ('pot' is short for potentiometer) will have its maximum level usually set at the 5 o'clock position and, the −10 dB point will be around the 2 o'clock

position. An even subjective attenuation of volume level is therefore produced by the log law as the control is gradually turned down. A linear law causes very little to happen subjectively until one reaches the lowest quarter of the range, at which point most of the effect takes place.

The linear law is, however, used where a symmetrical effect is required about the central position; for example, the cut and boost control of a tone control section will have a central zero position about which the signal is cut and boosted to an equal extent either side of this.

Electrical quality

There are two types of electrical track in use, along which a conductive 'wiper' runs as the fader is moved to vary its resistance. One type of track consists of a carbon element, and is cheap to manufacture. The quality of such carbon tracks is, however, not very consistent and the 'feel' of the fader is often scrapy or grainy, and as it is moved the sound tends to jump from one level to another in a series of tiny stages rather than in a continuous manner. The carbon track wears out rather quickly, and can become unreliable.

The second type employs a conductive plastic track. Here, an electrically conductive material is diffused into a strip of plastic in a controlled manner to give the desired resistance value and law (linear or log). Much more expensive than the carbon track, the conductive plastic track gives smooth, continuous operation and maintains this standard over a long period of time. It is the only serious choice for professional-quality equipment.

6.1.4 Miscellaneous features

Professional-quality microphones have an output impedance of around 200 ohms, and the balanced microphone inputs will have an input impedance of between 1000 and 2000 ohms ('2 kΩ', k = thousand). The outputs should have an impedance of around 200 ohms or lower. The headphone output impedance will typically be 100 ohms or so. Small mixers usually have a separate power supply which plugs into the

FACT FILE 6.2   Pan control

The pan control on a mixer is used for positioning a signal somewhere between left and right in the stereo mix image. It does this by splitting a *single* signal from the output of a fader into two signals (left and right), setting the position in the image by varying the level difference between left and right channels. It is thus not the same as the balance control on a stereo amplifier, which takes in a *stereo* signal and simply varies the relative levels between the two channels. A typical pan-pot law would look similar to that shown in the diagram, and ensures a roughly constant perceived level of sound as the source is panned from left to right in stereo. The output of the pan-pot usually feeds the left and right channels of the stereo mix bus (the two main summation lines which combine the outputs of all channels on the mixer), although on mixers with more than two mix buses the pan-pot's output may be

switched to pan between any pair of buses, or perhaps simply between odd and even groups (see Fact File 6.4).

On some older consoles, four way routing is provided to a quadraphonic mix bus, with a left–right pot and a front–back pot. These are rare now. Many stereo pan-pots use a dual-gang variable resistor which follows a law giving a 4.5 dB level drop to each channel when panned centrally, compared with the level sent to either channel at the extremes. The 4.5 dB figure is a compromise between the −3 dB and −6 dB laws. Pan-pots which only drop the level by 3 dB in the centre cause a rise in level of any centrally panned signal if a mono sum is derived from the left and right outputs of that channel, since two identical signals summed together will give a rise in level of 6 dB. A pot which gives a 6 dB drop in the centre results in no level rise for centrally panned signals in the mono sum. Unfortunately, the 3 dB drop works best for stereo reproduction, resulting in no perceived level rise for centrally panned signals.

Only about 18 dB of level difference is actually required between left and right channels to give the impression that a source is either fully left or fully right in a loudspeaker stereo image, but most pan-pots are designed to provide full attenuation of one channel when rotated fully towards the other. This allows for the two buses between which signals are panned to be treated independently, such as when a pan control is used to route a signal either to odd or even channels of a multitrack bus (see section 6.4.2).

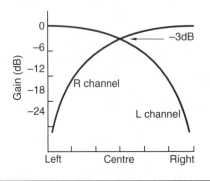

mains. This typically contains a mains transformer, rectifiers and regulating circuitry, and it supplies the mixer with relatively low DC voltages. The main advantage of a separate power supply is that the mains transformer can be sited well away from the mixer, since the alternating 50 Hz mains field around the former can be induced into the audio circuits. This manifests itself as 'mains hum' which is only really effectively dealt with by increasing the distance between the mixer and the transformer. Large mixers usually have separate rack-mounting power supplies.

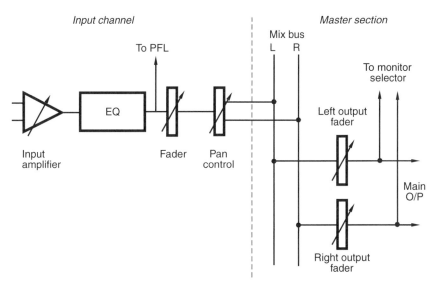

Figure 6.2 Block diagram of a typical signal path from channel input to main output on a simple mixer

The above-described mixer is very simple, offering few facilities, but it provides a good basis for the understanding of more complex models. A typical commercial example of a compact mixer is shown in Figure 6.3.

6.2 A multitrack mixer

6.2.1 Overview

The stereo mixer outlined in the previous section only forms half the story in a multitrack recording environment. Conventionally, popular music recording involves at least two distinct stages: the 'track-laying' phase, and the 'mixdown' phase. In the former, musical tracks are layed down on a multitrack tape recorder in stages, with backing tracks and rhythm tracks being recorded first, followed by lead tracks and vocals. In the mixdown phase, all the previously recorded tracks from the tape recorder are played back through the mixer and combined into stereo to form the finished product which goes to be made into a commercial release (see section 3.2). More recently, with the widespread adoption of electronic instruments and MIDI equipment (see Chapter 14), the multitrack tape recorder has begun to play a smaller rôle in some recording studios, because MIDI-sequenced sound sources are now played directly into the mix in the second stage.

For these reasons, as well as requiring mixdown signal paths from many inputs to a stereo bus the mixer also requires signal paths for routing many input signals to a multitrack tape recorder. Often it will be necessary to perform both of these functions simultaneously – that is, recording microphone signals to multitrack tape whilst also mixing the return from tape into stereo, so that the engineer and pro-

FACT FILE

6.3

Pre-fade
listen (PFL)

Pre-fade listen, or PFL, is a facility which enables a signal to be monitored without routing it to the main outputs of the mixer. It also provides a means for listening to a signal in isolation in order to adjust its level or EQ.

Normally, a separate mono mixing bus runs the length of the console picking up PFL outputs from each channel. A PFL switch on each channel routes the signal from before the fader of that channel to the PFL bus (see diagram), sometimes at the same time as activating internal logic which switches the mixer's monitor outputs to monitor the PFL bus. If no such logic exists, the mixer's monitor selector will allow for the selection of PFL, in which position the monitors will reproduce any channel currently with its PFL button pressed. On some broadcast and live consoles a separate small PFL loudspeaker is provided on the mixer itself, or perhaps on a separate output, in order that selected sources can be checked without affecting the main monitors.

Sometimes PFL is selected by 'over-pressing' the channel fader concerned at the bottom of its travel (i.e.: pushing it further down). This activates a microswitch which performs the same functions as above. PFL has great advantages in live work and broadcasting, since it allows the engineer to listen to sources before they are faded up (and thus routed to the main outputs which would be carrying the live programme). It can also be used in studio recording to isolate sources from all the others without cutting all the other channels, in order to adjust equalisation and other processing with greater ease.

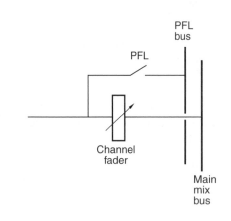

ducer can hear what the finished result will sound like, and so that any musicians who may be overdubbing additional tracks can be given a mixed feed of any previously recorded tracks in headphones. The latter is known as the *monitor mix* and this often forms the basis for the stereo mixdown when the tracklaying job is finished.

So there are two signal paths in this case: one from the microphone or line source to the multitrack tape recorder, and one from the multitrack recorder back to the stereo mix, as shown in Figure 6.4. The path from the microphone input which usually feeds the multitrack machine will be termed the *channel path*, whilst the path from the line input or tape return which usually feeds the stereo mix will be termed the *monitor path*.

It is likely that some basic signal processing such as equalisation will be required in the feed to the multitrack recorder (see below), but the more comprehensive signal processing features are usually applied in the mixdown path. The situation is somewhat different in the American market where there is a greater tendency to

Figure 6.3 A compact stereo mixer: the Seemix 'Seeport'. (Courtesy of Seemix Sound AS)

record on multitrack 'wet', that is with all effects and EQ, rather than applying the effects on mixdown.

6.2.2 In-line and split configurations

As can be seen from Figure 6.4, there are two complete signal paths, two faders, two sets of EQ, and so on. This takes up space, and there are two ways of arranging this physically, one known as the *split-monitoring*, or *European-style* console, the other as the *in-line* console. The split console is the more obvious of the two, and its physical layout is shown in Figure 6.5. It contains the input channels on one side (usually the left), a master control section in the middle, and the monitor mixer on the other side. So it really is two consoles in one frame. It is necessary to have as many monitor channels as there are tracks on the tape, and these channels are likely to need some signal processing. The monitor mixer is used during track laying for mixing a stereo version of the material that is being recorded, so that everyone can hear a rough mix of what the end result will sound like, and then on mixdown every

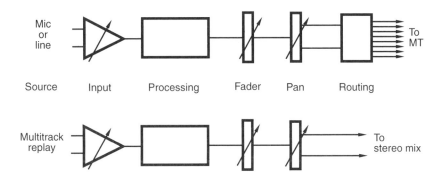

Figure 6.4 In multitrack recording two signal paths are needed – one from mic or line input to the multitrack recorder, and one returning from the recorder to contribute to a 'monitor' mix

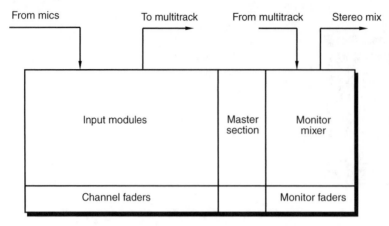

Figure 6.5 A typical 'split' or 'European-style' multitrack mixer has input modules on one side and monitor modules on the other: two separate mixers in effect

input to the console can be routed to the stereo mix bus so as to increase the number of inputs for outboard effects, etc. and so that the comprehensive facilities provided perhaps only on the left side of the console are available for the tape returns.

This layout has advantages in that it is easily assimilated in operation, and it makes the channel module less cluttered than the in-line design (described below), but it can make the console very large when a lot of tracks are involved. It can also increase the build cost of the console because of the near doubling in facilities and metalwork required, and it lacks flexibility, especially when switching over from track laying to remixing.

The in-line layout involves the translation of everything from the right-hand side of the split console (the monitor section) into the left side, rather as if the console were sawn in half and the right side merged with the left, as shown in Figure 6.6. In this process a complete monitor signal path is fitted into the same module as the same-numbered channel path, making it no more than a matter of a few switches to enable facilities to be shared between the two paths. In such a design each module will contain two faders (one for each signal path), but usually only one EQ section, one set of auxiliary sends (see below), one dynamics control section, and so on, with switches to swap facilities between paths. (A simple example showing only the switching needed to swap one block of processing is shown in Figure 6.7.) Usually this means that it is not possible to have EQ in both the multitrack recording path *and* the stereo mix path, but some more recent designs have made it possible to split the equaliser so that some frequency-band controls are in the channel path whilst others are in the monitor path. The band ranges are then made to overlap considerably which makes the arrangement quite flexible.

6.2.3 Further aspects of the in-line design

It has already been stated that there will be two main faders associated with each channel module in an in-line console: one to control the gain of each signal path.

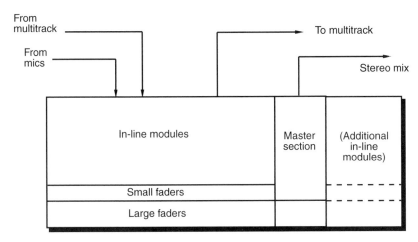

Figure 6.6 A typical 'in-line' mixer incorporates two signal paths in one module, providing two faders per module (one per path). This has the effect of reducing the size of the mixer for a given number of channels, when compared with a split design

Sometimes the small fader is not a linear slider but a rotary knob. It is not uniformly agreed as to whether the large fader at the bottom of the channel module should normally control the *monitor* level of the like-numbered tape track or whether it should control the *channel* output level to multitrack tape. Convention has it that American consoles make the large fader 'the monitor fader' in normal operation, while British consoles tend to make it 'the channel fader'. Normally their functions may be swapped over, depending on whether one is mixing down or track laying, either globally (for the whole console), in which case the fader swap will probably happen automatically when switching the console from 'recording' to 'remix' mode, or on individual channels, in which case the operation is usually performed using a control labelled something like 'fader flip', 'fader reverse' or 'changeover'. The process of fader swapping is mostly used for convenience, since

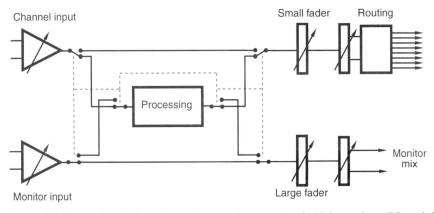

Figure 6.7 The in-line design allows for sound processing facilities such as EQ and dynamics to be shared or switched between the signal paths

more precise control can be exercised over a large fader near the operator than over a small fader which is further away, and thus the large fader is assigned to the function which is being used most in the current operation. This is coupled with the fact that in an automated console, it is almost invariably the large fader which is automated, and the automation is required most in the mixdown process.

Confusion can arise when operating in-line mixers, such as when a microphone signal is fed into, say, mic input 1 and is routed to track 13 on the tape, because the operator will control the monitor level of that track (and therefore the level of that microphone's signal in the stereo mix) on monitor fader 13, whilst the channel fader on module 1 will control the multitrack record level for that mic signal.

If a 24 track tape machine is in use with the mixer, then monitor faders higher than number 24 will not normally carry a tape return, but will be free for other sources. Remember that more than one microphone signal can be routed to each track on the tape, and so there will be a number of level controls which affect each source's level in the monitor mix, each of which has a different purpose:

- **MIC LEVEL TRIM** – adjusts the gain of the microphone pre-amplifier at the channel input. Usually located at the top of the module.

- **CHANNEL FADER** – comes next in the chain and controls the individual level of the mic (or line) signal connected to that module's input before it goes to tape. Located on the same-numbered module as the input. (May be switched to be either the large or small fader, depending on configuration.)

- **BUS TRIM or TRACK SUBGROUP** – will affect the overall level of all signals routed to a particular tape track. Usually located with the track routing buttons at the top of the module. Sometimes a channel fader can be made to act as a subgroup master.

- **MONITOR FADER** – is located in the return path from the multitrack recorder to the stereo mix. Does not affect the recorded level on the multitrack tape, but affects the level of this track in the mix. (May be switched to be either the large or small fader, depending on configuration.)

A typical in-line multitrack mixer is shown in the photograph in Figure 6.8.

6.3 Channel grouping

Grouping is a term which refers to the simultaneous control of more than one signal at a time. It usually means that one fader controls the levels of a number of slave channels. Two types of channel grouping are currently common: *audio grouping* and *'control' grouping*. The latter is often called VCA grouping, but there are other means of control grouping that are not quite the same as the direct VCA control method. The two approaches have very different results, although initially they may appear to be very similar due to the fact that one fader appears to control a number of signal levels. The primary reason for adopting group faders of any kind is in

Figure 6.8 A typical in-line mixer: the Soundcraft 'Sapphyre'. (Courtesy of Soundcraft Electronics Ltd)

order to reduce the number of faders which the engineer has to handle at a time, and is feasible in a situation where a number of channels are carrying audio signals which can be faded up and down together. These signals do not all have to be at the same initial level, and indeed one is still free to adjust levels individually within a group. A collection of channels carrying drum sounds, or carrying an orchestral string section, would be examples of suitable groups. The two approaches are described in Fact Files 6.4 and 6.5.

6.4 An overview of typical mixer facilities

Most mixing consoles provide a degree of sound signal processing on board, as well as routing to external processing devices. The very least of these facilities is some form of equalisation (a means of controlling the gain at various frequencies), and there are few consoles which do not include this. As well as signal processing, there will be a number of switches which make changes to

FACT FILE 6.4 Audio groups

Audio groups are so called because they create a single audio output which is the sum of a number of channels. A single fader controls the level of the summed signal, and there will be a group output from the console which is effectively a mix of the audio signals in that group, as shown in the diagram. The audio signals from each input to the group are fed via equal-value resistors to the input of a summing or virtual-earth amplifier.

The stereo mix outputs from an in-line console are effectively audio groups, one for the left, one for the right, as they constitute a sum of all the signals routed to the stereo output and include overall level control. In the

same way, the multitrack routing buses on an in-line console are also audio groups, as they are sums of all the channels routed to their respective tracks. More obviously, some smaller or older consoles will have routing buttons on each channel module for, say, four audio group destinations, these being really the only way of routing channels to the main outputs.

The master faders for audio groups will often be in the form of four or eight faders in the central section of the console. They may be arranged such that one may pan a channel between odd and even groups, and it would be common for two of these groups (an odd and an even one) to be used as the stereo output in mixdown. It is also common for perhaps eight audio group faders to be used as 'subgroups', themselves having routing to the stereo mix, so that channel signals can be made more easily manageable by routing them to a subgroup (or panning between two subgroups) and thence to the main mix via a single level control (the subgroup fader), as shown in the diagram. (Only four subgroups are shown in the diagram, without pan controls. Subgroups 1 and 3 feed the left mix bus, and 2 and 4 feed the right mix bus. Sometimes subgroup outputs can be panned between left and right main outputs.)

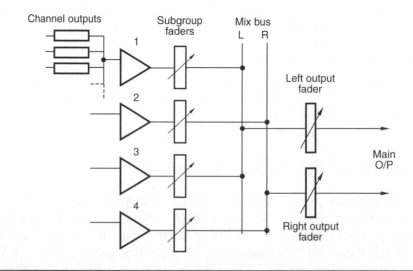

FACT FILE

6.5

Control groups

Control grouping differs from audio grouping primarily because it does not give rise to a single summed audio output for the group: the levels of the faders in the group are controlled from one fader, but their outputs remain separate. Such grouping can be imagined as similar in its effect to a large hand moving many faders at the same time, each fader maintaining its level in relation to the others.

The most common way of achieving control grouping is to use VCAs (Voltage-Controlled Amplifiers), whose gain can be controlled by a DC voltage applied to a control pin. In the VCA fader, audio is not passed through the fader itself but is routed through a VCA, whose gain is controlled by a DC voltage derived from the fader position, as shown in the diagram. So the fader now carries DC instead of audio, and the audio level is controlled indirectly.

Indirect gain control opens up all sorts of new possibilities. The gain of the channel could

be controlled externally from a variety of sources, either by combining the voltage from an external controller in an appropriate way with the fader's voltage so that it would still be possible to set the relative level of the channel, or by breaking the direct connection between the DC fader and the VCA so that an automation system could intervene, as discussed in section 7.3. It becomes possible to see that group faders could be DC controls which could be connected to a number of channel VCAs such that their gains would go up and down together. Further to this, a channel VCA could be assigned to any of the available groups simply by selecting the appropriate DC path: this is often achieved by means of thumbwheel switches on each fader, as shown in the diagram.

Normally, there are dedicated VCA group master faders in a non-automated system. They usually reside in the central section of a mixer and will control the overall levels of any channel faders assigned to them by the thumbwheels by the faders. In such a system, the channel audio outputs would normally be routed to the main mix directly, the grouping affecting the levels of the individual channels in this mix.

In an automated system grouping may be achieved via the automation processor which will allow any fader to be designated as the group master for a particular group. This is possible because the automation processor reads the levels of all the faders, and can use the position of the designated master to modify the data sent back to the other faders in the group (see section 7.3).

the signal path or operational mode of the console. These may operate on individual channels, or they may function globally (affecting the whole console at once). The following section is a guide to the facilities commonly found on multitrack consoles. Figure 6.9 shows the typical location of these sections on an in-line console module.

6.4.1 Input section

- *Input gain control*
  Sets the microphone or line input amplifier gain to match the level of the incoming signal. This control is often a coarse control in 10 dB steps, sometimes accompanied by a fine trim. Opinion varies as to whether this control should be in detented steps or continuous. Detented steps of 5 or 10 dB make for easy reset of the control to an exact gain setting, and precise gain matching of channels.

- *Phantom power*
  Many professional mics require 48 volts phantom powering (see section 4.9). There is sometimes a switch on the module to turn it on or off, although most balanced mics which do not use phantom power will not be damaged if it is accidentally left on. Occasionally this switch is on the rear of the console, by the mic input socket, or it may be in a central assignable switch panel. Other methods exist: for example, one console requires that the mic gain control is pulled out to turn on the phantom power.

- *MIC/LINE switch*
  Switches between the channel's mic input and line input. The line input could be the playback output from a tape machine, or another line level signal such as a synth or effects device.

- *PAD*
  Usually used for attenuating the mic input signal by something like 20 dB, for situations when the mic is in a field of high sound pressure. If the mic is in front of a kick drum, for example, its output may be so high as to cause the mic input to clip. Also, capacitor mics tend to produce a higher output level than dynamic mics, requiring that the pad be used on some occasions.

- *Phase reverse or 'ϕ'*
  Sometimes located after the mic input for reversing the phase of the signal, to compensate for a reversed directional mic, a mis-wired lead, or to create an effect. This is often left until later in the signal path.

- *HPF/LPF*
  Filters can sometimes be switched in at the input stage, which will usually just be basic high- and low-pass filters which are either in or out, with no frequency adjustment. These can be used to filter out unwanted rumble or perhaps hiss from noisy signals. Filtering rumble at this stage can be an advantage because it saves clipping later in the chain.

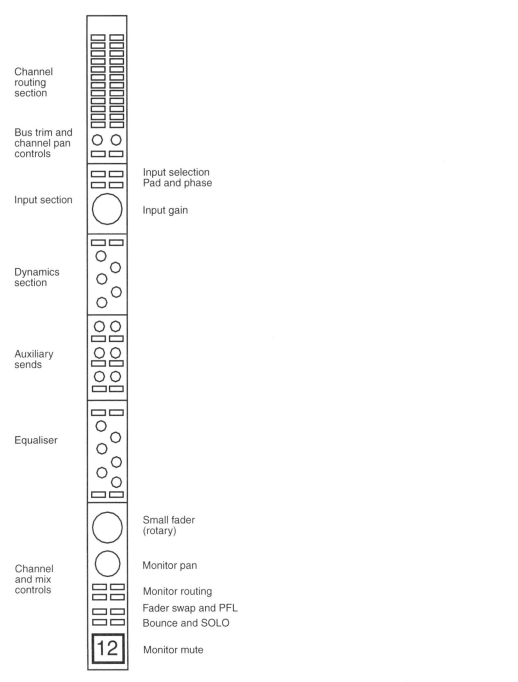

Channel
routing
section

Bus trim and
channel pan
controls

Input section

Dynamics
section

Auxiliary
sends

Equaliser

Channel
and mix
controls

Input selection
Pad and phase

Input gain

Small fader
(rotary)

Monitor pan

Monitor routing

Fader swap and PFL

Bounce and SOLO

Monitor mute

Figure 6.9 Typical layout of controls on an in-line mixer module (for description see text)

6.4.2 Routing section

- *Track routing switches*
  The number of routing switches depends on the console: some will have 24, some 32 and some 48. The switches route the channel path signal to the multitrack machine, and it is possible to route a signal to more than one track. The track assignment is often arranged as pairs of tracks, so that odd and even tracks can be assigned together, with a pan-pot used to pan between them as a stereo pair, e.g.: tracks 3 and 4 could be a stereo pair for background vocals, and each background vocal mic could be routed to 3 and 4, panned to the relevant place in the image. In an assignable console these controls may be removed to a central assignable routing section.

  It is common for there to be fewer routing switches than there are tracks, so as to save space, resulting in a number of means of assigning tracks. Examples are rotary knobs to select the track, one button per pair of tracks with 'odd/even/both' switch, and a 'shift' function to select tracks higher than a certain number. The multitrack routing may be used to route signals to effects devices during mixdown, when the track outputs are not being used for recording. In this case one would patch into the track output on the patchfield (see below) and take the relevant signal to an effects input somewhere else on the patchfield. In order to route monitor path signals to the track routing buses it may be necessary to use a switch which links the output of the monitor fader to the track assignment matrix.

  In theatre sound mixers it is common for output routing to be changed very frequently, and thus routing switches may be located close to the channel fader, rather than at the top of the module as in a music mixer. On some recent mixers, track routing is carried out on a matrix which resides in the central section above the main faders. This removes unnecessary clutter from the channel modules and reduces the total number of switches required. It may also allow the storing of routing configurations in memory for later recall.

- *Mix routing switches*
  Sometimes there is a facility for routing the channel path output signal to the main monitor mix, or to one of perhaps four output groups, and these switches will often be located along with the track routing.

- *Channel pan*
  Used for panning channel signals between odd and even tracks of the multitrack, in conjunction with the routing switches.

- *Bus trim*
  Used for trimming the overall level of the send to multitrack for a particular bus. It will normally trim the level sent to the track which corresponds to the number of the module.

- *Odd/Even/Both*

   Occasionally found when fewer routing buttons are used than there are tracks. When one routing button is for a pair of tracks, this switch will determine whether the signal is sent to the odd channel only, the even channel only, or to both (in which case the pan control is operative).

- *DIRECT*

   Used for routing the channel output directly to the corresponding track on the multitrack machine without going via the summing buses. This can reduce the noise level from the console since the summing procedure used for combining a number of channel outputs to a track bus can add noise. If a channel is routed directly to a track, no other signals can be routed to that track.

6.4.3 Dynamics section

Some advanced consoles incorporate dynamics control on every module, so that each signal can be treated without resorting to external devices. The functions available on the best designs rival the best external devices, incorporating compressor and expander sections which can act as limiters and gates respectively if required. One system allows the EQ to be placed in the side-chain of the dynamics unit, providing frequency-sensitive limiting, among other things, and it is usually possible to link the action of one channel's dynamics to the next in order to 'gang' stereo channels so that the image does not shift when one channel has a sudden change in level while the other does not.

When dynamics are used on stereo signals it is important that left and right channels have the same settings, otherwise the image may be affected. If dynamics control is not available on every module, it is sometimes offered on the central section with inputs and outputs on the patchbay. Dynamics control will not be covered further here, but is discussed in more detail in section 13.2.

6.4.4 Equaliser section

The EQ section is usually split into three or four sections, each operating on a different frequency band. As each band tends to have similar functions these will be described in general. The principles of EQ are described in greater detail in section 6.5.

- *HF, MID 1, MID 2, LF*

   A high-frequency band, two mid-frequency bands, and a low-frequency band are often provided. If the EQ is *parametric* these bands will allow continuous variation of frequency (over a certain range), 'Q', and boost/cut. If it is not parametric, then there may be a few switched frequencies for the mid band, and perhaps a fixed frequency for the LF and HF bands.

- *Peaking/shelving or BELL*

   Often provided on the upper and lower bands for determining whether the filter will provide boost/cut over a fixed band (whose width will be determined by the

Q), or whether it will act as a shelf, with the response rising or rolling off above or below a certain frequency (see Figure 6.14).

- *Q*

  The Q of a filter is defined as its centre frequency divided by its bandwidth (the distance between frequencies where the output of the filter is 3 dB lower than the peak output). In practice this affects the 'sharpness' of the filter peak or notch, high Q giving the sharpest response, and low Q giving a very broad response. Low Q would be used when boost or cut over a relatively wide range of frequencies is required, while high Q is used to boost or cut one specific region (see Fact File 6.6).

- *Frequency control*

  Sets the centre frequency of a peaking filter, or the turnover frequency of a shelf.

- *Boost/cut*

  Determines the amount of boost or cut applied to the selected band, usually up to a maximum of around ±15 dB.

- *HPF/LPF*

  Sometimes the high- and low-pass filters are located here instead of at the input, or perhaps in addition. They normally have a fixed frequency turnover point and a fixed roll-off of either 12 or 18 dB per octave. Often these will operate even if the EQ is switched out.

- *CHANNEL*

  The American convention is for the main equaliser to reside normally in the monitor path, but it can be switched so that it is in the channel path. Normally the whole EQ block is switched at once, but on some recent models a section of the EQ can be switched separately. This would be used to equalise the signal which is being *recorded* on multitrack tape. If the EQ is in the monitor path then it will only affect the replayed signal. The traditional European convention is for EQ to reside normally in the *channel* path, so as to allow recording with EQ.

- *IN/OUT*

  Switches the EQ in or out of circuit. Equalisation circuits can introduce noise and phase distortion, so they are best switched out when not required.

6.4.5 Channel and mix controls

- *Pan*

  See Fact File 6.2.

- *Fader reverse*

  Swaps the faders between mix and channel paths, so that the large fader can be made to control either the mix level or the channel level. Some systems defeat any fader automation when the large fader is put in the channel path. Fader reverse can often be switched globally, and may occur when the console mode is changed from recording to mixdown.

FACT FILE
6.6   Variable Q

Some EQ sections provide an additional control whereby the Q of the filter can be adjusted. This type of EQ section is termed a parametric EQ since all parameters, cut/ boost, frequency, and Q can be adjusted. The diagram below illustrates the effect of varying the Q of an EQ section. High Q settings affect very narrow bands of frequencies, low Q settings affect wider bands. The low Q settings

sound 'warmer' because they have gentle slopes and therefore have a more gradual and natural effect on the sound. High Q slopes are good for a rather more overt emphasis of a particular narrow band, which of course can be just as useful in the appropriate situation. Some EQ sections are labelled parametric even though the Q is not variable. This is a misuse of the term, and it is wise to check whether or not an EQ section is truly parametric even though it may be labelled as such.

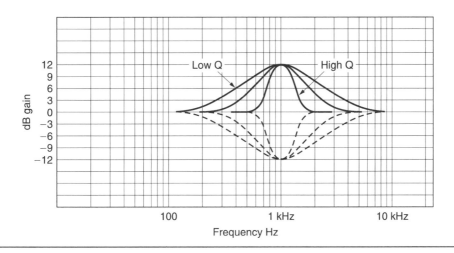

- *Line/Tape or Bus/Tape*
  Switches the source of the input to the monitor path between the line *output* of the same-numbered channel and the return from multitrack tape. Again it is possible that this may be switched globally. In 'line' or 'bus' mode the monitor paths are effectively 'listening to' the line output of the console's track assignment buses, while in 'tape' mode the monitor paths are listening to the off-tape signal (unless the tape machine's monitoring is switched to monitor the line input of the tape machine, in which case 'line' and 'tape' will effectively be the same thing!). If a problem is suspected with the tape machine, switching to monitor 'line' will bypass the tape machine entirely and allow the operator to check if the console is actually sending anything.

- *Broadcast, or 'mic to mix', or 'simulcast'*
  Used for routing the mic signal to both the channel and monitor paths simulta-

neously, so that a multitrack recording can be made while a stereo mix is being recorded or broadcasted. The configuration means that any alterations made to the channel path will not affect the stereo mix, which is important when the mix output is live (see Figure 6.10).

- *BUS or 'monitor-to-bus'*
  Routes the output of the monitor fader to the input of the channel path (or the channel fader) so that the channel path can be used as a post-fader effects send to any one of the multitrack buses (used in this case as aux sends), as shown in Figure 6.11. If a BUS TRIM control is provided on each multitrack output this can be used as the master effects-send level control.

- *DUMP*
  Incorporated (rarely) on some consoles to route the stereo panned mix output of a track (i.e.: after the monitor path pan-pot) to the multitrack assignment switches. In this way, the mixed version of a group of tracks can be 'bounced down' to two tracks on the multitrack, panned and level-set as in the monitor mix (see Figure 6.11).

- *BOUNCE*
  A facility for routing the output of the monitor fader to the multitrack assignment matrix, before the pan control, in order that tracks can be 'bounced down' so as to free tracks for more recording by mixing a group of tracks on to a lower number of tracks. BOUNCE is like a mono version of DUMP (see Figure 6.11).

- *MUTE or CUT*
  Cuts the selected track from the mix. There may be two of these switches, one for cutting the channel signal from the multitrack send, the other for cutting the mix signal from the mix.

Figure 6.10 A 'broadcast mode' switch in an in-line console allows the microphone input to be routed to both signal paths, such that a live stereo mix may be made independent of any changes to multitrack recording levels

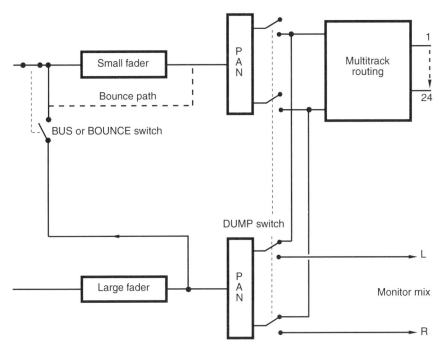

Figure 6.11 Signal routings for 'bounce', 'bus' and 'dump' modes (see text)

- *PFL*
  See Fact File 6.3.

- *AFL*
  After fade listen is similar to PFL, except that it is taken from after the fader. This is sometimes referred to as *SOLO*, which routes a panned version of the track to the main monitors, cutting everything else. These functions are useful for isolating signals when setting up and spotting faults. On many consoles the AFL bus will be stereo. Solo functions are useful when applying effects and EQ, in order that one may hear the isolated sound and treat it individually without hearing the rest of the mix. Often a light is provided to show that a solo mode is selected, because there are times when nothing can be heard from the loudspeakers due to a solo button being down with no signal on that track. A *solo safe* control may be provided centrally, which prevents this feature from being activated.

- *In-place solo*
  On some consoles, solo functions as an 'in-place' solo, which means that it actually changes the mix output, muting all tracks which are not solo'ed and picking out all the solo'ed tracks. This may be preferable to AFL as it reproduces the exact contribution of each channel to the mix, at the presently set master mix level. Automation systems often allow the solo functions to be automated in groups, so that a whole section can be isolated in the mix. In certain designs,

the function of the automated mute button on the monitor fader may be reversed so that it becomes solo.

6.4.6 Auxiliary sends

The number of aux sends depends on the console, but there can be up to ten on an ordinary console, and sometimes more on assignable models. Aux sends are 'take-off points' for signals from either the channel or mix paths, and they appear as outputs from the console which can be used for foldback to musicians, effects sends, cues, and so on. Each module will be able to send to auxiliaries, and each numbered auxiliary output is made up of all the signals routed to that aux send. So they are really additional mix buses. Each aux will have a master gain control, usually in the centre of the console for adjusting the overall gain of the signal sent from the console, and may have basic EQ. Aux sends are often a combination of mono and stereo buses. Mono sends are usually used as routes to effects, while stereo sends may have one level control and a pan control per channel for mixing a fold-back source.

- *Aux sends 1–n*
  Controls for the level of each individual channel in the numbered aux mix.

- *Pre/post*
  Determines whether the send is taken off before or after the fader. If it is before then the send will still be live even when the fader is down. Generally, 'cue' feeds will be pre-fade, so that a mix can be sent to foldback which is independent of the monitor mix. Effects sends will normally be taken post-fade, in order that the effect follows a track's mix level.

- *Mix/channel*
  Determines whether the send is taken from the mix or channel paths. It will often be sensible to take the send from the channel path when effects are to be recorded on to multitrack rather than on to the mix. This function has been labelled 'WET' on some designs.

- *MUTE*
  Cuts the numbered send from the aux mix.

6.4.7 Master control section

The master control section usually resides in the middle of the console, or near the right-hand end. It will contain some or all of the following facilities:

- *Monitor selection*
  A set of switches for selecting the source to be monitored. These will include tape machines (stereo), aux sends, the main stereo mix, and perhaps some miscella-neous external sources like CD players, cassette machines, etc. They only select the signal going to the loudspeakers, not the mix outputs. This may be duplicated

to some extent for a set of additional studio loudspeakers, which will have a separate gain control.

- *DIM*
  Reduces the level sent to the monitor loudspeakers by a considerable amount (usually around 40 dB), for quick silencing of the room.

- *MONO*
  Sums the left and right outputs to the monitors into mono so that mono compatibility can be checked.

- *Monitor phase reverse*
  Phase reverses one channel of the monitoring so that a quick check on suspected phase reversals can be made.

- *TAPE/LINE*
  Usually a global facility for switching the inputs to the mix path between the tape returns and the console track outputs. Can be reversed individually on modules.

- *FADER REVERSE*
  Global swapping of small and large faders between mix and channel paths.

- *Record/Overdub/Mixdown*
  Usually globally configures mic/line input switching, large and small faders and auxiliary sends depending on mode of operation. (Can be overridden on individual channels.)

- *Auxiliary level controls*
  Master controls for setting the overall level of each aux send output.

- *Foldback and Talkback*
  There is often a facility for selecting which signals are routed to the stereo foldback which the musicians hear on their headphones. Sometimes this is as comprehensive as a cue mixer which allows mixing of aux sends in various amounts to various stereo cues, while often it is more a matter of selecting whether foldback consists of the stereo mix, or one of the aux sends. Foldback level is controllable, and it is sometimes possible to route left and right foldback signals from different sources. *Talkback* is usually achieved using a small microphone built into the console, which can be routed to a number of destinations. These destinations will often be aux sends, multitrack buses, mix bus, studio loudspeakers and foldback.

- *Oscillator*
  Built-in sine-wave oscillators vary in quality and sophistication, some providing only one or two fixed frequencies, while others allow the generation of a whole range. If the built-in oscillator is good it can be used for lining up the tape machine, as it normally can be routed to the mix bus or the multitrack outputs. The absolute minimum requirement is for accurate 1 kHz and 10 kHz tones, the 10 kHz being particularly important for setting the bias of an analogue tape machine. The oscillator will have an output level control.

- *Slate*
  Provides a feed from the console talkback mic to the stereo output, often super-imposing a low-frequency tone (around 50 Hz) so that the slate points can be heard when winding a tape at high speed. Slate would be used for recording take information on to tape.

- *Master faders*
  There may be either one stereo fader or left and right faders to control the overall mix output level. Often the group master faders will reside in this section.

6.4.8 Effects returns

Effects returns are used as extra inputs to the mixer, supplied specifically for inputs from external devices such as reverberation units. These are often located in the central section of the console and may be laid out like reduced-facility input channels. Returns sometimes have EQ, perhaps more basic than on channels, and they may have aux sends. Normally they will feed the mix, although sometimes facilities are provided to feed one or more returns to the multitrack via assignment switches. A small fader or rotary level control is provided, as well as a pan-pot for a mono return. Occasionally, automated faders may be assigned to the return channels so as to allow automated control of their levels in the mix.

6.4.9 Patchfield or jackfield

Most large consoles employ a built-in jackfield or patchbay for routing signals in ways which the console switching does not allow, and for sending signals to and from external devices. Just about every input and output on every module in the console comes up on the patchbay, allowing signals to be cross-connected in virtually any configuration. The jackfield is usually arranged in horizontal rows, each row having an equal number of jacks. Vertically, it tries to follow the signal path of the console as closely as possible, so the mic inputs are at the top and the multitrack outputs are nearer the bottom. In between these there are often *insert points* which allow the engineer to 'break into' the signal path, often before or after the EQ, to insert an effects device, compressor, or other external signal processor. Insert points usually consist of two rows, one which physically breaks the signal chain when a jack is inserted, and one which does not. Normally it is the lower row which breaks the chain, and should be used as inputs. The upper row is used as an output or send. *Normalling* is usually applied at insert points, which means that unless a jack is inserted the signal will flow directly from the upper row to the lower.

At the bottom of the patchfield will be all the master inputs and outputs, playback returns, perhaps some parallel jacks, and sometimes some spare rows for connection of one's own devices. Some consoles bring the microphone signals up to the patchbay, but there are some manufacturers who would rather not do this unless absolutely necessary as it is more likely to introduce noise, and phantom power may be present on the jackfield. Jackfields are covered in further detail in section 12.12.

6.5 EQ explained

The tone control or EQ (= equalisation) section provides mid-frequency controls in addition to bass and treble. A typical comprehensive EQ section may have firstly an HF (High-Frequency) control similar to a treble control but operating only at the highest frequencies. Next would come a hi-mid control, affecting frequencies from around 1 kHz to 10 kHz, the centre frequency being adjusted by a separate control. Lo-mid controls would come next, similar to the hi-mid but operating over a range of say 200 Hz to 2 kHz. Then would come an LF (Low-Frequency) control. Additionally, high- and low-frequency filters can be provided. The complete EQ section looks something like Figure 6.12. An EQ section takes up quite a bit of space, and so it is quite common for dual concentric controls to be used. For instance, the cut/boost controls of the hi- and lo-mid sections can be surrounded by annular skirts which select the frequency. Console area is therefore saved.

6.5.1 Principal EQ bands

The HF section affects the highest frequencies and provides up to 12 dB of boost or cut. This type of curve is called a shelving curve because it gently boosts or cuts the frequency range towards a shelf where the level remains relatively constant (see Figure 6.13(a)). Next comes the hi-mid section. Two controls are provided here, one to give cut or boost, the other to select the desired centre frequency. The latter is commonly referred to as a 'swept mid' because one can sweep across the frequency range.

Figure 6.13(b) shows the result produced when the frequency setting is at the 1 kHz position, termed the centre frequency. Maximum boost and cut affects this frequency the most, and the slopes of the curve are considerably steeper than those of the previous shelving curves. This is often referred to as a 'bell' curve due to the upper portion's resemblance to the shape of a bell. It has a fairly high 'Q', that is its sides are steep. Q is defined as:

$$Q = \text{centre frequency} \div \text{bandwidth}$$

where the bandwidth is the distance in hertz between the two points at which the response of the filter is 3 dB lower than that at the centre frequency. In the example shown the centre frequency is 1 kHz and the bandwidth is 400 Hz, giving $Q = 2.5$.

MF EQ controls are often used to hunt for trouble-spots; if a particular instrument (or microphone) has an emphasis in its spectrum somewhere which does not sound very nice, some mid cut can be introduced, and the frequency control can be used to search for the precise area in the frequency spectrum where the trouble lies. Similarly, a dull sound can be given a lift in an appropriate part of the spectrum which will bring it to life in the overall mix. Figure 6.13(c) shows the maximum cut and boost curves obtained with the frequency selector at either of the three settings of 1, 5 and 10 kHz. The high Q of the filters enables relatively narrow bands to be affected. Q may be varied in some cases, as described in Fact File 6.6.

Figure 6.12 Typical layout of an EQ section

Figure 6.13 (a) Typical HF and LF shelf EQ characteristics shown at maximum settings

Figure 6.13 (b) Typical MF peaking filter characteristic

The lo-mid section is the same as the hi-mid section except that it covers a lower band of frequencies. Note though that the highest frequency setting overlaps the lowest setting of the hi-mid section. This is quite common, and ensures that no 'gaps' in the frequency spectrum are left uncovered.

6.5.2 Filters

High- and low-cut filters provide fixed attenuation slopes at various frequencies. Figure 6.13(d) shows the responses at LF settings of 80, 65, 50, 35 and 20 Hz. The slopes are somewhat steeper than is the case with the HF and LF shelving curves, and slope rates of 18 or 24 dB per octave are typical. This enables just the lowest, or highest, frequencies to be rapidly attenuated with minimal effect on the mid band.

Figure 6.13 (c) MF peaking filter characteristics at 1, 5 and 10 kHz

Figure 6.13 (d) High-pass filters with various turnover frequencies

Very low traffic rumble could be removed by selecting the 20 or 35 Hz setting. More serious low-frequency noise may require the use of one of the higher turnover frequencies. High-frequency hiss from, say, a noisy guitar amplifier or air escaping from a pipe organ bellows can be dealt with by selecting the turnover frequency of the HF section which attenuates just sufficient HF noise without unduly curtailing the HF content of the wanted sound.

6.6 Stereo line input modules

In broadcast situations it is common to require a number of inputs to be dedicated to stereo line level sources, such as cart machines, CD players, tapes, etc. Such modules are sometimes offered as an option for multitrack consoles, acting as replacements for conventional I/O modules and allowing two signals to be faded up and down together with one fader. Often the EQ on such modules is more limited, but the module may provide for the selection of more than one stereo source, and routing to the main mix as well as the multitrack. It is common to require that stereo modules always reside in special slots on the console, as they may require special wiring. Such modules may also provide facilities for handling LP turntable outputs, offering RIAA equalisation (see section A2.2).

With the advent of stereo television, the need for stereo *microphone* inputs is also becoming important, with the option for MS (middle and side) format signals as well as AB (conventional left and right) format (see section 4.7).

6.7 Dedicated monitor mixer

A dedicated monitor mixer is often used in live sound reinforcement work to provide a separate monitor mix for each musician, in order that each artist may specify his or her precise monitoring requirements. A comprehensive design will have, say, 24 inputs containing similar facilities to a conventional mixer, except that below the EQ section there will be a row of rotary or short-throw faders which individually send the signal from that channel to the group outputs, in any combination of relative levels. Each group output will then provide a separate monitor mix to be fed to headphones or amplifier racks.

6.8 Basic operational techniques

6.8.1 Level setting

If one is using a microphone to record speech or classical music then normally a fairly high input gain setting will be required. If the microphone is placed up against a guitar amplifier then the mic's output will be high and a much lower input gain setting can be used. There are essentially three ways of setting the gain control to the optimum position. Firstly, using PFL or prefade listen (see Fact File 6.3).

PFL is pressed, or the fader over-pressed (i.e.: pressed beyond the bottom of its travel against a sprung microswitch), on the input module concerned and the level read on either a separate PFL meter or with the main meters switched to monitor the PFL bus. The channel input gain should be adjusted to give a meter reading of, say, PPM 5, or 0 VU (although bearing in mind that VUs tend to give a lower reading than PPMs on material with a large transient content, thus making it easy to over-record when using VUs, as discussed in section 7.2). This gain-setting procedure must be carried out at a realistic input level from the source. It is frequently the case during rehearsals that vocalists and guitarists will produce a level that is rather lower than that which they will use when they actually begin to play.

The pan control should be set next (see Fact File 6.2) to place the source in the stereo image. The main output faders will normally be set to 0 dB on their calibration, which is usually at the top. The channel faders can then be set to give both the desired subjective sound balance and appropriate output meter readings.

The second way of setting the gain is a good way in its own right, and it has to be used if PFL facilities are not provided. First of all both the channel fader and the output faders need to be positioned to the 0 dB point. This will be either at the top of the faders' travels or at a position about a quarter of the way down from the top of their travel. If no 0 dB position is indicated then the latter position should be set. After the pan control and faders have been positioned, the input gain may then be adjusted to give the desired reading on the output level meters. When several incoming signals need to be balanced the gain controls should all be positioned to give both the desired sound balance between them and the appropriate meter readings – normally PPM 6 or just over 0 VU during the loudest passages.

These two gain-setting methods differ in that with the former method the channel fader positions will show a correspondence to the subjective contribution each channel is making towards the overall mix, whereas the latter method places all the channel faders at roughly the same level.

The third way is similar to the second way, but one channel at a time is set up, placing channel and output faders at 0 dB and adjusting the gain for a peak meter reading. That channel fader is then turned completely down and the next channel is set up in a similar way. When all the channels which are to be used have been set up, the channel faders can then be advanced to give both the desired subjective balance and peak meter readings.

Use of the EQ controls often necessitates the resetting of the channel's input gain. For example, if a particular instrument requires a bit of bass boost, applying this will also increase the level of signal and so the gain will often need to be reduced a little to compensate. Applying bass or treble cut will sometimes require a small gain increase.

6.8.2 Using auxiliary sends

Aux facilities were described in section 6.4.6. The auxiliaries are configured either 'pre-fade' or 'post-fade'. Pre-fade aux sends are useful for providing a monitor mix for musicians, since this balance will be unaffected by movements of the faders which control the main mix. The engineer then retains the freedom to experiment in the control room without disturbing the continuity of feed to the musicians.

Post-fade sends are affected by the channel fader position. These are used to send signals to effects devices and other destinations where it is desirable to have the aux level under the overall control of the channel fader. For example, the engineer may wish to add a little echo to a voice. Aux 2, set to post-fade, is used to send the signal to an echo device, probably positioning the aux 2 control around the number 6 position and the aux 2 master at maximum. The output of the echo device is returned to another input channel or an echo return channel, and this fader can be adjusted to set the amount of echo. The level of echo will then rise and fall with the fader setting for the voice.

The post-fade aux could also be used simply as an additional output to drive separate amplifiers and speakers in another part of a hall, for example.

6.8.3 Using audio groups

The group outputs (see section 6.3) or multitrack routing buses (see section 6.4.2) can be used for overall control of various separate groups of instruments, depending on whether mixing down or track laying. For example, a drum kit may have eight microphones on it. These eight input channels can be routed to groups 1 and 2 with appropriate stereo pan settings. Groups 1 and 2 would then be routed to stereo outputs left and right respectively. Overall control of the drum kit level is now achieved simply by moving group faders 1 and 2.

When feeding a multitrack tape machine it is normally desirable to use the highest possible recording level on every track regardless of the final required balance, in order to achieve the best noise performance, and each multitrack group output will usually have an output level meter to facilitate this.

Recommended further reading

See *General further reading* at the end of this book.

Chapter 7

Mixers 2

7.1 Technical specifications of analogue mixers

7.1.1 Input noise

The output from a microphone is in the millivolt range, and so needs considerable amplification to bring it up to line level. Amplification of the signal also brings with it amplification of the microphone's own noise output (discussed in section 4.8.2), which one can do nothing about, and amplification of the mixer's own input noise. The latter must therefore be as low as possible so as not to compromise the noise performance unduly. A 200 ohm source resistance on its own generates 0.26 μV of noise (20 kHz bandwidth). Referred to the standard line level of 775 mV (0 dBu) this is −129.6 dBu. A microphone amplifier will add its own noise to this, and so manufacturers quote an 'equivalent input noise' (EIN) value which should be measured with a 200 ohm source resistance across the input.

An amplifier with a noise contribution equal to that of the 200 ohm resistor will degrade the theoretically 'perfect' noise level by 3 dB, and so the quoted equivalent input noise will be −129.6 + 3 = −126.6 dBm. (Because noise contributions from various sources sum according to their power content, not their voltage levels, dBm is traditionally used to express input noise level.) This value is quite respectable, and good-quality mixers should not be noisier than this. Values of around −128 dBm are sometimes encountered which are excellent, indicating that the input resistance is generating more noise than the amplifier is. Make sure that the EIN is quoted with a 200 ohm source, and a bandwidth up to 20 kHz, unweighted. A 150 ohm source, sometimes specified, will give an apparently better EIN simply because this resistor is itself quieter than a 200 ohm one, resistor noise being proportional to ohmic value. Also, weighting gives a flattering result, so one always has to check the measuring conditions. Make sure that EIN is quoted in dBm or dBu. Some manufacturers quote EIN in dBV (i.e.: ref. 1 volt) which gives a result 2.2 dB better.

An input should have high common mode rejection as well as low noise, as discussed in Fact File 7.1.

FACT FILE 7.1	Common mode rejection

As discussed in section 12.4, common mode rejection is the ability of a balanced input to reject interference which can be induced into the signal lines. A microphone input should have a CMRR (Common Mode Rejection Ratio) of 70 dB or more; i.e.: it should attenuate the interference by 70 dB. But look at how this measurement is made. It is relatively easy to achieve 70 dB at, say, 500 Hz, but rejection is needed most at high frequencies – between 5 and 20 kHz – and

so a quoted CMRR of '70 dB at 15 kHz' or '70 dB between 100 Hz and 10 kHz' should be sought. Line level CMRR can be allowed to be rather lower since the signal voltage level is a lot higher than in microphone cabling. CMRRs of as low as 30 dB at 10 kHz are deemed to be adequate.

Common mode rejection is a property of a balanced input, and so it is not applicable to a balanced output. However, output balance is sometimes quoted which gives an indication of how closely the two legs of a balanced output are matched. If the two legs were to be combined in antiphase total cancellation would ideally be achieved. In practice, around 70 dB of attenuation should be looked for.

7.1.2 Output noise

The output residual noise of a mixer, with all faders at minimum, should be at most −90 dBu. There is no point in having a very quiet microphone amplifier if a noisy output stage ruins it. With all channels routed to the output, and all faders at the 'zero' position, output noise (or 'mixing' noise) should be at least −80 dBu with the channel inputs switched to 'line' and set for unity gain. Switching these to 'mic' inevitably increases noise levels because this increases the gain of the input amplifier. It underlines the reason why all unused channels should be switched out, and their faders brought down to a minimum. Make sure that the aux outputs have a similarly good output noise level.

7.1.3 Impedance

A microphone input should have a minimum impedance of 1 kΩ. A lower value than this degrades the performance of many microphones. A line level input should have a minimum impedance of 10 kΩ. Whether it is balanced or unbalanced should be clearly stated, and consideration of the type of line level equipment that the mixer will be partnered with will determine the importance of balanced line inputs. All outputs should have a low impedance, below 200 ohms, balanced (600 ohms sounds nice and professional, but it is much too high, as described in section 12.9). Check that the aux outputs are also of very low impedance. Sometimes they are not. If insert points are provided on the input channels and/or outputs, these also should have very low output and high input impedances.

7.1.4 Frequency response

A frequency response which is within 0.2 dB between 20 Hz and 20 kHz for all combinations of input and output is desirable. The performance of audio transformers varies slightly with different source and load impedances, and a specification should state the range of loads between which a 'flat' frequency response will be obtained. Above 20 kHz, and probably below 15 Hz or so, the frequency response should fall away so that unwanted out-of-band frequencies are not amplified, for example radio-frequency breakthrough or subsonic interference.

7.1.5 Distortion

Distortion should be quoted at maximum gain through the mixer and a healthy output level of, say, +10 dBu or more. This will produce a typical worst case, and should normally be less than 0.1 per cent THD (Total Harmonic Distortion). The distortion of the low-gain line level inputs to outputs can be expected to be lower: around 0.01 per cent. The outputs should be loaded with a fairly low impedance which will require more current from the output stages than a high impedance will, this helping to reveal any shortcomings. A typical value is 600 ohms.

Clipping and overload margins are discussed in Fact File 7.2.

FACT FILE

7.2

Clipping

A good mixer will be designed to provide a maximum electrical output level of at least +20 dBu. Many will provide +24 dBu. Above this electrical level clipping will occur, where the top and bottom of the audio waveform are chopped off, producing sudden and excessive distortion (see diagram). Since the nominal reference level of 0 dBu usually corresponds to

Max. +ve output voltage

Max. –ve output voltage

a meter indication of PPM 4 or –4 VU, it is very difficult to clip the output stages of a mixer. The maximum meter indication on a PPM would correspond in this case to an electrical output of around +12 dBu, and thus one would have to be severely bending the meter needles to cause clipping.

Clipping, though, may occur at other points in the signal chain, especially when large amounts of EQ boost have been added. If, say, 12 dB of boost has been applied on a channel, and the fader is set well above the 0 dB mark, clipping on the mix bus may occur, depending on overload margins here. Large amounts of EQ boost should not normally be used without a corresponding overall gain reduction of the channel for this reason.

An input pad or attenuator is often provided to prevent the clipping of mic inputs in the presence of high-level signals (see section 6.4.1).

7.1.6 Crosstalk

A signal from one input may induce a small signal in another channel, and this is termed 'crosstalk'. Crosstalk from adjacent channels should be well below the level of the legitimate output signal, and a figure of −80 dB or more should be looked for. Crosstalk performance tends to deteriorate at high frequencies due to capacitive coupling in wiring harnesses for instance, but a crosstalk of at least −60 dB at 15 kHz should still be sought. Similarly, very low-frequency crosstalk often deteriorates due to the power supply source impedance rising here, and a figure of −50 dB at 20 Hz is reasonable.

Ensure that crosstalk between all combinations of input and output is of a similarly good level. Sometimes crosstalk between channel auxiliaries is rather poorer than that between the main outputs.

7.2 Metering systems

Metering systems are provided on audio mixers to indicate the levels of audio signals entering and leaving the mixer. Careful use of metering is vital to the recording of the correct audio level on tape (and thus the avoidance of distortion and noise). In this section the merits of different metering systems are examined.

7.2.1 Mechanical metering

Two primary types of mechanical meters are in existence today: the VU (Volume Unit) meter (Figure 7.1) and the PPM (Peak Program Meter), as shown in Figure 7.2. These are very different to each other, the only real similarity being that they both have swinging needles. The British, or BBC-type, PPM is distinctive in styling in that it is black with numbers ranging from 1 to 7 equally spaced across its scale, there being 4 dB level difference between each gradation, except between 1 and 2 where there is usually a 6 dB change in level. The EBU PPM (also shown in Figure 7.2) has a scale calibrated in decibels. The VU, on the other hand, is usually white or cream, with a scale running from −20 dB up to +3 dB, ranged around a zero point which is usually the studio's electrical reference level.

Although this section is not intended to be a tutorial on tape machine line-up and reference levels, it is impossible to cover the subject of metering without reference to such topics, as they are inextricably intertwined. It is important to know how meter readings relate to the line-up standard in use in a particular environment, and to understand that these standards may vary between establishments. Fact File 7.3 discusses the relationship between meter indication and recording level on a tape recorder.

7.2.2 Problems with mechanical meters

PPMs respond well to signal peaks, that is they have a fast rise-time, whereas VUs are quite the opposite: they have a very slow rise-time. This means that VUs do not give a true representation of the peak level going on to tape, especially in cases when

Figure 7.1 Typical VU meter scale

a signal with a high transient content, such as a harpsichord, is being recorded, often showing as much as 10–15 dB lower than a peak-reading meter. This can result in overmodulation of the recording, especially with digital recorders where the system is very sensitive to peak overload. None the less, many people are used to working with VUs, and have learned to interpret them. They are good for measuring continuous signals such as tones, but their value for monitoring programme material is dubious in the age of digital recording.

VUs have no control over the fall-time of the needle, which is much the same as the rise-time, whereas PPMs are engineered to have a fast rise-time and a longer fall-time, which tends to be more subjectively useful. The PPM was designed to indicate peaks which would cause audible distortion, but does not measure the absolute peak level of a signal. Mechanical meters take up a lot of space on a console, and it can be impossible to find space for one meter per channel in the case of a multitrack console. In this case there are often only meters on the main outputs, and perhaps measuring some auxiliary signals, these being complemented on more expensive consoles by electronic bargraph meters, usually consisting of LED or liquid crystal displays, or some form of 'plasma' display.

Figure 7.2 (a) BBC-type peak programme meter (PPM). (b) European-type PPM

FACT FILE
7.3
Metering
and
distortion

Within a studio there is usually a 'reference level' and a 'peak recording level'. The reference level usually relates to the level at which a 1 kHz line-up tone from an analogue test tape, recorded at a reference flux level of 200, 250 or 320 nanowebers per metre, should play back on the console's meters. In analogue mixers this may correspond to PPM 4 on a BBC-type PPM or 'Test' on a European PPM. This may in turn correspond to −4 dB on a VU meter (the relationship between VUs and PPMs depends on the standard in use as described in section 7.2.4). Electrically PPM 4 usually corresponds to a level of 0 dBu. In the digital domain, line-up level usually corresponds to either −20 dBFS or −18 dBFS, depending on the area of the world and standard concerned. A relationship is therefore established between meter reading and signal level in each domain.

With analogue recording it is the magnetic flux level that mostly governs distortion and compression. Any distortion introduced by the

tape recorder's electronics will be minimal in comparison unless the level is excessively high (above, say, +20 dBu, which would be bending the needles on most normal console meters). High-quality analogue tape typically reaches its maximum output level (the level at which harmonic distortion reaches 3 per cent at 1 kHz) at 10–12 dB above 320 nWb m^{-1}, corresponding to around PPM7 if PPM4 is aligned for 320 nWb m^{-1}. This is therefore an advisable maximum for analogue tape recording unless a particular effect is desired.

In broadcasting it is normal to peak no more than 8–9 dB above line-up level (PPM 6 in the UK) as higher levels than this can have serious effects on analogue transmitter distortion. In digital audio systems it is possible to peak anywhere up to 0 dBFS without incurring increases in distortion, and many recording engineers use all this 'headroom' in order to maximise dynamic range unless the recordings are made to broadcasting standards that limit peak levels. Digital systems clip hard at 0 dBFS whereas analogue tape tends to give rise to gradually increasing distortion and level compression as levels rise.

7.2.3 Electronic bargraph metering

Unlike mechanical meters, electronic bargraphs have no mechanical inertia to be overcome, so they can effectively have an infinitely fast rise-time although this may not be the ideal in practice. Cheaper bargraphs are made out of a row of LEDs (Light Emitting Diodes), and the resolution accuracy depends on the number of LEDs used. This type of display is sometimes adequate, but unless there are a lot of gradations it is difficult to use them for line-up purposes. Plasma and liquid crystal displays look almost continuous from top to bottom, and do not tend to have the glare of LEDs, being thus more comfortable to work with for any period of time. Such displays often cover a dynamic range far greater than any mechanical meter, perhaps from −50 dB up to +12 dB, and so can be very useful in showing the presence of signals which would not show up on a mechanical PPM. Such a meter is illustrated in Figure 7.3.

Figure 7.3 Typical peak-reading bargraph meter with optional VU scale

There may be a facility provided to switch the peak response of these meters from PEAK to VU mode, where they will imitate the scale and ballistic response of a VU meter. On more up-market designs it may be possible to use the multitrack bargraphs as a spectrum analyser display, indicating perhaps a one-third octave frequency-band analysis of the signal fed to it. Occasionally, bargraph displays incorporate a peak-hold facility. A major advantage of these vertical bargraphs is that they take up very little horizontal space on a meter bridge and can thus be used for providing one meter for every channel of the console: useful for monitoring the record levels on a multitrack tape machine. In this case, the feed to the meter is usually taken off at the input to the monitor path of an in-line module.

Miscellaneous meters may also be provided on the aux send outputs for giving some indication of the level being sent to auxiliary devices such as effects. These are commonly smaller than the main meters, or may consist of LED bargraphs with lower resolution. A phase meter or correlation meter is another option often available, this usually being connected between the left and right main monitor outputs to indicate the degree of phase correlation between these signals. This can be either mechanical or electronic. In broadcast environments, sum and difference (or M and S) meters may be provided to show the level of the mono-compatible and stereo difference signals in stereo broadcasting. These often reside alongside a stereo meter for left and right output levels.

7.2.4 Correlation between different metering standards

The correlation between meter indication and electrical output level varies depending on the type of meter and the part of the world concerned. Figure 7.4 shows a number of common meter scales and the relationship between these scales and

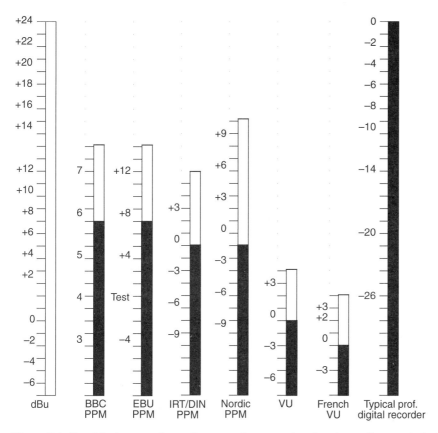

Figure 7.4 Graphical comparison of commonly encountered meter scalings and electrical levels in dBu. (After David Pope, with permission)

electrical output level of the mixer. As introduced in Fact File 7.3, there is a further correlation to be concerned with, this being the relationship between the electrical output level of the mixer and the recording level on an analogue or digital tape machine. This is discussed in greater detail in Fact File 8.5.

7.2.5 Meter take-off point

Output level meter-driving circuits should normally be connected directly across the outputs so that they register the real output levels of the mixer. This may seem self-evident but there are certain models in which this is not the case, the meter circuit taking its drive from a place in the circuit just before the output amplifiers. In such configurations, if a faulty lead or piece of equipment the mixer is connected to places, say, a short-circuit across the output the meter will nevertheless read normal levels, and lack of signal reaching a destination will be attributed to other causes. The schematic circuit diagrams of the mixer can be consulted to ascertain whether such an arrangement has been employed. If it is not clear, a steady test tone can be sent to the mixer's output, giving a high meter reading. Then a short-circuit can be deliberately applied across the output (the output amplifier will not normally be harmed by several seconds of short-circuit) and the meter watched. If the indicated level drastically reduces then the meter is correctly registering the real output. If it stays high then the meter is taking its feed from elsewhere.

7.3 Automation

7.3.1 Background

The original, and still most common form of mixer automation is a means of storing fader positions dynamically against time for reiteration at a later point in time, synchronous with recorded material. The aim of automation has been to assist an engineer in mixdown when the number of faders that need to be handled at once become too great for one person. Fader automation has resulted in engineers being able to concentrate on sub-areas of a mix at each pass, gradually building up the finished product and refining it.

MCI first introduced VCA (voltage controlled amplifier) automation for their JH500 series of mixing consoles in the mid-seventies, and this was soon followed by imitations with various changes from other manufacturers. Moving fader automation systems, such as Neve's NECAM, were introduced slightly later and tended to be more expensive than VCA systems. During the mid 1980s, largely because of the falling cost of microprocessor hardware, console automation enjoyed further advances resulting in developments such as snapshot storage, total dynamic automation, retrofit automation packages, and MIDI-based automation. It is now possible to install basic fader automation on a console for only a few thousand pounds, whereas previously one might have been contemplating tens of thousands. The rise of digital mixers and digitally controlled analogue mixers with integral automation is likely to continue the trend towards total automation of most mixer controls as a standard feature of new products.

In the following sections a number of different approaches to console automation will be presented and discussed.

7.3.2 Fader automation

There are two common means of memorising and controlling the gain of a channel: one which stores the positions of the fader and uses this data to control the gain of a VCA or digital attenuator, the other which also stores fader movements but uses this information actually to drive the fader's position using a motor. The former is cheaper to implement than the latter, but is not so ergonomically satisfactory because the fader's physical position may not always correspond to the gain of the channel.

It is possible to combine elements of the two approaches in order that gain control can be performed by a VCA but with the fader being moved mechanically to display the gain. This allows for rapid changes in level which might be impossible using physical fader movements, and also allows for dynamic gain offsets of a stored mix whilst retaining the previous gain profile (see below). In the following discussion the term 'VCA faders' may be taken to refer to any approach where indirect gain control of the channel is employed.

With VCA faders it is possible to break the connection between a fader and the corresponding VCA, as was described in Fact File 6.5. It is across this break point that an automation system will normally be connected. The automation processor then reads a digital value corresponding to the position of the fader and can return a value to the VCA to control the gain of the channel (see Figure 7.5). The information sent back to the VCA would depend on the operational mode of the system at the time, and might or might not correspond directly to the fader position. Common operational modes are:

- WRITE: VCA gain corresponds directly to the fader position
- READ: VCA gain controlled by data derived from a previously stored mix
- UPDATE: VCA gain controlled by a combination of previously stored mix data and current fader position
- GROUP: VCA gain controlled by a combination of the channel fader's position and that of a group master

The fader position is measured by an analogue-to-digital convertor (see Chapter 10), which turns the DC value from the fader into a binary number (usually eight or ten bits) which the microprocessor can read. An eight bit value suggests that the fader's position can be represented by one of 256 discrete values, which is usually enough to give the impression of continuous movements, although professional

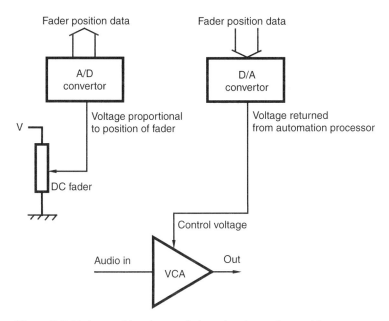

Figure 7.5 Fader position is encoded so that it can be read by an automation computer. Data returned from the computer is used to control a VCA through which the audio signal flows

systems tend to use ten bit representation for more precise control (1024 steps). The automation computer 'scans' the faders many times a second and reads their values. Each fader has a unique address and the information obtained from each address is stored in a different temporary memory location by the computer. A generalised block diagram of a typical system is shown in Figure 7.6.

The disadvantage of such a system is that it is not easy to see what the level of the channel is. During a read or update pass the automation computer is in control of the channel gain, rather than the fader. The fader could be half way to the bottom of its travel while the gain of the VCA was near the top. Sometimes a mixer's bargraph meters can be used to display the value of the DC control voltage which is being fed from the automation to the VCA, and a switch is sometimes provided to change their function to this mode. Alternatively a separate display is provided for the automation computer, indicating fader position with one marker and channel gain with another.

VCA faders are commonly provided with 'null' LEDs: little lights on the fader package which point in the direction that the fader must be moved to make its position correspond to the gain of the VCA. When the lights go out (or when they are both on), the fader position is correct. This can sometimes be necessary when

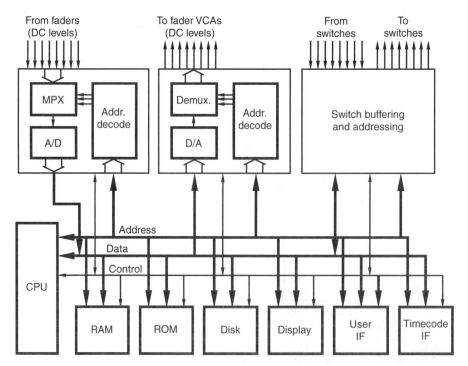

Figure 7.6 Generalised block diagram of a mixer automation system handling switches and fader positions. The fader interfaces incorporate a multiplexer (MPX) and demultiplexer (Demux) to allow one convertor to be shared between a number of faders. RAM is used for temporary mix data storage, ROM may hold the operating software program. The CPU is the controlling microprocessor

modifying a section of the mix by writing over the original data. If the data fed to the VCA from the automation is different to the position of the fader, then when the mode is switched from read to write there will be a jump in level as the fader position takes over from the stored data. The null lights allow the user to move the fader towards the position dictated by the stored data, and most systems only switch from read to write when the null point is crossed, to ensure a smooth transition. The same procedure is followed when coming out of rewrite mode, although it can be bypassed in favour of a sudden jump in level.

Update mode involves using the *relative* position of the fader to modify the stored data. In this mode, the fader's absolute position is not important because the system assumes that its starting position is a point of unity gain, thereafter adding the *changes* in the fader's position to the stored data. So if a channel was placed in update mode and the fader moved up by 3 dB, the overall level of the updated passage would be increased by 3 dB (see Figure 7.7). For fine changes in gain the fader can be preset near the top of its range before entering update mode, whereas larger changes can be introduced nearer the bottom (because of the gain law of typical faders).

Some systems make these modes relatively invisible, anticipating which mode is most appropriate in certain situations. For example, WRITE mode is required for the first pass of a new mix, where the absolute fader positions are stored, whereas subsequent passes might require all the faders to be in UPDATE.

A moving fader system works in a similar fashion, except that the data which is returned to the fader is used to set the position of a drive mechanism which physically moves the fader to the position in which it was when the mix was written. This has the advantage that the fader is its own means of visual feedback from the automation system and will always represent the gain of the channel.

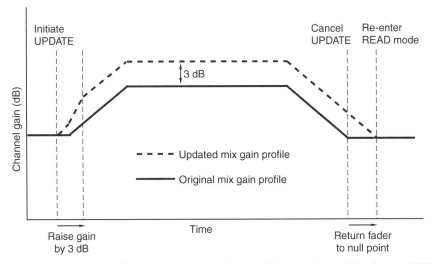

Figure 7.7 Graphical illustration of stages involved in entering and leaving an UPDATE or RELATIVE mode on an automated VCA fader

If the fader was permanently driven, there would be a problem when both the engineer and the automation system wanted to control the gain. Clutches or other forms of control are employed to remove the danger of a fight between fader and engineer in such a situation, and the fader is usually made touch-sensitive to detect the presence of a hand on it.

Such faders are, in effect, permanently in update mode, as they can at any time be touched and the channel gain modified, but there is usually some form of relative mode which can be used for offsetting a complete section by a certain amount. The problem with relative offsets and moving faders is that if there is a sudden change in the stored mix data while the engineer is holding the fader, it will not be executed. The engineer must let go for the system to take control again. This is where a combination of moving fader and VCA-type control comes into its own.

7.3.3 Grouping automated faders

Conventional control grouping (Fact File 6.5) is normally achieved by using dedicated VCA master faders. In an automated console it may be possible to do things differently. The automation computer has access to data representing the positions of all the main faders on the console, so it may allow any fader to be designated a group master for a group of faders assigned to it. It can do this by allowing the user to set up a fader as a group master (either by pressing a button on the fader panel, or from a central control panel). It will then use the level from this fader to modify the data sent back to all the other VCAs in that group, taking into account their individual positions as well. This idea means that a master fader can reside physically within the group of faders to which it applies, although this may not always be the most desirable way of working.

Sometimes the computer will store automation data relating to groups in terms of the motions of the individual channels in the group, without storing the fact that a certain fader was the master, whereas other systems will store the data from the master fader, remembering the fact that it was a master originally.

7.3.4 Mute automation

Mutes are easier to automate than faders because they only have two states. Mute switches associated with each fader are also scanned by the automation computer, although only a single bit of data is required to represent the state of each switch. A simple electronic switch can be used to effect the mute, and this often takes the form of a FET (field effect transistor) in the signal path, which has very high attenuation in its 'closed' position (see Figure 7.8). Alternatively, some more basic systems effect mutes by a sudden change in VCA gain, pulling it down to maximum attenuation.

7.3.5 Storing the automation data

Early systems converted the data representing the fader positions and mute switches into a modulated serial data stream which could be recorded alongside the audio to which it related on a multitrack tape. In order to allow updates of the data, at least

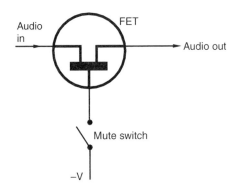

Figure 7.8 Typical implementation of a FET mute switch

two tracks were required: one to play back the old data, and one to record the updated data, these usually being the two outside tracks of the tape (1 and 24 in the case of a 24 track machine). This was limiting, in that only the two most recent mixes were ever available for comparison (unless more tracks were set aside for automation), whole tracks had to be mixed at a time (because otherwise the updated track would be incomplete), and at least two audio tracks were lost on the tape. Yet it meant that the mix data was always available alongside the music, eliminating the possibility of losing a disk with the mix data stored separately.

More recent systems use computer hardware to store mix data, in RAM and on disks (either hard or floppy). Data is synchronised to the audio by recording time-code on one track of the tape which uniquely identifies any point in time, this being read by the automation system and used to relate tape position to stored data. This method gives almost limitless flexibility in the modification of a mix, allowing one to store many versions, of which sections can be joined together 'off-line' (that is, without the tape running) or on-line, to form the finished product. The finished mix can be dumped to a disk for more permanent storage, and this disk could contain a number of versions of the mix.

It is becoming quite common for cheaper automation systems to use MIDI for the transmission of fader data. A basic automation computer associated with the mixer converts fader positions into MIDI information using a device known as a UART which generates and decodes serial data at the appropriate rate for the MIDI standard, as shown in Figure 7.9. MIDI data can then be stored on a conventional sequencer or using dedicated software. This proves adequate for a small number of channels or for uncomplicated mixes, especially if a dedicated sequencer is used for the storage of automation data, but it is possible that a single MIDI interface would be overloaded with information if large mixes with much action were attempted. One recent system overcomes these limitations by using a multiport MIDI interface (see Chapter 14) and a non-standard implementation of MIDI to carry ten bit fader data for a large number of channels.

7.3.6 Integrating machine control

Control of a tape machine or machines is a common feature of modern mixers. It may only involve transport remotes being mounted in the centre panel somewhere,

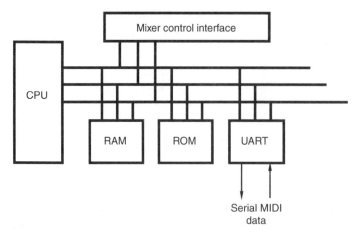

Figure 7.9 A UART is used to route MIDI data to and from the automation computer

or it may involve a totally integrated autolocator/synchroniser associated with the rest of the automation system. On top-flight desks, controls are provided on the channel modules for putting the relevant tape track into record-ready mode, coupled with the record function of the transport remotes. This requires careful interfacing between the console and the tape machine, but means that it is not necessary to work with a separate tape machine remote unit by the console.

It is very useful to be able to address the automation in terms of the mix in progress: in other words, 'go back to the second chorus', should mean something to the system, even if abbreviated. The alternative is to have to address the system in terms of timecode locations. Often, keys are provided which allow the engineer to return to various points in the mix, both from a mix data point-of-view and from the tape machines' point-of-view, so that the automation system locates the tape machine to the position described in the command, ready to play. This facility is often provided using an integral synchroniser with which the automation computer can communicate, and this can be a semi-customised commercial synchroniser from another manufacturer.

MIDI Machine Control (MMC) is also becoming a popular means of remote control for modular multitrack recording equipment, and is another way of interfacing a tape recorder to an automation system.

7.3.7 Retrofitting automation

Automation can usually be retrofitted into existing consoles which do not have any automation. These systems usually control only the faders and the mutes, as anything else requires considerable modification of the console's electronics, but the relatively low price of some systems makes them attractive, even on a modest budget. Fitting normally involves a modification or replacement of the fader package, to incorporate VCAs in consoles which don't have them, or to break into the control path between fader and VCA in systems which do. This job can normally be achieved in a day. It is also possible to retrofit moving fader automation.

A separate control panel may be provided, with buttons to control the modes of operation, as well as some form of display to show things like VCA gains, editing modes, and set up data. The faders will be interfaced to a processor rack which would reside either in a remote bay, or under the console, and this will normally contain a disk drive to store the final mixes. Alternatively a standard desktop computer will be used as the control interface.

7.3.8 Total automation systems

This title means different things to different people. SSL originally coined the term 'Total Recall' for its system, which was not in fact a means of completely resetting every control on the console; rather it was a means of telling the operator where the controls should be and leaving him to reset them himself. None the less, this saved an enormous amount of time in the resetting of the console in between sessions, because it saved having to write down the positions of every knob and button.

True Total Reset is quite a different proposition and requires an interface between the automation system and every control on the console, with some means of measuring the position of the control, some means of resetting it, and some means of displaying what is going on. A number of options exist, for example one could:

- motorise all the rotary pots
- make all the pots continuously rotating and provide a display
- make the pots into up/down-type incrementers with display
- provide assignable controls with larger displays

Of these, the first is impractical in most cases due to the space that motorised pots would take up, the reliability problem, and the cost, although it does solve the problem of display. The second would work, but again there is the problem that a continuously rotating pot would not have a pointer because it would merely be a means of incrementing the level from wherever it was at the time, so extra display would be required and this takes up space. None the less some ingenious solutions have been developed, including incorporating the display in the head of rotary controls (see Figure 7.10). The third is not ergonomically very desirable, as the human prefers analogue interfaces rather than digital ones, and there is no room on a conventional console for all the controls to be of this type with their associated displays. Most of the designs which have implemented total automation have adopted a version of the fourth option: that is to use fewer controls than there are functions, and to provide larger displays.

The concept of total automation is inherent in the principles of an assignable mixing console. In such a console, few of the controls carry audio directly as they are only interfaces to the control system, so one knob may control the HF EQ for any channel to which it is assigned, for example. Because of this indirect control,

(a)   (b)

Figure 7.10 Two possible options for positional display with continuously rotating knobs in automated systems. (a) Lights around the rim of the knob itself; (b) lights around the knob's base

usually via a microprocessor, it is relatively easy to implement a means of storing the switch closures and settings in memory for reiteration at a later date.

7.3.9 Dynamic and static systems

Many analogue assignable consoles use the modern equivalent of a VCA: the digitally controlled attenuator, to control the levels of various functions such as EQ, aux sends, and so on. If every function on the desk is to be dynamically memorised (that is, changes stored and replayed in real time, and continuously) then a lot of data is produced and must be dealt with quickly to ensure suitably seamless audible effects. Complex software and fast processors are required, along with a greater requirement for memory space.

Static systems exist which do not aim to store the continuous changes of all the functions, but they will store 'snapshots' of the positions of controls which can be recalled either manually or with respect to timecode. This can often be performed quite regularly (many times a second) and in these cases we approach the dynamic situation, but in others the reset may take a second or two which precludes the use of it during mixing. Changes must be silent to be useful during mixing.

Other snapshot systems merely store the settings of switch positions, without storing the variable controls, and this uses much less processing time and memory. Automated routing is of particular use in theatre work where sound effects may need to be routed to a complex combination of destinations. A static memory of the required information is employed so that a single command from the operator will reset all the routing ready for the next set of sound cues.

7.4 Digital mixers

In a digital mixer incoming analogue signals are converted to a digital signal as early as possible so that all the functions are performed entirely in the digital domain. Digital inputs and outputs can be provided to connect recording devices and other digital equipment (see Chapter 10) without conversion to analogue. The advantage of this is that once the signal is in the digital domain it is inherently more robust than its analogue counterpart: it is virtually immune from crosstalk, and is un-affected by lead capacitance, electromagnetic fields from mains wiring, additional circuit distortion and noise, and other forms of interference.

Functions such as gain, EQ, delay, phase, routing, and effects such as echo, reverb, compression and limiting, can all be carried out in the digital domain precisely and repeatably using digital signal processing as described in Chapter 10. The microphone amplifiers can be sited remotely from the mixer, close to the microphones themselves, the gain still being adjusted from the mixer, so that their line level outputs can now be converted to digital format before being input to the mixer. Operationally a digital mixer can remain similar to its analogue counterpart, although the first commercial examples have tended at least partially to follow the assignable route described above. The more fully assignable digital mixing console is ergonomically quite different from its analogue counterpart, and a brief description follows.

Because many of the controls of the traditional console such as pan, eq, aux send and group assign are either absent entirely from the assignable console's control surface or present only as single assignable sections or multi-function controls, many facilities can be packed into a unit of modest dimensions and quite modest cost, the signal remaining in the digital domain with as much as 32-bit internal processing resolution to cope with extremes of signal level, eq settings and other effects. Inputs can be a mixture of analogue and digital (the latter configurable via plug-in modules for Tascam, ADAT, Yamaha and AES/EBU formats, for example) with digital and analogue main and monitoring outputs.

Typically, the control surface consists of many input channel faders and/or rotary knobs, each channel having 'active' and 'select' buttons. Much smaller areas of the control surface are given over to single sections of eq, routing (aux and group) and processing: these sections are automatically assigned to one particular channel when its 'select' button is active before adjustments can take place. Thus most processes which are affecting the signals are not continuously on view or at the fingertips of the operator as is the case with the traditional analogue desk. The assignable console is therefore more suitable to recording work (particularly post-session mixdowns) where desk states can be built up gradually and saved to scene memories, rather than to live performance and primary recording work where continuous visual indication of and access to controls remains desirable.

Facilities such as channel delay, effects processing, moving fader automation and fader ganging, scene memories offering total recall of all settings, MIDI (including memory load and dump via separate MIDI data filers), and timecode interface are typically offered, and a display screen shows the status of all controls: either in simple global formats for the whole console for parameters such as routing, channel delay, scene memory details and the like, or in much greater detail for each individual channel. Metering can also be shown. Cursors facilitate both navigation around the screen displays and adjustments of the various parameters.

Digital mixing has now reached the point where it can be implemented cost effectively, and Yamaha has released a number of low cost digital mixers with full automation. An example is pictured in Figure 7.11. At the other end of the scale companies are manufacturing large-scale studio mixers with an emphasis on ultra-high sound quality and an ergonomically appropriate control interface. An example of such a mixer with a multi-purpose assignable control surface is pictured

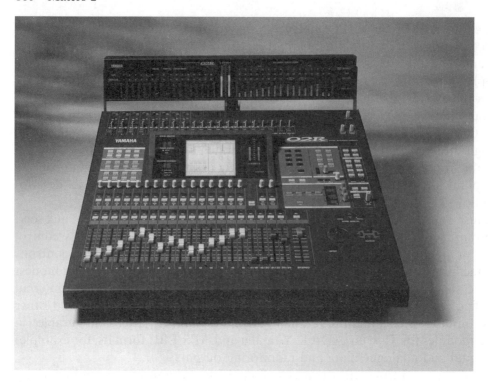

Figure 7.11 Yamaha 02R digital mixing console. (Courtesy of Sound PR)

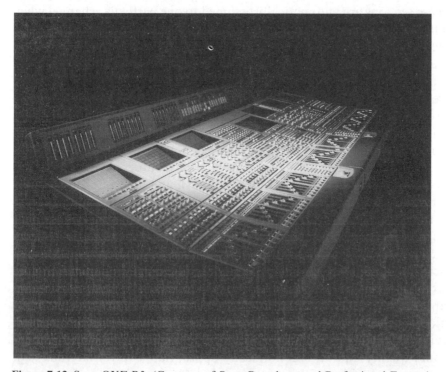

Figure 7.12 Sony OXF-R3. (Courtesy of Sony Broadcast and Professional Europe)

Figure 7.13 Solid State Logic *OmniMix*. (Courtesy Solid State Logic)

in Figure 7.12, and the alternative knob per function 'in-line' design of Solid State Logic's *MT Production* console is shown in Figure 7.13. At the low cost end of the scale, digital mixers are implemented within computer-based workstations and represented graphically on the computer display. Faders and other controls are moved using a mouse.

Recommended further reading

See *General further reading* at the end of this book.

Chapter 8

Analogue tape recording

Analogue sound recording is the process by which a continuously variable electrical waveform, representing an acoustic sound signal (see Chapter 1), is converted into an analogous pattern of modulation on a physical medium, which may be stored and subsequently replayed. In tape recording, this pattern of modulation is magnetic, such that the pattern of magnetisation (strength and direction of field) on the tape after recording is analogous to the sound waveform which was recorded. Analogue recording differs from digital recording (see Chapter 10) in that the recorded signal is physically and temporally continuous, whereas in digital recording it is time discrete (sampled) and stored in the form of data. A magnified analysis of the pattern of modulation of an analogue recording would show up clear similarities with the causatory sound waveform, whereas one would be unlikely to see any such similarity in a visual analysis of the recorded signal on a digital tape.

8.1 Magnetic tape

8.1.1 Structure

Magnetic tape consists of a length of plastic material which is given a surface coating capable of retaining magnetic flux rather in the manner that, say, an iron rod is capable of being magnetised (see Figure 8.1). The earliest recorders actually used a length of iron wire as the recording medium. In practice all modern tape has a polyester base which was chosen, after various trials with other formulations which proved either too brittle (they snapped easily) or too plastic (they stretched), for its good strength and dimensional stability. It is used throughout the tape industry from the dictation microcassette to the 2 inch (5 cm) multitrack variety. The coating is of a metal oxide, or metal alloy particles.

Magnetic layer

Plastic substrate

Figure 8.1 Cross-section through a magnetic recording tape

The most common coating used is of gamma-ferric oxide, a kind of purified rust with specially shaped particles (signified by 'gamma'). This formulation is used in cassettes along with the alternative formulations of chromium dioxide and its substitutes, and metal particles. It is also almost exclusively used for open-reel analogue master tapes of all widths – quarter inch, half inch, 1 inch, 2 inch – although a particular brand and formulation may not always be available in all widths.

8.1.2 Recent history

The BASF company of Germany introduced a chromium dioxide formulation in the early 1970s, claiming better high-frequency performance and improved signal-to-noise ratios. The patent on this formulation spurred other manufacturers to experiment with alternatives, and a substitute which consisted of cobalt-enriched ferric oxide emerged. Then so-called 'ferri-chrome' tape was introduced which consisted of a layer of ferric oxide over which was applied a layer of chromium dioxide or a substitute. High frequencies tend to be recorded close to the surface of an oxide layer, lower frequencies being recorded more deeply. The dual layer was therefore intended to exploit this, taking advantage of the top layer's good performance at high frequencies together with the lower layer's strengths in the areas of distortion and good lower-frequency output capability. This type of formulation is infrequently used today, partially because it tended to display appalling middle-frequency distortion characteristics, and also because of improvements in standard ferric and chrome formulations.

During the 1980s pure metal tape was developed for the cassette medium. Initially this posed great difficulties for the manufacturers due to the fact that very tiny particles of pure iron or iron alloy of the size required for recording tape tended to oxidise rapidly – they caught fire! This problem was, however, overcome and metal cassettes were launched which were rather more expensive than the oxide types and additionally required specially equipped cassette machines with record heads and circuitry capable of effectively magnetising the high-coercivity metal tape. Many cheaper cassette machines, although intended to cope with this tape and labelled as such, could not fully exploit it. The coercivity of the tape – its willingness to accept and retain magnetic flux – was such that cheap tape heads often saturated magnetically before the tape did, due to the extremely high bias currents required to magnetise it (see section 8.2.3). Improvements in distortion, output level and signal-to-noise ratios were obtainable with metal tape from good cassette machines, at a cost.

The metal formulation came into its own in a modified form when its high recording density capability proved ideal for digital tape recording, in special cassette form for R-DAT (see section 10.8) and also for hand-held domestic camcorder machines.

8.1.3 Cassette tape

Cassette tape comes in various lengths to give appropriate playing times for given applications – 'C5' gives 2.5 minutes playing time per side, C90 gives 45 minutes per side and so on. The longer-playing tapes are thinner in order that more tape can be loaded on to the spools. The C120 cassette has very thin tape which causes problems with many machines since their transports are unable to handle the delicate and very flimsy tape adequately. The tape will tend to stick on to the rubber pinch roller as it passes between the roller and the capstan, causing the transport to chew the tape up in bad cases, wrapping it around the capstan and pinch roller. Excessively thin tape does not sit comfortably in the tape guides in the cassette housing, and it is also not pulled across the tape heads very evenly, causing poor head-to-tape contact with loss of output and frequency response. It would be difficult to find a cassette machine manufacturer who actually sanctioned the use of C120 cassettes.

8.1.4 Open-reel tape

Open-reel quarter-inch tape intended for analogue recorders has been available in a variety of thicknesses. Standard Play tape has an overall thickness of 50 microns (micrometres), and a playing time (at 15 inches (38 cm) per second) of 33 minutes is obtained from a 10 inch (25 cm) reel. Long Play tape has an overall thickness of 35 microns giving a corresponding 48 minutes of playing time, which is very useful for live recording work. In the past 'Double Play' and even 'Triple Play' thicknesses have been available, these being aimed at the domestic open-reel market. These formulations are prone to snapping or stretching, as well as offering slightly poorer sound quality, and should not really be considered for professional use.

Standard Play tape is almost always 'back coated'. A rough coating is applied to the back of the tape during manufacture which produces neater and more even winding on a tape machine, by providing a certain amount of friction between layers which holds the tape in place. Also, the rough surface helps prevent air being trapped between layers during fast spooling which can contribute to uneven winding. Long Play tape is also available with a back coating, but as often as not it will be absent. It is worth noting that the flanges of a tape spool should only serve to protect the tape from damage. The 'pancake' of tape on the spool should not touch these flanges. Metal spools are better than plastic spools because they are more rigid and they do not warp. Professional open-reel tape can be purchased either on spools or in 'pancake' form on hubs without flanges. The latter is of course cheaper, but considerable care is needed in its handling so that spillage of the unprotected tape does not occur. Such pancakes are either spooled on to empty reels before use, or they can be placed on top of a special reel with only a lower flange. Professional tape machines are invariably operated with their decks horizontal. Half inch, 1 inch and 2 inch tape intended for multitrack recorders always comes on spools, is always of Standard Play thickness, and is always back coated.

Open-reel tape should have a batch number printed on the box. This number ensures that a given delivery of tape was all manufactured at the same time and so will have virtually identical magnetic properties throughout the batch. Different batches can have slightly different properties and a studio may wish to realign its machines when a new batch is started. It can be said, however, that variations from batch to batch are minimal these days.

8.2 The magnetic recording process

8.2.1 Introduction

Since tape is magnetic, the recording process must convert an electrical audio signal into a magnetic form. On replay the recorded magnetic signal must be converted back into electrical form. The process is outlined in Fact File 8.1. Normally a professional tape recorder has three heads, as shown in Figure 8.2, in the order erase–record–replay. This allows for the tape to be first erased, then re-recorded, and then monitored by the third head. The structure of the three heads is similar, but the gap of the replay head is normally smaller than that of the record head. It is possible to use the same head for both purposes, but usually with a compromise in performance. Such a two-head arrangement is often found in cheaper cassette machines which do not allow off-tape monitoring whilst recording. A simplified block diagram of a typical tape recorder is shown in Figure 8.3.

The magnetisation characteristics of tape are by no means linear, and therefore a high-frequency signal known as *bias* is added to the audio signal at the record head, generally a sine wave of between 100 and 200 kHz, which biases the tape towards a more linear part of its operating range. Without bias the tape retains very little magnetisation and distortion is excessive. The bias signal is of too high a frequency to be retained by the tape, so does not appear on the output during replay. Different types of tape require different levels of bias for optimum recording conditions to be achieved, and this will be discussed in section 8.2.3.

8.2.2 Equalisation

'Pre-equalisation' is applied to the audio signal before recording. This equalisation is set in such a way that the replayed short-circuit flux in an ideal head follows a standard frequency response curve (see Figure 8.4). A number of standards exist for different tape speeds, whose time constants are the same as those quoted for replay EQ in Table 8.1. Although the replayed *flux level* must conform to these curves, the electrical pre-EQ may be very different, since this depends on the individual head and tape characteristics. Replay equalisation (see Figure 8.5) is used to ensure that a flat response is available at the tape machine's output. It compensates for losses incurred in the magnetic recording/replay process, the rising output of the replay head with frequency, the recorded flux characteristic, and the fall-off in HF response where the recorded wavelength approaches the head gap width (see Fact File 8.2). Table 8.1 shows the time constants corresponding to the turnover frequencies of replay equalisers at a number of tape speeds. Again a number of standards exist. Time constant (normally quoted in microseconds) is the product

FACT FILE 8.1 A magnetic recording head

When an electrical current flows through a coil of wire a magnetic field is created. If the current only flows in one direction (DC) the electromagnet thus formed will have a north pole at one end and a south pole at the other (see diagram). The audio signal to be recorded on to tape is alternating current (AC), and when this is passed through a similar coil the result is an alternating magnetic field whose direction changes according to the amplitude and phase of the audio signal.

basically an electromagnet with a small gap in it. The tape passes across the gap, as shown in the diagram. The electrical audio signal is applied across the coil and an alternating magnetic field is created across the gap. Since the gap is filled with a non-magnetic material it appears as a very high 'resistance' to magnetic flux, but the tape represents a very low resistance in comparison and thus the flux flows across the gap via the tape, leaving it magnetised.

Magnetic flux is rather like the magnetic equivalent of electrical current, in that it flows from one pole of the magnet to the other in invisible 'lines of flux'. For sound recording it is desirable that the tape is magnetised with a pattern of flux representing the sound signal. A recording head is used which is

On replay, the magnetised tape moves across the head gap of a similar or identical head to that used during recording, but this time the magnetic flux on the tape flows through the head and thus *induces* a current in the coil, providing an electrical output.

of resistance and capacitance *(RC)* in the equivalent equalising filter, and the turnover frequency corresponding to a particular time constant can be calculated using:

$$f = 1/(2\pi RC)$$

The LF time constant of 3180 μs was introduced in the American NAB standard to reduce hum in early tape recorders, and has remained. HF time constants resulting in low turnover frequencies tend to result in greater replay noise, since HF is boosted over a wider band on replay, thus amplifying tape noise considerably. This is mainly why Type I cassette tapes (120 μs EQ) sound noisier than Type II tapes (70 μs EQ). Most professional tape recorders have switchable EQ to allow the

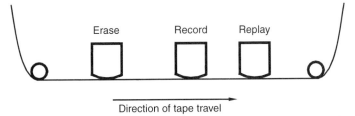

Figure 8.2 Order of heads on a professional analogue tape recorder

replay of NAB- and IEC/CCIR-recorded tapes. EQ switches automatically with tape speed in most machines.

Additional adjustable HF and LF EQ is provided on many tape machines, so that the recorder's frequency response may be optimised for a variety of operational conditions, bias levels and tape types.

8.2.3 Bias requirements

Higher bias levels enable the audio signal to be recorded more deeply into the oxide layer than would be the case with lower bias levels. High-coercivity tapes require a higher bias than low-coercivity tapes in order to magnetise the tape suitably. Such tapes can, however, display slightly poorer print-through performance (see below). The correct setting of bias level is vital for obtaining optimum performance from an analogue tape, and in professional systems the setting of bias is part of the day-to-day alignment of the tape machine (see section 8.7).

Bias requirements vary from tape to tape as has been said, and cassette tapes are grouped into four 'slots': Type I (ferric), Type II ('chrome' or CrO_2), Type III (ferri-chrome), and Type IV (metal). Cassette formulations are expected to conform closely to one of the above groups with respect to required bias level so that good results are achieved when various different brands are used with a given machine. This is important because variations in bias level cause significant variations in

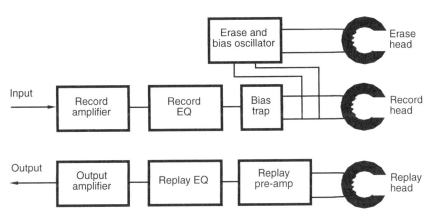

Figure 8.3 Simplified block diagram of a typical analogue tape recorder. The bias trap is a filter which prevents the HF bias signal feeding back into an earlier stage

Figure 8.4 Examples of standardised recording characteristics for short-circuit flux. (N.B: this is not equivalent to the electrical equalisation required in the record chain, but represents the resulting flux level replayed from tape, measured using an ideal head)

Figure 8.5 Examples of replay equalisation required to correct for the recording characteristic (see Figure 8.4), replay-head losses, and the rising output of the replay head with frequency

FACT FILE
8.2

Replay
head
effects

The output level of the replay head coil is proportional to the *rate of change of flux*, and thus the output level increases by 6 dB per octave as frequency rises (assuming a constant flux recording). Replay equalisation is used to correct for this slope.

At high frequencies the recorded wavelength on tape is very short (in other words the distance between magnetic flux reversals is very short). The higher the tape speed, the longer the recorded wavelength. At a certain high frequency the recorded wavelength will equal the replay-head gap width (see diagram) and the net flux in the head will be zero, thus no current will be induced. The result of this is that there is an upper cut-off frequency on replay (the extinction frequency), which is engineered to be as high as possible. Gap effects are noticeable below the cut-off frequency, resulting in a gradual roll-off in the frequency response as the wavelength approaches the gap length. Clearly, at low tape speeds (in which case the recorded wavelength is short) the cut-off frequency will be lower than at high tape speeds for a given gap width.

At low frequencies, the recorded wavelength approaches the dimensions of the length of tape in contact with the head, and various additive and cancellation effects occur when not all of the flux from the tape passes through the head, or when flux takes a 'short-circuit' path through the head. This results in low-frequency 'head bumps' or 'woodles' in the frequency response. The diagram below summarises these effects on the output of the replay head.

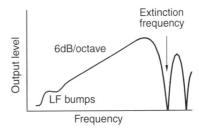

high-frequency performance. The domestic user cannot be expected to optimise his or her cassette machine for a particular type of tape as is routine in professional open-reel usage, although some cassette machines do offer fine bias adjustment, either manual or automatic. It is usually found that a given cassette machine will perform better with one brand of tape than with another depending on exactly how the manufacturer or distributor has aligned the machine, and it is worth experimenting with different types in order to find the one which gives the best performance on that machine.

8.2.4 Print-through

Print-through is caused by a modulated area of tape inducing its magnetism into the adjacent layer of tape on the spool during storage, rather in the manner that a pin, when left stuck to a magnet, will itself become magnetised. This manifests itself as pre-echo or post-echo, depending on which way the tape is wound off the machine, meaning that one can hear the beginning of a movement or programme after a pause at a very low level a second or so before it actually starts. At the end of the programme, faint traces of the last second or so can be heard repeated, although this is usually effectively masked by decaying reverberation or, if no significant reverberation is present, some leader tape will be spliced on immediately after the programme material has ceased.

The storing of master tapes 'tail out' – standard industry practice, indicated by red leader tape which should always be present on the end of every master tape – helps to avoid pre-echo because now a silent section will lie adjacent to the previous programme material rather than adjacent to the beginning of the following section. Tail-out storage of master tapes is also desirable because tape machines invariably give neater spooling on replay than on fast wind or rewind, and so a tape which has just been played and is now tail out will be stored in this neat state which helps to maintain the tape in good condition. Post-echo is preferable because it is at least more natural to hear a faint echo of what one has just heard than to hear a preview of a following section. Dying reverberation also helps to mask it as has been said.

Noise reduction (see Chapter 9) helps to reduce the consequences of print-through because on replay the decoding process pushes low-level signals further down in level as part of the expansion process. Since print-through signals are introduced in between noise reduction encoding and decoding they are therefore reduced in level.

8.3 The tape recorder

8.3.1 Studio recorder

Professional open-reel recorders fall into two categories: console mounted and portable. The stereo console recorder, intended for permanent or semi-permanent installation in a recording studio, outside broadcast truck or whatever generally sports rather few facilities, but has balanced inputs and outputs at line level (no microphone inputs), transport controls, editing modes, possibly a headphone socket, a tape counter (often in real time rather than in arbitrary numbers or revs), tape speed selector, reel size selector, and probably (though not always) a pair of level meters. It is deliberately simple because its job is to accept a signal, store it as faithfully as possible, and then reproduce it on call. It is also robustly built, stays aligned for long periods without the need for frequent adjustment, and will be expected to perform reliably for long periods. A typical example is pictured in Figure 8.6.

The inputs of such a machine will be capable of accepting high electrical levels – up to at least +20 dBu or around 8 volts – so that there is virtually no possibility of electrical input overload. The input impedance will be at least 10 kΩ. The outputs will be capable of driving impedances down to 600 ohms, and will have a source impedance of below 100 ohms. A facility will be provided for connecting a remote control unit so that the transport can be controlled from the mixing console, for instance. Often the real-time tape counter can also be remotely displayed so that the machine itself can virtually be ignored during a recording session. A noise reduction system can often be located within the housing of the machine, and the record button can send a DC control voltage to this which will automatically switch the noise reduction to encode during recording. Input and output level, bias and EQ

Figure 8.6 A typical professional open-reel two-track analogue tape recorder: the Studer A807-TC. (Courtesy of FWO Bauch Ltd)

controls will be provided, but tucked away so that they cannot be accidently misaligned. Often, small screwdriver holes are employed for these.

Depending on the machine, record 'safe' and 'ready' switches are normally provided for each track, as well as the ability to switch between sync replay and normal replay if it is a multitrack machine (see below). Varispeed controls often exist too, being a means of finely adjusting the speed of the machine either side of a standard play speed. There may also be the option to lock the machine to an external speed reference for synchronisation purposes.

8.3.2 Semi-professional recorder

Its semi-professional counterpart will be capable at its best of a performance that is little inferior, and in addition to being smaller and lighter will sport rather more facilities such as microphone inputs and various alternative input and output options. Headphone outlets will be provided along with record-level meters, source/tape monitor switching, variable output level, and perhaps 'sound on sound'-type facilities for simple overdub work. A typical example is shown in Figure 8.7. The semi-professional machine will not usually be as robustly constructed, this being of particular concern for machines which are to be transported since rough treatment can easily send a chassis askew, causing misalignment of the

Figure 8.7 A typical semi-professional two-track recorder: the Revox PR99. (Courtesy of Revox UK Ltd)

tape transport system which will be virtually impossible to correct. Some chassis are constructed of pressed steel which is not very rigid. A casting is much better.

Input and output connectors on a semi-pro machine will generally be of the domestic type – phono and unbalanced jack – and voltage levels may well be referenced around −10 dBV instead of +4 dBu. Tape guides may well not include the rollers and low-friction ball races of the professional machine. Facilities for interfacing with noise reduction systems will probably be absent. Such a machine is considerably cheaper than its fully professional counterpart, but the best examples of such machines do, however, return excellent performance consistently and reliably.

8.3.3 The portable machine

The professional portable tape machine, unlike its console equivalent, needs to offer a wide range of facilities since it will be required to provide such things as balanced outputs and inputs, both at line and microphone level, phantom and A–B mic powering, metering, battery operation which allows usefully long recording times, the facility to record timecode and pilot tone for use in TV and film work, illumination of the important controls and meters, and possibly even basic mixing facilities. It must be robust to stand up to professional field use, and small enough to be carried easily. Nevertheless, it should also be capable of accepting professional 10 inch (25 cm) reels, and adaptors are usually available to facilitate this. A lot has to be provided in a small package, and the miniaturisation necessary does not come cheap. The audio performance of such machines is at least as good as that of a studio recorder. A typical commercial example is pictured in Figure 8.8.

Figure 8.8 A typical professional portable two-track recorder: the Nagra IV-S. (Courtesy of Nagra Kudelski (GB) Ltd)

8.3.4 The multitrack machine

Multitrack machines come in a variety of track configurations and quality levels. The professional multitrack machine tends to be quite massively engineered and is designed to give consistent, reliable performance on a par with the stereo mastering machine. The transport needs to be particularly fine so that consistent performance across the tracks is achieved. A full reel of 2 inch tape is quite heavy, and powerful spooling motors and brakes are required to keep it under control. Apart from the increased number of tracks, multitrack machines are basically the same as their stereo counterparts and manufacturers tend to offer a range of track configurations within a given model type. Alignment of course takes a lot longer, and computer control of this is most welcome when one considers that 24 tracks implies 168 separate adjustments! A typical 24 track tape machine is pictured in Figure 8.9.

A useful feature to have on a multitrack recorder is an automatic repeat function or autolocate. The real-time counter can be programmed so that the machine will repeat a section of the tape over and over again within the specified start and end points to facilitate mixdown rehearsals. Multitrack recorders will be equipped with a number of unique features which are vital during recording sessions. For example, sync replay (see Fact File 8.3), gapless, noiseless punch-in (allowing any track to be

Figure 8.9 A typical professional multitrack recorder: the Saturn 824. (Courtesy of Saturn Research)

FACT FILE

8.3

Sync replay

The overdubbing process used widely in multitrack recording requires that musicians can listen to existing tracks on the tape whilst recording others. If replay was to come from the replay head and the new recording was to be made on to the record head, a recorded delay would arise between old and new material due to the distance between the heads. Sync replay allows the record head to be used as a replay head on the tracks which are not currently recording, thus maintaining synchronisation. The sound quality coming off the record head (called the sync head in this mode) is not always as good as that coming off the replay head, because the gap is larger, but it is adequate for a cue feed. Often separate EQ is provided for sync replay to optimise this. Mixdown should always be performed from the replay head.

Some manufacturers have optimised their head technology such that record and replay heads are exactly the same, and thus there is no difference between true replay and sync replay.

---

dropped into record at any point without introducing a gap or a click) and spot erasure (allowing a track to be erased manually over a very small portion of tape).

Various semi-professional multitrack machines have appeared over the years, making use of decreased track widths in the interests of cheaper transports and lower tape running costs at the expense of quality. Quarter-inch tape has been used for four, eight and even sixteen tracks; 1 inch tape has also been used for eight and sixteen tracks. Some models incorporate domestic-type noise reduction systems such as Dolby C, and a recent machine makes use of the Dolby S system, derived from professional SR (see section 9.2.5).

Cheap multitrack machines will have domestically orientated input and output sockets, and will be unbalanced. Comprehensive alignment facilities may not be provided. Phase errors between the tracks can be high, and the transport will not be first class. Crosstalk between tracks can be poor. The machine may not last as long as its professional counterpart, nor may it stay in alignment for as long.

8.4 Track formats

8.4.1 Mono, two-track and stereo formats

The professional stereo format is known as half track, because each track occupies approximately half of the tape width. Full-track mono recorders record the signal across the whole of the tape width (see Figure 8.10). Domestic open-reel two-track machines conformed to the quarter-track format, in which left and right channels were recorded in the first (upper) and third quarters of the tape's width. The tape could then be turned over so that fresh material could be recorded on the other two quarters of the tape. Twice the recording time could therefore be accommodated. The 'other side' of the tape, as it is conveniently called, is actually the same side but

Figure 8.10 Track patterns for quarter-inch recording. (a) Mono, full-track. (b) Stereo, half-track. (c) Four track, or stereo quarter-track (tracks 1 and 3 in one direction and tracks 2 and 4 in the other)

uses a different part of the tape's area. The disadvantages are that reduced track widths mean higher distortion and poorer signal-to-noise ratios; greater possibility of 'drop-out' (momentary loss of signal) due to head-to-tape contact being more critical; and the fact that the editing of a programme will also chop up any material recorded on the other side. Some semi-professional machines are quarter track, and one occasionally finds such a machine in a studio to cater for quarter-track tapes which may sometimes be brought in. Alternatively, some half-track machines have a second replay head in quarter-track form.

FACT FILE

8.4

NAB and DIN formats

It is important to differentiate between a 'stereo' machine and a 'two-track' machine. With the latter, it is possible to record on the two tracks at separate times if required. Synchronised recordings are also possible if the record head can be switched to perform the replay function instead, thus enabling monitoring of an existing recording on, say, track 1, whilst recording on to track 2. Because entirely separate and unconnected sounds may sometimes need to be recorded on the two tracks, crosstalk between the two needs to be kept to a minimum. The unmodulated band in between the tracks needs to be wider

therefore, and so the head guard bands are wider than is the case with the 'stereo' machine. The wide-guard-band format is called the NAB format (which is not necessarily related to NAB equalisation). The narrow-guard-band stereo format is called the DIN format.

This has consequences with respect to compatibility between the two types of machine. A recording made on a DIN machine will occupy more of the guard band than will be the case with the NAB machine. If a recording made on the stereo machine is erased on the two track, the latter's erase head will not completely remove all the signal and traces of it will still be heard on replay. The stereo machine's erase head is full track. Also, if NAB tapes are replayed on DIN heads there will be a marginal increase in noise of 1–2 dB.

Two different, professional two-track formats exist, NAB and DIN, as described in Fact File 8.4.

8.4.2 Multitrack formats

The professional standard for tape width adheres as far as possible to the scaling derived from the quarter-inch two-track mastering machine. A four-track machine therefore uses half-inch-width tape. Next comes the eight-track which makes use of 1 inch tape. Sixteen track utilises 2 inch tape, and 24 tracks are also accommodated across 2 inches. Comparable quality levels of all the tracks across all the formats are therefore achieved, with no audio degradation as one moves, say, from four track to eight track.

Tracks are normally numbered from 1 at the top of the tape to the highest number at the bottom.

8.5 Magnetic recording levels

It has already been said that the equivalent of electrical current in magnetic terms is magnetic flux, and it is necessary to understand the relationship between electrical levels and magnetic recording levels on tape (a little was said about this in Fact File 7.3). The performance of an analogue tape recorder depends very much on the magnetic level recorded on the tape, since at high levels one encounters distortion and saturation, whilst at low levels there is noise (see Figure 8.11). A window exists, between the noise and the distortion, in which the audio signal must be recorded, and the recording level must be controlled to lie optimally within this region. For this reason the relationship between the electrical input level to the tape machine and the flux level on tape must be established so that the engineer knows what meter indication on a mixer corresponds to what magnetic flux level. Once a relationship has been set up it is possible largely to forget about magnetic flux levels and concentrate on the meters. Fact File 8.5 discusses magnetic flux reference levels.

Figure 8.11 The available dynamic range on an analogue tape lies between the noise floor and the MOL. Precise figures depend on tape and tape machine

FACT FILE 8.5   Magnetic reference levels

Magnetic flux density on tape is measured in nanowebers per meter ($nWb\ m^{-1}$), the weber being the unit of magnetic flux. Modern tapes have a number of important specifications, probably the most significant being maximum output level (MOL), HF saturation point and noise level. (These parameters are also discussed in Appendix 1.) The MOL is the flux level at which third-harmonic distortion reaches 3 per cent of the fundamental's level, measured at 1 kHz (or 5 per cent and 315 Hz for cassettes), and can be considered as a sensible peak recording level unless excessive distortion is required for some reason. The MOL for a modern high-quality tape lies at a magnetic level of around $1000\ nWb\ m^{-1}$, or even slightly higher in some cases, and thus it is wise to align a tape machine such that this magnetic level corresponds fairly closely to the peak level indication on a mixer's meters.

A common reference level in electrical terms is 0 dBu, which often lines up with PPM 4 or −4 VU on a mixer's meter. This must be aligned to correspond to a recognised *magnetic* reference level on the tape, such as $320\ nWb\ m^{-1}$. Peak recording level, in this case, would normally be around +8 dBu if the maximum allowed PPM indication was to be 6, as is conventional. This would in turn correspond to a magnetic recording level of $804\ nWb\ m^{-1}$, which is close to the MOL of the tape and would probably result in around 2 per cent distortion.

There are a number of accepted magnetic reference levels in use worldwide, the principal ones being 200, 250 and $320\ nWb\ m^{-1}$. There is 4 dB between 200 and $320\ nWb\ m^{-1}$, and thus a $320\ nWb\ m^{-1}$ test tape should replay 4 dB higher in level on a meter than a $200\ nWb\ m^{-1}$ test tape. American test tapes often use $200\ nWb\ m^{-1}$ (so-called NAB level), whilst German tapes often use $250\ nWb\ m^{-1}$ (sometimes called DIN level). Other European tapes tend to use $320\ nWb\ m^{-1}$ (sometimes called IEC level). Test tapes are discussed further in the main text.

There is currently a likelihood in recording studios that analogue tapes are being under-recorded, since the performance characteristics of modern tapes are now good enough to allow higher peak recording levels than before. A studio which aligned PPM 4 to equal 0 dBu, in turn to correspond to only $200\ nWb\ m^{-1}$ on tape, would possibly be leaving 4–6 dB of headroom unused on the tape, sacrificing valuable signal-to-noise ratio.

8.6 What are test tapes for?

A test tape is a reference standard recording containing pre-recorded tones at a guaranteed magnetic flux level. A test tape is the only starting point for aligning a tape machine, since otherwise there is no way of knowing what magnetic level will end up on the tape during recording. During alignment, the test tape is replayed, and a 1 kHz tone at the specified magnetic flux level (say $320\ nWb\ m^{-1}$) produces a certain electrical level at the machine's output. The output level would then be adjusted for the desired electrical level, according to the studio's standard (say 0 dBu), to read at a standard meter indication (say PPM 4). It is then absolutely clear that if the output level of the tape machine is 0 dBu then the magnetic level on tape is $320\ nWb\ m^{-1}$. After this relationship has been set up it is then possible to *record* a signal on tape at a known magnetic level – for example, a 1 kHz tone at 0 dBu could be fed to the input of the tape machine, and the *input* level adjusted until the output read 0 dBu also. The 1 kHz tone would then be recording at a flux level of 320 nWb m^{-1}.

Test tapes also contain tones at other frequencies for such purposes as azimuth alignment of heads and for frequency response calibration of replay EQ (see below). A test tape with the required magnetic reference level should be used, and it should also conform to the correct EQ standard (NAB or CCIR, see section 8.2.2). Tapes are available at all speeds, standards and widths, with most being recorded across the full width of the tape.

8.7 Tape machine alignment

8.7.1 Head inspection and demagnetisation

Heads and tape guides must be periodically inspected for wear. Flats on guides and head surfaces should be looked for; sometimes it is possible to rotate a guide so that a fresh portion contacts the tape. Badly worn guides and heads cause sharp angles to contact the tape which can damage the oxide layer. Heads have been made of several materials. Mu-metal heads have good electromagnetic properties, but are not particularly hard wearing. Ferrite heads wear extremely slowly and their gaps can be machined to tight tolerances. The gap edges can, however, be rather brittle and require careful handling. Permalloy heads last a long time and give a good overall performance, and are often chosen. Head wear is revealed by the presence of a flat area on the surface which contacts the tape. Slight wear does not necessarily indicate that head replacement is required, and if performance is found to be satisfactory during alignment with a test tape then no action need be taken.

Replay-head wear is often signified by exceptionally good high-frequency response, requiring replay EQ to be reduced to the lower limit of its range. This seems odd but is because the replay gap on many designs gets slightly narrower as the head wears down, and is at its narrowest just before it collapses!

Heads should be cleaned regularly using either isopropyl alcohol and a cotton bud, or a freon spray. They should also be demagnetised fairly regularly, since heads can gradually become slightly permanently magnetised, especially on older machines, resulting in increased noise and a type of 'bubbling' modulation noise in the background on recordings. A demagnetiser is a strong AC electromagnet which should be switched on well away from the tape machine, keeping it clear of anything else magnetic or metal. This device will erase a tape if placed near one! Once turned on the demagger should be drawn smoothly and slowly along the tape path (without a tape present), across the guides and heads, and drawn away gently on the far side. Only then should it be turned off.

8.7.2 Replay alignment

Replay alignment should be carried out before record alignment, as explained above. The method for setting replay and record levels has already been covered in the previous section. HF tones for azimuth adjustment normally follow (see Fact File 8.6). The test tape will contain a sequence of tones for replay frequency response alignment, often at 10 or 20 dB below reference level so that tape saturation is avoided at frequency extremes, starting with a 1 kHz reference followed by, say, 31.5 Hz, 63 Hz, 125 Hz, 250 Hz, 500 Hz, 2 kHz, 4 kHz, 8 kHz and 16 kHz. Spoken identification of each section is provided. As the tape runs, the replay

---

FACT FILE 8.6   Azimuth alignment

Azimuth

Azimuth describes the orientation of the head gap with respect to the tape. The gap should be exactly perpendicular to the edge of the tape otherwise two consequences follow. Firstly, high frequencies are not efficiently recorded or replayed because the head gap becomes effectively wider as far as the tape is concerned, as shown in the diagram (B is wider than A). Secondly, the relative phase between tracks is changed.

The high-frequency tone on a test tape (8, 10, or 16 kHz) can be used with the outputs of both channels combined, adjusting replay azimuth so as to give maximum output level which indicates that both channels are in phase. Alternatively, the two channels can be displayed separately on a double-beam oscilloscope, one wave being positioned above the other on the screen, where it can easily be seen if phase errors are present. Azimuth is adjusted until the two sine waves are in step. It is advisable to begin with a lower-frequency tone than 8 kHz if a large azimuth error is suspected, since there is a danger of ending up with tracks a multiple of 360° out of phase otherwise.

In multitrack machines a process of trial and error is required to find a pair of tracks which

most closely represents the best phase alignment between all the tracks. Head manufacturing tolerances result in gaps which are not perfectly aligned on all tracks. Cheap multitrack machines display rather wider phase errors between various tracks than do expensive ones.

Azimuth of the replay head is normally adjusted regularly, especially when replaying tapes made on other machines which may have been recorded with a different azimuth. Record-head azimuth is not modified unless there is reason to believe that it may have changed.

Height

Absolute height of the head should be such that the centre of the face of the head corresponds with the centre of the tape. Height can be adjusted using a test tape that is *not* recorded across the full width of the tape but with two discrete tracks. The correct height gives both equal output level from both channels and minimum crosstalk between them. It is also possible to buy tapes which are only recorded in the guard band, allowing the user to adjust height for minimum breakthrough onto the audio tracks. It can also sometimes be adjusted visually.

Zenith

Zenith is the vertical orientation of the head with respect to the surface of the tape. The head should neither lean forwards towards the tape, nor lean backwards, otherwise uneven wrap of the tape across the surface of the head results causing inconsistent tape-to-head contact and uneven head wear. Zenith is not normally adjusted unless the head has been changed or there is reason to believe that the zenith has changed.

Wrap
Wrap is the centrality of the head gap in the area of tape in contact with the head. The gap should be exactly in the centre of that portion, so that the degree of approach and recede contact of the tape with respect to the gap is exactly equal. Uneven frequency response can be caused if this is not the case. Wrap can be adjusted by painting the head surface with a removable dye and running the tape across it. The tape will remove the dye over the contact area, and adjustments can be made accordingly.

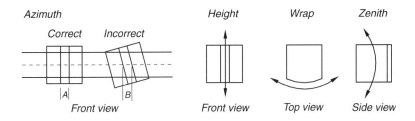

Azimuth — *Correct* — *Incorrect* — |A| — |B| — *Front view*
Height — *Front view*
Wrap — *Top view*
Zenith — *Side view*

equalisation is adjusted so as to achieve the flattest frequency response. Often both LF and HF replay adjustment is provided, sometimes just HF, but normally one should only adjust HF response on replay, since LF can suffer from the head bumps described in Fact File 8.2 and a peak or dip of response may coincide with a frequency on the test tape, leading to potential misalignment. Also full-track test tapes can cause 'fringing' at LF, whereby flux from the guard band leaks on to adjacent tracks. (Although it seems strange, replay LF EQ is normally adjusted during recording, to obtain the flattest record–replay response.)

8.7.3 Record alignment

The frequency response of the machine during recording is considerably affected by bias adjustment, and therefore bias is aligned first before record equalisation. The effects and alignment of bias are described in Fact File 8.7. It is wise to set a roughly correct input level before adjusting bias, by sending a 1 kHz tone at reference level to the tape machine and adjusting the input gain until it replays at the same level.

After bias levels have been set, record azimuth can be adjusted if necessary (see Fact File 8.6) by recording an HF tone and monitoring the now correctly aligned replay output. It may also be necessary to go back and check the 1 kHz record level if large changes have been made to bias.

Record equalisation can now be aligned. Normally only HF EQ is available on record. A 1 kHz tone is recorded at between 10 and 20 dB below reference level and the meter gain adjusted so that this can be seen easily on replay. Spot frequencies are then recorded to check the machine's frequency response, normally only at the

FACT FILE

8.7

Bias adjustment

Bias level affects the performance of the recording process and the correct level of bias is a compromise between output level, distortion, noise level and other factors. The graph below shows a typical tape's performance with increasing bias, and it can be seen that output level increases up to a point,

after which it falls off. Distortion and noise go down as bias increases, but unfortunately the point of minimum noise and distortion is not quite the same as the point of maximum output level. Typically the optimum compromise between all the factors, offering the best dynamic range, is where the bias level is set just slightly higher than the point giving peak output. In order to set bias, a 10 kHz tone is recorded at, say, 10 dB below reference level, whilst bias is gradually increased from the minimum. The output level from the tape machine gradually rises to a peak and then begins to drop off as bias continues to increase. Optimum bias is set for a number of decibels of fall-off in level after this peak – the so-called 'overbias' amount.

   The optimum bias point depends on tape speed and formulation, but is typically around 3 dB of overbias at a speed of 15 ips (38 cm s^{-1}). At 7.5 ips the overbias increases to 6 dB and at 30 ips it is only around 1.5 dB. If bias is adjusted at 1 kHz there is much less change of output level with variation in bias, and thus only between 0.5 and 0.75 dB of overbias is required at 15 ips. This is difficult to read on most meters.

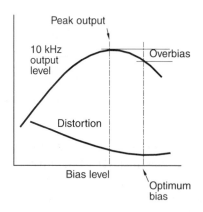

extremes of the range. A 5 kHz tone, followed by tones at 10 kHz and 15 kHz can be recorded and monitored off tape. The HF EQ is adjusted for the flattest possible response. The LF *replay* EQ (see above) can similarly be adjusted, sweeping the oscillator over a range of frequencies from, say, 40 Hz to 150 Hz, and adjusting for the best compromise between the upper and lower limits of the 'head bumps'.

   Some machines have a built-in computer which will automatically align it to any tape. The tape is loaded and the command given, and the machine itself runs the tape adjusting bias, level and EQ as it goes. This takes literally seconds. Several settings can be stored in its memory so that a change of tape type can be accompanied simply by telling the machine which type is to be used, and it will automatically set its bias and EQ to the previously stored values. This is of particular value when aligning multitrack machines!

   Once the tape machine has been correctly aligned for record and replay, a series of tones should be recorded at the beginning of every tape made on the machine. This allows the replay response of any machine which might subsequently be used for replaying the tape to be adjusted so as to replay the tape with a flat frequency response. The minimum requirement should be a tone at 1 kHz at reference level, followed by tones at HF and LF (say 10 kHz and 63 Hz) at either reference level (if the tape can cope) or at -10 dB. The levels and frequencies of these tones must be marked on the tape box (e.g.: 'Tones @ 1 kHz, 320 nWb m^{-1} (= 0 dB); 10 kHz and 63 Hz @ -10 dB). Designations on the tape box such as '1 kHz @ 0 VU' mean almost nothing, since 0 VU is not a magnetic level. What the engineer means in this case is that he sent a tone from his desk to the tape machine, measuring 0 VU on the meters, but this gives no indication of the magnetic level that resulted on the tape. Noted on the box should also be an indication of where peak recording level lies in relation to the 1 kHz reference level (e.g.: 'peak recording level @ 8 dB above 320 nWb m^{-1}), in order that the replay chain can be set up to accommodate the likely signal peaks. In broadcasting, for example, it is most important to know where the peak signal level will be, since this must be set to peak at PPM 6 on a program meter, corresponding to maximum transmitter modulation.

When this tape comes to be replayed, the engineer will adjust the replay level and EQ controls of the relevant machine, along with replay azimuth, to ensure that the recorded magnetic reference level replays at his or her studio's electrical reference level, and to ensure a flat response. *This is the only way of ensuring that a tape made on one machine replays correctly on another day or on another machine.*

8.8 Mechanical transport functions

Properly, mechanical alignment of the tape transport should be looked at before electrical alignment, because the electromagnetic performance is affected by it, but the converse is not the case. Mechanical alignment should be required far less frequently than electrical adjustments, and sometimes it also requires rather specialised tools. Because most mechanical alignments are fairly specialised, and because they differ with each tape machine, detailed techniques will not be covered further here. The manual for a machine normally details the necessary procedures. Looking at the diagram in Figure 8.12, it can be seen that the tape unwinds from the reel on the left, passes through various guides on its way to the head block, and then through various further guides and on to the take-up reel on the right. Some tape guides may be loaded with floppy springs which give on the instant of start-up, then slowly swing back in order to control the tension of the tape as the machine starts. The capstan is the shaft of a motor which pokes up through the deck of the machine by a couple of centimetres or so (more of course for multitrack machines with their increased tape widths) and lies fairly close to the tape when the tape is at rest, on the right-hand side of the head block. A large rubber wheel will be located close to the capstan but on the opposite side of the tape. This is called the pinch roller or pinch wheel. The capstan motor rotates at a constant and carefully controlled speed, and its speed of rotation defines the speed at which the tape runs. When record or play is selected the pinch roller rapidly moves towards the capstan, firmly sandwiching the tape in between the two. The rotation of the capstan now controls the speed of tape travel across the heads.

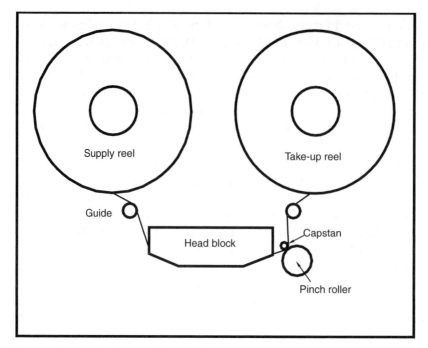

Figure 8.12 Typical layout of mechanical components on the deckplate of an analogue open-reel recorder

The take-up reel is controlled by a motor which applies a low anticlockwise torque so that the tape is wound on to it. The supply reel on the left is also controlled by a motor, which now applies a low clockwise torque, attempting to drag the tape back in the opposite direction, and this 'back tension' keeps the tape in firm contact with the heads. Different reel sizes require different degrees of back tension for optimum spooling, and a reel size switch will usually be provided although this is sometimes automatic. One or two transports have been designed without pinch rollers, an enlarged diameter capstan on its own providing speed control. The reel motors need to be rather more finely controlled during record and replay so as to avoid tape slippage across the capstan. Even capstanless transports have appeared, the tape speed being governed entirely by the reel motors.

When fast wind or rewind is selected the tape is lifted away from the heads by tape lifters, whilst spooling motors apply an appropriately high torque to the reel which is to take up the tape and a low reverse torque to the supply reel to control back tension. The tape is kept away from the heads so that its rapid movement does not cause excessive heating and wear of the tape heads. Also, very high-level, high-frequency energy is induced into the playback head if the tape is in contact with it which can easily damage speakers, particularly tweeters and HF horns. Nevertheless, a facility for moving the tape into contact with the heads during fast spooling is provided so that a particular point in the tape can be listened for.

Motion sensing and logic control is an important feature of a modern open-reel machine. Because the transport controls are electronically governed on modern machines, one can go straight from, say, rewind to play, leaving the machine itself to store the command and bring the tape safely to a halt before allowing the pinch wheel to approach the capstan. Motion sensing can be implemented by a number of means, often either by sensing the speed of the reel motors using tachometers, or by counting pulses from a roller guide.

The tape counter is usually driven by a rotating roller between the head block and that reel. Slight slippage can be expected, this being cumulative over a complete reel of tape, but remarkably accurate real-time counters are nevertheless to be found.

8.9 The Compact Cassette

8.9.1 Background

The Compact Cassette was invented by Philips, and was launched in 1963. It was originally intended as a convenient low-quality format suitable for office dictation machines and the like. It was envisaged that domestic tape recording would be open-reel, and a boom in this area was predicted. Pre-recorded open-reel tapes were launched. The expected boom never really materialised, however, and the sheer convenience of the cassette medium meant that it began to make inroads into the domestic environment. The format consists of tape one-eighth of an inch wide (3 mm), quarter track, running at a speed of 1.875 ips. Such drastically reduced dimensions and running speed compared with open-reel returned a poor level of audio performance, and if it was to be used for reasonable-quality music reproduction considerable development was needed.

Leaving the standards which had been set for the Compact Cassette unaltered, tape and machine manufacturers worked hard to develop the format, and the level of performance now available from this medium is quite impressive given its humble beginnings. It is worth mentioning that in the mid 1970s Sony introduced a rival called the Elcaset. The cassette housing was larger to accommodate 0.25 inch (6 mm) wide tape and the tape speed was 3.75 ips, promising rather better quality. But the format came too late, the Compact Cassette already being well established particularly since Dolby B noise reduction had been widely exploited since the early 1970s.

8.9.2 Cassette housing and transport

The cassette housing incorporates a felt pressure pad on the opposite side of the tape's recording surface which maintains the tape in intimate contact with the head during record and replay, this being particularly important when dealing with such small dimensions. During record and replay the heads are moved towards the cassette housing, and through appropriate access holes to contact the tape. Originally specified as a two-head format, an erase and a single record/replay head, no provision was made for a third head which would allow record and replay to be carried out by separate dedicated heads optimised for each purpose. In the continuing process of development manufacturers wished to add a third head and some chose to incorporate the record and playback heads in one housing, to be positioned in the normal record/replay-head position. Others chose to use a separate housing locating a third head elsewhere, using a separate access hole in the cassette. No pressure pad is provided for this, and the machine's transport needs to be up to the task of providing optimum back tension of the tape to give good head-to-tape contact. Three heads of course means that off-tape monitoring is possible and 'Double Dolby', 'Double dbx', etc. noise reduction circuits are employed which encode the incoming signal and simultaneously decode the off-tape signal for monitoring.

Dual-capstan drives have been offered with a capstan and pinch roller placed at each end of the cassette. The capstan which is placed at the end which supplies the tape to the heads is engineered to run at a marginally slower speed than the other one, the latter defining the actual tape speed. Such an arrangement ensures very consistent tensioning of the tape across the heads and the mechanical performance of the cassette reels no longer affects the results. Some machines push the pressure pad away from the head so that it does not influence the performance. Auto-reverse has been seen, as has automatic 'preview' whereby the machine senses the silences between recorded items on the tape and then plays the first few seconds of each item before fast winding on to the next.

8.9.3 Tape selection

The different tape formulations were discussed in section 8.1.3. A number of machines automatically select EQ and bias for the tape type which is loaded, sensing special holes in the cassette housing which are provided for this purpose (see Figure 8.13). Internal bias adjustment is usually provided in a cassette machine, but record and replay EQ are usually fixed. If it is desired to bias a machine for a particular tape, a good way to do this in the absence of EQ adjustments is to record pink noise at a level 20 dB below zero on the meters from a test record, or 'white' noise provided by FM tuner interstation noise, increasing the bias from a low level until the sound coming off tape is as near as possible indistinguishable from the sound going on. Too much bias produces a dull sound; too little gives an overbright sound. With a three-head machine this is very easy since the source/tape switch can be flicked to and fro for instant comparison, but two-head machines require regular rewinding and comparison.

Figure 8.13 Holes in the top edge of a compact cassette may be uncovered to signify different tape types and to prevent recording

It is very important to set the same replay level for the noise source and the off-tape noise, otherwise the ear hears frequency response differences which may simply be a subjective result caused by the different levels. The process should be carried out with noise reduction switched off, as this will exaggerate response errors. After alignment the noise reduction should be switched on and another section of noise recorded. The noise reduction should not introduce significant degradation of the signal off tape. In many cheaper recorders the bias control either affects all tape types equally, or simply works for ferric tapes which have the widest range of requirements.

8.9.4 Other alignments

Cassette test tapes are available, enabling frequency response and azimuth checks. Thorough cleaning and demagnetisation of the machine should be carried out before one is used. Small azimuth adjustments can bring particularly worthwhile improvements in cassette performance, especially when replaying tapes recorded on another machine. Azimuth can simply be adjusted for maximum subjective HF response using the little spring-loaded screw on one side of the record/replay head. Some machines incorporate computer systems similar to those found in certain professional open-reel models which automatically align the machine for a particular type of tape. Settings for several types can be stored in the computer's memory.

Automatic replay azimuth adjustment is also possible. The two channels of the stereo output are filtered, converted into square waves and then fed to a comparator. Phase differences produce an output control voltage and this drives a small low-speed motor which adjusts the azimuth setting of the replay head. When azimuth is correct no control voltage is produced and the azimuth is left alone. The system is continuously active throughout the replay process and it is designed to extract the best performance from pre-recorded musicassettes and recordings made on other machines.

8.9.5 Multitrack cassette recorders

In the late 1970s the Japanese TEAC company introduced a machine called the Portastudio. It was a four-channel multitrack cassette recorder with mixing facilities and multiple inputs built in. The tape ran at twice normal speed, 3.75 ips, and the four tracks were recorded across the full width of the tape. Each track could be recorded on separately, sync facilities were provided, and 'bounce down' could be achieved in the manner of a professional multitrack machine whereby signals recorded on, say, tracks 1, 2 and 3 could be mixed and recorded on to the fourth track, freeing the other three tracks for further use. The final four-track tape could then be mixed down into stereo, these stereo outputs being fed to a conventional cassette recorder (or even open-reel).

One mixer company even offered an eight-track cassette-based system which incorporated a mixing section that offered facilities such as multiband EQ and auxiliary sends.

8.9.6 Cassette duplication

Cassette duplication is an important area of activity, pre-recorded musicassettes being very popular. The vast majority of such cassettes are produced on special machines which run the tape at 16, 32 or 64 times normal speed so that a length of tape for a 20 minute cassette can be duplicated in a few seconds. In one system, used mainly for short runs, banks of cassettes are duplicated in one go by each particular duplicator, and the labelling and packaging is carried out automatically. Master tapes to be duplicated are copied on to a half-inch 'loop-bin master' tape which is fed from a vacuum bin where the tape is stored in loose loops to allow for high-speed repeated reproduction. Tones are recorded on to the tape during duplication which tell the cassette loading machine where the beginning and end of each section is.

A speed 32 times normal for duplication requires record bias frequencies in the megahertz range and record-head gaps need to be rather less than a micron in width for high frequencies to be adequately recorded. Level and frequency response errors will be magnified by Dolby encode/decode processes. Consistent head-to-tape contact is difficult to maintain at such high speeds, high frequency loss normally being the result. Such duplicating equipment therefore needs to be maintained in first-class working order if the final cassette is to stand comparison with an equivalent home recorded example. Many musicassettes are found to be recorded at too low a level, failing to exploit the maximum dynamic range of the medium. Many sound dull, and several actually sound rather better with Dolby switched out during replay. Some manufacturers use higher-quality tape than others, and several chromium dioxide examples are available, but recorded using 120 µs equalisation.

An alternative to high-speed duplication is real-time copying. Instead of the highspeed duplicators a bank of carefully maintained cassette recorders are used which all chug away in real time, programme signal being provided by an open-reel machine or even a digital recorder, linked to a distribution amplifier to feed the cassette machines. Such a setup is ideal for the production of relatively low numbers of cassettes when it may not be economical to prepare the special production tapes necessary for the high-speed duplication process. Sound quality can of course be as good as the medium is capable of.

Recommended further reading

Jorgensen, F. (1995) *The Complete Handbook of Magnetic Recording*. 4th Edition. McGraw-Hill

Mallinson, J. C. (1987) *The Foundations of Magnetic Recording*. Academic Press

See also *General further reading* at the end of this book.

Chapter 9

Noise reduction

Noise reduction techniques have been applied to analogue tape machines of all formats, radio microphones, radio transmission and reception, land lines, satellite relays, gramophone records, and even some digital tape machines. The general principles of operation will be outlined, followed by a discussion of particular well-known examples. Detailed descriptions of some individual systems are referred to in the Further reading list at the end of this chapter.

9.1 Why is noise reduction required?

A noise reduction system, used correctly, reduces the level of unwanted signals introduced in a recording–replay or transmission–reception process (see Figure 9.1). Noise such as hiss, hum, and interference may be introduced, as well as, say, print-through in analogue recording, due to imperfections in the storage or transmission process. In communications, a signal sent along a land line may be prone to interference from various sources, and will therefore emerge with some of this interference signal mixed with it. A signal recorded on a cassette machine replays with high-frequency hiss. Unwanted noise already present in a signal before recording or transmitting, though, is very difficult to remove without also removing

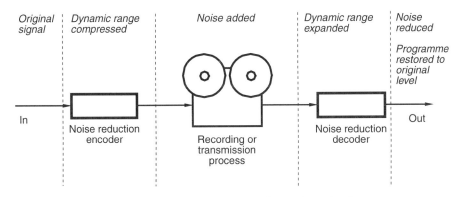

Figure 9.1 Graphical representation of a companding noise reduction process

a part of the wanted signal. One could roll off the treble of a cassette recording to reduce the hiss, but one would also lose the high-frequency information from the sound, causing it to sound muffled and 'woolly'.

9.2 Methods of reducing noise

9.2.1 Variable pre-emphasis

Pre-emphasis (see Fact File 9.1) is a very straightforward solution to the problem of noise reduction, but is not a panacea. Many sound sources, including music, have a falling energy content at high frequencies, so lower-level HF signals can be boosted to an extent without too much risk of saturating the tape. But tape tends to saturate more easily at HF than at LF (see the previous chapter), so high levels of distortion and compression would result if too much pre-emphasis were applied at the recording stage. What is needed is a circuit which senses the level of the signal on a continuous basis, controlling the degree of pre-emphasis so as to be non-existent at high signal levels but considerable at low signal levels (see Figure 9.2). This can be achieved by incorporating a filter into a side-chain which passes only high-frequency, low-level signals, adding this component into the un-pre-emphasised signal. On replay, a reciprocal de-emphasis circuit could then be used. The lack of

FACT FILE	Pre-emphasis
9.1	

One approach to the problem of reducing the apparent level of noise could be to pre-condition the incoming signal in some way so as to raise it further above the noise. Hiss is most annoying at high frequencies, so one

could boost HF on recording. On replay, HF signals would therefore be reproduced with unnatural emphasis, but if the same region is now attenuated to bring the signal down to its original level any hiss in the same band will also be attenuated by a corresponding amount, and so a degree of noise reduction can be achieved without affecting the overall frequency balance of the signal. This is known as *pre-emphasis* (on record) and *de-emphasis* (on replay), as shown in the diagram.

Figure 9.2 A simple complementary noise reduction system could boost high frequencies at low signal levels during encoding, and cut them on decoding (encoding characteristic shown)

noise reduction at high signal levels does not matter, since high-level signals have a masking effect on low-level noise (see Fact File 2.3).

Such a process may be called a *compansion* process, in other words a process which compresses the dynamic range of a signal during recording and expands it on replay. The variable HF emphasis described above is an example of selective compansion, acting only on a certain band of frequencies. It is most important to notice that the decoding stage is an exact mirror image of the encoding process, and that it is not possible to use one without the other. Recordings not *encoded* by a noise reduction system cannot simply be passed through a decoder to reduce their noise. Similarly, encoded tapes sound unusual unless properly decoded, normally sounding overbright and with fluctuations in HF level.

9.2.2 Dolby B

The above process is used as the basis for the Dolby B noise reduction system, found in most cassette decks. Specifically, the threshold below which noise reduction comes into play is around 20 dB below a standard magnetic reference level known as 'Dolby level' (200 nWb m^{-1}). The maximum HF boost of the Dolby B system is 10 dB above 8 kHz, and therefore a maximum of 10 dB of noise reduction is provided. A high-quality cassette deck, without noise reduction, using a good ferric tape, will yield a signal-to-noise ratio of about 50 dB ref. Dolby level. When Dolby B noise reduction is switched in, the 10 dB improvement brings this up to 60 dB (which is more adequate for good-quality music and speech recording). The quoted improvement is seen when noise is measured according to the CCIR 468-2 weighting curve (see section A1.5) and will not be so great when measured unweighted.

Dolby B incorporates a sliding band over which pre-emphasis is applied, such that the frequency above which compansion occurs varies according to the nature of the signal. It may slide as low as 400 Hz. This aims to ensure that maximum masking of low-level noise always occurs, and that high-level signals at low frequencies do not result in 'noise pumping' (a phenomenon which arises when a high level signal in one band causes less overall noise reduction, causing the noise in another band to rise temporarily, often not masked by the high-level signal due to the difference in frequency of the signal and the noise).

The Dolby process, being level dependent, requires that the reproduced signal level on decoding is exactly the same with respect to Dolby level as on encoding. This means that a particular cassette machine must be set up internally so that Dolby-encoded tapes recorded on it or other machines will replay into the decoder at the correct electrical level for proper decoding. This is independent of the actual output level of the machine itself, which varies from model to model. If the replay level, for instance, is too low, the decoder applies too much treble cut because the −20 dB threshold level will have moved downwards, causing recorded signal levels above this to be de-emphasised also. Frequency response error will therefore be the result. Similarly, if the frequency response of a cassette machine shows significant errors at HF, these will be exaggerated by the Dolby record/replay process.

A so-called MPX (multiplex) filter is mandatory with Dolby B systems, and removes the 19 kHz pilot tone present in FM stereo radio broadcasts. This is needed because the pilot tone may still be present in the output of an FM tuner, artificially affecting the encoded level of HF signals on a recording from the radio. Since the frequency response of many cassette machines does not extend to 20 kHz the tone would not be reproduced on replay, and thus the decoder would not track the encoded signal correctly, leading to noise pumping and response errors. On some recorders the filter is switchable. On cheaper machines the filter simply rolls off everything above 15 kHz, but on better machines it is a notch at 19 kHz.

9.2.3 Dolby C

Dolby B became widely incorporated into cassette players in the early 1970s, but by the end of the 1970s competition from other companies offering greater levels of noise reduction prompted Dolby to introduce Dolby C, which gives 20 dB of noise reduction. The system acts down to a lower frequency than Dolby B (100 Hz), and incorporates additional circuitry (known as 'anti-saturation') which reduces HF tape squashing when high levels of signal are present. Most of the noise reduction action takes place between 1 kHz and 10 kHz, and less action is taken on frequencies above 10 kHz (where noise is less noticeable) in order to desensitise the system to HF response errors from such factors as azimuth misalignment which would otherwise be exaggerated (this is known as 'spectral skewing'). Dolby C, with its greater compression/expansion ratio compared with Dolby B, will exaggerate tape machine response errors to a correspondingly greater degree, and undecoded Dolby C tapes will sound extremely bright.

9.2.4 Dolby A

Dolby A was introduced in 1965, and is a professional noise reduction system. In essence there is a similarity to the processes described above, but in the Dolby A encoder the noise reduction process is divided into four separate frequency bands, as shown in Figure 9.3. A low-level 'differential' component is produced for each band, and the differential side-chain output is then recombined with the main signal. The differential component's contribution to the total signal depends on the input level, having maximum effect below −40 dB ref. Dolby level (see Figures 9.4(a) and (b)).

The band splitting means that each band acts independently, such that a high-level signal in one band does not cause a lessening of noise reduction effort in another low-level band, thus maintaining maximum effectiveness with a wide

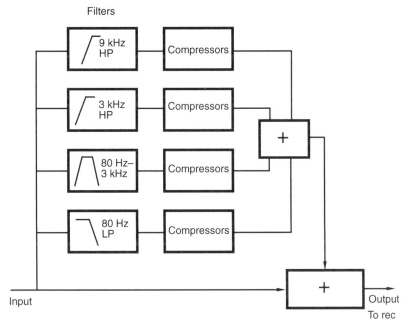

Figure 9.3 In the Dolby A system a low-level 'differential' signal is added to the main signal during encoding. This differential signal is produced in a side-chain which operates independently on four frequency bands. The differential signal is later subtracted during decoding

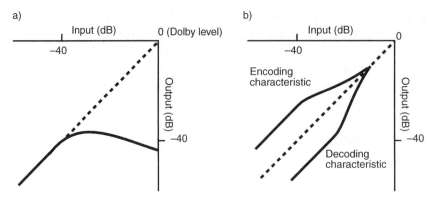

Figure 9.4 (a) Differential signal component produced in a Dolby A side-chain. (b) Input level plotted against output level of Dolby A unit after adding or subtracting differential component

range of programme material. The two upper bands are high pass and overlap, offering noise reduction of 10 dB up to around 5 kHz, rising to 15 dB at the upper end of the spectrum.

The decoder is the mirror image of the encoder, except that the differential signal produced by the sidechain is now *subtracted* from the main signal, restoring the signal to its original state and reducing the noise introduced between encoding and decoding.

9.2.5 Dolby SR

The late 1980s saw the introduction of Dolby SR – Spectral Recording – which gives greater noise reduction of around 25 dB. It has been successful in helping to prolong the useful life of analogue tape machines, both stereo mastering and multi-track, in the face of the coming of digital tape recorders. Dolby SR differs from Dolby A in that whereas the latter leaves the signal alone until it drops below a certain threshold, the former seeks to maintain full noise reduction (i.e.: maximum signal boost during recording) across the whole frequency spectrum until the incoming signal rises above the threshold level. The band of frequencies where this happens is then subject to appropriately less boost. This is rather like looking at the same process from opposite directions, but the SR system attempts to place a comparably high recording level on the tape across the whole frequency spectrum in order that the dynamic range of the tape is always used optimally.

This is achieved by ten fixed and sliding-band filters with gentle slopes. The fixed-band filters can vary in gain. The sliding-band filters can be adjusted to cover different frequency ranges. It is therefore a fairly complex multiband system, requiring analysis of the incoming signal to determine its energy at various frequencies. Spectral skewing and anti-saturation are also incorporated (see section 9.2.3). Dolby SR is a particularly inaudible noise reduction system, more tolerant of level mismatches and replay speed changes than previous systems. A simplified 'S'-type version has been introduced for the cassette medium, and is also used on some semi-professional multitrack recorders.

9.2.6 dbx

dbx is another commonly encountered system. It offers around 30 dB of noise reduction and differs from the various Dolby systems as follows. dbx globally compresses the incoming signal across the whole of the frequency spectrum, and in addition gives pre-emphasis at high frequencies (treble boost). It is not level dependent, and seeks to compress an incoming signal with, say, a 90 dB dynamic range into one with a 60 dB dynamic range which will now fit into the dynamic range capabilities of the analogue tape recorder. On replay, a reciprocal amount of expansion is applied together with treble de-emphasis.

Owing to the two factors of high compansion ratios and treble pre- and de-emphasis, frequency response errors can be considerably exaggerated. Therefore, dbx type 1 is offered which may be used with professional equipment and type 2 is to be used with domestic equipment such as cassette decks where the noise reduction at high frequencies is relaxed somewhat so as not to exaggerate response errors unduly. The degree of compression/expansion is fixed, that is it does not depend on the level of the incoming signal. There is also no division of noise reduction between frequency bands. These factors sometimes produce audible modulation of background hiss with critical programme material such as wide dynamic range classical music, and audible 'pumping' noises can sometimes be heard. The system does, however, offer impressive levels of noise reduction, particularly welcome with the cassette medium, and does not require accurate level alignment.

9.2.7 telcom c4

The ANT telcom c4 noise reduction system arrived somewhat later than did Dolby and dbx, in 1978. Capitalising on the experience gained by those two systems, the telcom c4 offers a maximum noise reduction of around 30 dB, is level dependent like Dolby, and also splits the frequency spectrum up into four bands which are then treated separately. The makers claim that the c4 system is less affected by record/ replay-level errors than is Dolby A. The system works well in operation, and side-effects are minimal.

There is another system offered by the company, called 'hi-com', which is a cheaper, simpler version intended for home studio setups and domestic cassette decks.

9.3 Line-up of noise reduction systems

In order to ensure unity gain through the system on recording and replay, with correct tracking of a Dolby decoder, it is important to align the noise reduction signal chain. Many methods are recommended, some more rigorous than others, but in a normal studio operation for everyday alignment, the following process should be satisfactory. It should be done after the tape machine has been aligned (this having been done with the NR unit bypassed).

196   Noise reduction

For a Dolby A encoder, a 1 kHz tone should be generated from the mixer at +4 dBu (usually PPM 5), and fed to the input of the NR unit. The unit should be in 'NR out' mode, and set to 'record'. The input level of the NR unit should normally be adjusted so that this tone reads on the 'NAB' level mark on the meter (see Figure 9.5). The output of the unit should then be adjusted until its electrical level is also +4 dBu. (If the tape machine has meters then the level can be read here, provided that these meters are reliable and the line-up is known.)

It is customary to record a passage of 'Dolby tone' (in the case of Dolby A) or Dolby Noise (in the case of Dolby SR) at the beginning of a Dolby-encoded tape, along with the other line-up tones (see section 8.7.3). During record line-up, the Dolby tone is generated by the Dolby unit itself, and consists of a frequency-modulated 700 Hz tone at the Dolby's internal line-up reference level, which is easily recognised and distinguished from other line-up tones which may be present on a tape. Once the output level of the record Dolby has been set then the Dolby tone button on the relevant unit should be pressed, and the tone recorded at the start of the tape.

To align the replay Dolby (set to 'NR out', 'replay' mode), the recorded Dolby tone should be replayed and the input level adjusted so that the tone reads at the NAB mark on the internal meter. The output level should then be adjusted for +4 dBu, or so that the mixer's meter reads PPM 5 when switched to monitor the tape machine replay.

For operation, the record and replay units should be switched to 'NR in'.

Dolby SR uses pink noise instead of Dolby tone, to distinguish tapes recorded with this system, and it is useful because it allows for line-up of the replay Dolby in cases where accurate level metering is not available. Since level misalignment will result in response errors the effects will be audible on a band of pink noise. A facility is provided for automatic switching between internally generated pink noise and off-tape noise, allowing the user to adjust replay-level alignment until there appears to be no audible difference between the spectra of the two. In normal circumstances Dolby SR systems should be aligned in a similar way to Dolby A, except that a noise band is recorded on the tape instead of a tone. Most systems use LED meters to indicate the correct level, having four LEDs as shown in Figure 9.5.

Figure 9.5 Dolby level is indicated on Dolby units using either a mechanical meter (shown left), or using red and green LEDs (shown right). The meter is normally aligned to the '18.5 NAB' mark or set such that the two green LEDs are on together

9.4 Operational considerations

A word may be said about varispeed. It is not uncommon for the replay speed of a tape to need to be adjusted slightly to alter pitch, or total playing time. In creative work massive amounts of speed change are sometimes employed. The pitch change means that Dolby decoding will be inaccurate since the frequency bands will not now correspond to those during the recording process, and Dolby mistracking will result.

Professional noise reduction systems are available as single-channel units, stereo packages, and conveniently grouped multiples of 8, 16 and 24 for multitrack work. They generally fit into standard 19 inch (48 cm) racks. Certain models are designed to fit straight into multitrack recorders so that the complete recorder plus noise reduction combination is conveniently housed in one unit.

Each noise reduction channel is manually switchable between encode for record and decode for replay, and in addition a special input is usually provided which accepts a remote DC signalling voltage, which will switch the unit into encode. Removal of the DC causes the unit to revert to decode ready for replay. Professional tape machines can usually provide this DC requirement, linking it to the record status of each track. Those tracks which are switched to record will now automatically switch the appropriate noise reduction channels to encode ready for recording. The system enables the selection of correct noise reduction status to be left to the recorder itself which is a very convenient feature particularly when a large number of channels are in use.

9.5 Single-ended noise reduction

9.5.1 General systems

Several companies offer so-called 'single-ended' noise reduction systems, and these are intended to 'clean up' an existing noisy recording or signal. They operate by sensing the level of the incoming signal, and as the level falls below a certain threshold the circuit begins to roll off the treble progressively, thereby reducing the level of hiss. The wanted signal, being low in level, in theory suffers less from this treble reduction than would a high-level signal due to the change in response of the ear with level (see Fact File 2.2). High-level signals are left unprocessed. The system is in fact rather similar to the Dolby B decoding process, but of course the proper reciprocal Dolby B encoding is absent. The input level controls of such systems must be carefully adjusted so as to bring in the effect of the treble roll-off at the appropriate threshold for the particular signal being processed so that a suitable compromise can be achieved between degree of hiss reduction and degree of treble loss during quieter passages. Such single-ended systems should be judiciously used – they are not intended to be left permanently in circuit – and value judgements must always be made as to whether the processed signal is in fact an improvement over the unprocessed one.

If a single-ended system is to be used on a stereo programme, units which are capable of being electronically 'ganged' must be employed so that exactly the same degree of treble cut is applied to each channel; otherwise varying frequency balance between channels will cause stereo images to wander.

9.5.2 Noise gates

The noise gate can be looked upon as another single-ended noise reduction system. It operates as follows. A threshold control is provided which can be adjusted such that the output of the unit is muted (the gate is 'closed') when the signal level falls below the threshold. During periods when signal level is very low (possibly consisting of tape or guitar amplifier noise only) or absent the unit shuts down. A very fast attack time is employed so that the sudden appearance of signal opens up the output without audible clipping of the initial transient. The time lapse before the gate closes, after the signal has dropped below the chosen threshold level, can also be varied. The close threshold is engineered to be lower than the open threshold (known as *hysteresis*) so that a signal level which is on the borderline does not confuse the unit as to whether it should be open or closed, which would cause 'gate flapping'.

Such units are useful when, for instance, a noisy electric guitar setup is being recorded. During passages when the guitarist is not playing the output shuts down so that the noise is removed from the mix. They are sometimes also used in a similar manner during multitrack mixdown where they mute outputs of the tape machine during the times when the tape is unmodulated, thus removing the noise contribution from those tracks.

The noise gate is frequently heard in action during noisy satellite link broadcasts and long-distance telephone-line operation. An impressive silence reigns when no-one is talking, but when speech begins a burst of noise abruptly appears and accompanies the speaker until he or she stops talking. This can sometimes be disconcerting for the speaker at the other end of the line because he or she gets the impression that the line has been cut off when the noise abruptly disappears.

Noise gates can also be used as effects in themselves, and the 'gated snare drum' is a common effect on pop records. The snare drum is given a heavy degree of gated reverb, and a high threshold level is set on the gate so that around half a second or so after the drum is hit the heavy 'foggy' reverb is abruptly cut off. Drum machines can mimic this effect, as can some effects processors.

9.5.3 Digital noise extraction

Extremely sophisticated single-ended computer-based noise reduction systems have been developed. They are currently expensive, and are offered as a service by a limited number of facilities. A given noisy recording will normally have a short period somewhere in which only the noise is present without any programme, for instance the run-in groove of an old 78 rpm shellac disc recording provides a sample of that record's characteristic noise. This noise is analysed by a computer and can subsequently be recognised as an unwanted constituent of the signal, and then extracted electronically from it. Sudden discontinuities in the programme caused by scratches and the like can be recognised as such and removed. The gap is filled by new material which is made to be similar to that which exists either side of the gap. Not all of these processes are currently 'real time', and it may take several times longer than the progamme's duration for the process to be carried out, but as the speed of digital signal processing increases more operations become possible in real time.

Recommended further reading

Dolby, R. (1967) An audio noise reduction system. *J. Audio Eng. Soc.*, vol. 15, pp. 383–388

Dolby, R. (1970) A noise reduction system for consumer tape applications. Presented at the *39th AES Convention. J. Audio Eng. Soc. (Abstracts)*, vol. 18, p. 704

Dolby, R. (1983) A 20 dB audio noise reduction system for consumer applications. *J. Audio Eng. Soc.*, vol. 31, pp. 98–113

Dolby, R. (1986) The spectral recording process. Presented at the *81st AES Convention*. Preprint 2413 (C-6). Audio Engineering Society

See also *General further reading* at the end of this book.

Chapter 10

Digital recording

10.1 Digital and analogue recording contrasted

In analogue recording, as described in the previous chapters, sound is recorded by converting continuous variations in sound pressure into continuous variations in electrical voltage, using a microphone. This varying voltage is then converted into a varying pattern of magnetisation on a tape, or, alternatively, into a pattern of light and dark areas on an optical-film sound track, or a groove of varying deviation on an LP.

Because the physical characteristics of analogue recordings relate closely to the sound waveform, replaying them is a relatively simple matter. Variations in the recorded signal can be converted directly into variations in sound pressure using a suitable collection of transducers and amplifiers. The problem, though, is that the replay system is unable to tell the difference between *wanted* signals and *unwanted* signals. Unwanted signals might be distortions, noise and other forms of interference introduced by the recording process. For example, a record player cannot distinguish between the stylus movement it experiences because of a scratch on a record (unwanted) and that caused by a loud transient in the music (wanted). Imperfections in the recording medium are reproduced as clicks, crackles and other noises.

Digital recording, on the other hand, converts the electrical waveform from a microphone into a series of binary numbers, each of which represents the amplitude of the signal at a unique point in time, recording these numbers in a coded form which allows the system to detect whether the replayed signal is correct or not. A reproducing device is then able to distinguish between the wanted and the unwanted signals introduced above, and is thus able to reject all but the wanted original information in most cases. Digital audio is much more tolerant of a poor recording channel than analogue audio, and distortions and imperfections in the storage or transmission process need not affect the sound quality of the signal provided that they remain within the design limits of the system. These issues are given further coverage in Fact File 10.1.

Digital audio has made it possible for sound engineers to take advantage of developments in the computer industry, and this is particularly beneficial because

FACT FILE 10.1   Analogue and digital information

Analogue information is made up of a continuum of values, which at any instant may have any value between the limits of the system. For example, a rotating knob may have one of an infinite number of positions – it is therefore an analogue controller (see the diagram).

(a)   (b)

Analogue controller   Switch (binary)

A simple switch, on the other hand, can be considered as a digital controller, since it has only two positions – off or on. It cannot take any value in between. The brightness of light which we perceive with our eyes is analogue information, and as the sun goes down the brightness falls gradually and smoothly, whereas a household light without a dimmer may be either on or off – its state is binary (that is it has only two possible states).

Electrically, analogue information may be represented as a varying voltage or current. If a rotary knob is connected to a variable resistor and a voltage supply, its position will affect the output voltage as shown here. The switch may be connected to a similar voltage supply, and in this case the output voltage can only be either zero volts or +V. In other words the electrical information which results is binary.

The high (+V) state could be said to correspond to a binary one, and the low

state to binary zero (although in many real cases it is actually the other way around).

Binary information is inherently more resilient to noise and interference than analogue information, as shown below. If noise is added to an analogue signal then it becomes very difficult to tell at any later stage in the signal chain what is the wanted signal and what is the unwanted noise, since there is no means of distinguishing between the two. If noise is added to a digital signal it is possible to extract the important information at a later stage, since only two states matter – the one and zero states. By comparing the signal amplitude with a fixed decision point it is possible for a receiver to decide that everything above the decision point is 'high', whilst everything below it is 'low'. For any noise to influence the state of a digital signal it must be at least large enough in amplitude to cause a high level to be interpreted as 'low', or vice versa.

The timing of digital signals may also be corrected to some extent, using a similar method, which gives digital signals another advantage over analogue. If the timing of bits in a digital message becomes unstable, such as after having been passed over a long cable, resulting in timing 'jitter' the signal may be reclocked at a stable rate ensuring that the timing stability of the information is restored.

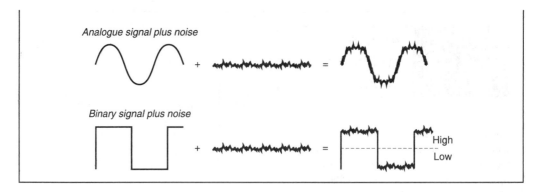

the size of that industry results in mass production (and therefore cost savings) on a scale not possible for audio products alone. Today it is common for sound to be recorded, processed and edited on relatively low cost desktop computer equipment, and this is a trend likely to continue.

This chapter contains an introduction to the main principles of digital recording, described in as non-technical a manner as possible. Further reading recommendations at the end of this chapter are given for those who want to study the subject in more depth.

10.2 The digital audio signal chain

Figure 10.1 shows the signal chain involved in any digital recording or broadcasting system. First the analogue audio signal (a time-varying electrical voltage) is passed through an analogue-to-digital (A/D) convertor where it is transformed from a continuously varying voltage into a series of 'samples', which are 'snapshots' of the analogue signal taken many thousand times per second. Each sample is represented by a number. This series of samples is coded into a form which makes it suitable for recording or broadcasting (a process known as coding or channel coding), and the signal is then recorded or transmitted. Upon replay or reception the signal is decoded and subjected to error correction, and it is this latter process which works out what damage has been done to the signal since it was coded. Any errors in timing or value of the samples are corrected if possible, and the result is fed to a digital-to-analogue (D/A) convertor which turns the numerical data back into a time-continuous analogue audio signal.

In the following sections each of the main processes involved in this chain will be explained, followed by a discussion of the implementation of this technology in real audio systems. Only the most straightforward principles are described here, since much of the theoretical matter involved in digital audio is above the intended level of this book.

Figure 10.1 Block diagram of the typical digital recording or broadcasting signal chain

10.3 Analogue-to-digital (A/D) conversion

10.3.1 Sampling

The sampling process employed in an A/D convertor involves the measurement or 'sampling' of the amplitude of the audio waveform at regular intervals in time (see Figure 10.2). From this diagram it will be clear that the sample pulses represent the instantaneous amplitudes of the signal at each point in time. The samples can be considered as like instantaneous 'still frames' or 'snapshots' of the audio signal which together and in sequence form a representation of the continuous waveform, rather as the still frames which make up a movie film give the impression of a continuously moving picture when played in quick succession.

In order to represent the fine detail of the signal it is necessary to take a large number of these samples per second, and the mathematical sampling theorem proposed by Shannon indicates that at least two samples must be taken per audio cycle if the necessary information about the signal is to be conveyed. In other words, the sampling frequency must be at least twice as high as the highest frequency of audio to be handled by the system (this is often called the Nyquist criterion). Another way of visualising this issue is presented in Fact File 10.2, which considers sampling as a modulation process.

10.3.2 Filtering and aliasing

A simple example in Figure 10.3 shows that if too few samples are taken per cycle of the audio signal then the samples may be interpreted as representing a wave other than that originally sampled, and this is one way of understanding the phenomenon known as aliasing. An 'alias' is an unwanted by-product of the original signal. It arises when the signal is reconstructed as an analogue signal during D/A conversion. If the sample values in Figure 10.3(b) were presented to a D/A convertor the lower frequency (dotted) wave could be reconstructed, which is not correct. There is

Figure 10.2 An arbitary audio signal is sampled at regular intervals of time t to create short sample pulses whose amplitudes represent the instantaneous amplitude of the audio signal at each point in time

FACT FILE
10.2

Sampling – frequency domain

One way of visualising the sampling process is to consider it in terms of modulation. The continuous audio waveform is used to modulate a regular chain of pulses. The frequency of these pulses is the sampling frequency (f_s). Before modulation all these pulses have the same amplitude (height), but after modulation the amplitude of the pulses is modified according to the instantaneous amplitude of the audio signal at that point in time. This process is called pulse amplitude modulation (PAM).

The frequency spectrum of the modulated signal is as shown in the diagram. The shape of the spectrum is not significant – it is just an example of a complex audio signal such as music. It will be seen that in addition to the 'baseband' audio signal (the original spectrum before sampling) there are now a number of additional spectra, each centred on multiples of the sampling frequency. Sidebands have been produced either side of the sampling frequency and its multiples, as a result of the amplitude modulation, and these extend above and below the sampling frequency and its multiples to the extent of the base bandwidth. In other words these sidebands are pairs of mirror images of the audio band.

a)

b)

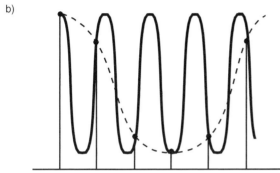

Figure 10.3 (a) Wave sampled at an adequate rate. (b) Wave sampled at an inadequate rate leading to aliasing

no information in the samples to indicate the rapid changes in the wave that take place in between.

To understand the issue in more detail one must consider the signals in the frequency domain. It is relatively easy to see why the sampling frequency must be at least twice the highest audio frequency from Figure 10.4. Here an extension of the audio signal's bandwidth (base-band) above the Nyquist frequency (half the sampling frequency) results in the lower sideband of the first spectral repetition overlapping the upper end of the baseband, resulting in an alias region where they overlap. Two further examples are shown to illustrate the point – the first in which a sine wave tone is sampled, having a low enough frequency for the sampled sidebands to lie above the audio frequency range, and the second in which a much higher frequency tone (above the Nyquist frequency) causes a sampled sideband to fall well within the baseband. This would be audible as an alias of the original tone.

The aliasing phenomenon can also be witnessed visually in the case of the well-known 'spoked-wheel' effect on films, since moving pictures are also an example of a sampled signal. In film, still pictures (image samples) are normally taken at a rate of 24 per second. If a rotating wheel with a marker on it is filmed it will appear to move round in a forward direction as long as the rate of rotation is much slower than the rate of the still photographs, but as its rotation rate increases it will appear to slow down, stop, and then appear to start moving backwards. The virtual impression of backwards motion gets faster as the rate of rotation of the wheel gets faster,

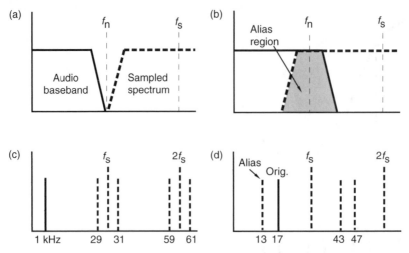

Figure 10.4 Aliasing viewed in the frequency domain. In (a) the audio baseband extends up to half the sampling frequency (the Nyquist frequency f_n) and no aliasing occurs. In (b) the audio baseband extends above the Nyquist frequency and consequently overlaps the lower sideband of the first spectral repetition, giving rise to aliased components in the shaded region. In (c) a tone at 1 kHz is sampled at a sampling frequency of 30 kHz, creating sidebands at 29 and 31 kHz (and at 59 and 61 kHz, etc.). These are well above the normal audio frequency range, and will not be audible. In (d) a tone at 17 kHz is sampled at 30 kHz, putting the first lower sideband at 13 kHz – well within the normal audio range. The 13 kHz sideband is said to be an alias of the original wave

and this backwards motion is the aliased result of sampling at too low a rate. Clearly the wheel is not really rotating backwards, it just appears to be – it is an incorrect reconstruction of the original information, resulting from sampling the picture at too slow a rate.

If audio signals are allowed to alias in digital recording one hears the audible equivalent of the backwards-rotating wheel – that is, sound components in the audible spectrum which were not there in the first place, moving downwards in frequency as the original frequency of the signal increases. In basic convertors, therefore, it is necessary to filter the baseband audio signal before the sampling process, as shown in Figure 10.5, so as to remove any components having a frequency higher than half the sampling frequency.

In real systems, and because filters are not perfect, the sampling frequency is made slightly higher than twice the highest audio frequency to be represented, allowing for the filter to roll off slightly more gently. Choice of sampling frequency is discussed further in Fact File 10.3.

10.3.3 Quantisation

Once the audio signal has been sampled, the samples must be converted into a series of numbers in a process known as quantisation. The result of sampling the sound signal is that a train of pulses of varying amplitude is produced, and the quantisation process assigns a numerical value to each of these samples according to its amplitude. It is rather like taking a tape measure to each of the sample pulses and

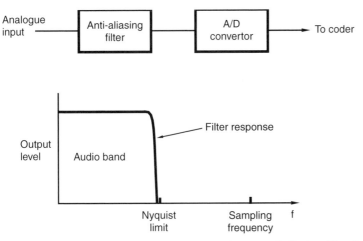

Figure 10.5 In simple A/D convertors an analogue anti-aliasing filter is used prior to conversion, which removes input signals with a frequency above the Nyquist limit

measuring its height, above or below the 0 volt line (see Figure 10.6). Each sample is assigned a value from a range of fixed possibilities, and in the example a ruler scale from 1 to 10 is used for both positive and negative ranges. Each sample must be given one of these integer values, and no fractions or decimal places are allowed; thus each sample is rounded to the nearest whole number during quantisation. The resulting sequence of numbers is shown in the figure.

In a digital audio system, the number system used is *binary* rather than decimal. This has great advantages, since it allows numbers to be represented in two-state form (on/off, high/low, true/false, voltage/no voltage, etc.), and it

---

FACT FILE 10.3 Sampling rates

The choice of sampling rate determines the maximum audio bandwidth available. The table here is an attempt to summarise the variety of sampling frequencies in existence and their applications. A wide range of rates is used in desktop PCs and sound cards, and only the most popular are shown here. Arguments have been presented for the standardisation of higher sampling rates such as 88.2 and 96 kHz, quoting evidence from sources claiming that information above 20 kHz is important for sound quality. It is certainly true that the ear's frequency response does not cut off completely at 20 kHz, but there is very little properly supported evidence that listeners can reliably distinguish between signals containing higher frequencies and those which do not. Doubling the sampling frequency leads to a doubling in the overall data rate of a digital audio system, and a consequent halving in storage time.

Low sampling frequencies such as those below 30 kHz are sometimes encountered in PC workstations for lower quality sound applications such as the storage of speech samples, the generation of internal sound effects and so forth.

Commonly encountered sampling frequencies

f_s (kHz)	Application
8	Used in telephony. Poor audio quality. CCITT G711 standard. IMA RP rate*
~ 11.025	One quarter of the CD sampling rate, used in the sound hardware of some desktop computers, particularly the Apple Macintosh, for low quality applications. IMA RP rate*
16	Used in some telephony applications. G722 data reduction
18.9	CD-ROM/XA and CD-I standard for low–moderate quality audio using ADPCM to extend playing time
~ 22.05	Half the CD rate is 22.05 kHz. The original Apple Macintosh audio sampling rate was 22254.5454.... IMA RP rate*
32	Used in some broadcast systems, e.g. NICAM 3, NICAM 728, DAT long play mode
37.8	CD-ROM/XA and CD-I sampling rate for intermediate quality audio using ADPCM
44.056	A slight modification of the 44.1 rate used in some older equipment to align digital audio with the NTSC television frame rate of 29.97 frames per second. Occasionally still encountered in the USA
44.1	CD sampling rate. Used widely for professional audio recording in many formats. Some professionally modified DAT machines will operate at this rate from analogue inputs. IMA RP rate*
47.952	Occasionally encountered when 48 kHz equipment is used in NTSC video operations. To be avoided
48	'Professional' rate, as specified in AES5-1984, and encountered mainly in digital video recorder sound tracks. Many DAT machines will only sample at this rate through analogue inputs
88.2, 96, 176.4 and 192	2× and 4× sampling rates used in some high-end equipment and standardised as options in the DVD-Audio format
2.8224	A highly oversampled rate used in Super Audio CD

* IMA RP rates were selected in the International Multimedia Association Recommended Practice for Enhancing Digital Audio Compatibility in Multimedia Systems, Rev. 3.00, Oct. 1992, for sound file transfer between workstations

allows audio information to be stored in the same way as computer data is stored. In the decimal number system, each digit of a number represents a power of ten. In a binary system each digit represents a power of two, as shown in Figure 10.7. Each *binary* digi*t* is called a bit. It is possible to calculate the decimal equivalent of a binary integer (whole number) by using the method shown. A number made up of more than one bit is called a binary 'word', and an 8 bit word is called a 'byte' (from 'by eight'). Four bits is called a 'nibble'. The more bits there are in a word the larger the number of states it can represent, with eight bits allowing 256 (2^8) states and sixteen bits allowing

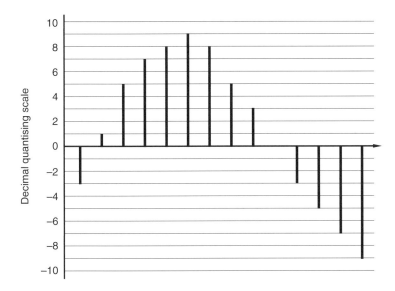

Resulting output data sequence = –3, 1, 5, 7, 8, 9, 8, 5, 3, 0, –3, –5, –7, –9

Figure 10.6 Sample pulses must be quantised to give them numerical values (decimal scale shown). Each sample is forced to the closest quantising level

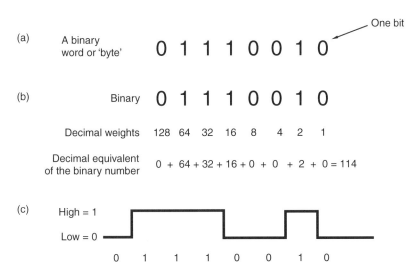

Figure 10.7 (a) A binary number (word or 'byte') consists of a number of bits. (b) Each bit represents a power of two. (c) Binary numbers can be represented electrically in pulse-code modulation (PCM) by a string of high and low voltages

65 536 (2^{16}). The bit with the lowest weight (2^0) is called the least significant bit or LSB, and that with the greatest weight is called the most significant bit or MSB. The term kilobyte or kbyte is used to mean 1024 or 2^{10} bytes, and the term megabyte or Mbyte represents 1024 kbytes.

A binary number (word) with 3 bits is capable of representing 2^3 (8) different values, and an example is shown in Figure 10.8 of a quantising scale using 3 bit binary instead of decimal values. The numbers have been arranged so that binary zero is at 0 volts (as explained in Fact File 10.4).

Because it is rare for samples to lie precisely at the amplitudes represented by each quantising step, there will usually be a slight difference between the quantised sample and its original amplitude. This is called the quantising error. The maximum value of the quantising error is plus or minus half of one quantising interval, since if a sample's amplitude is more than half of a quantising interval above a step it is rounded to the next one. If more steps are used to cover the same quantising range, the potential quantising error is smaller, since the individual quantising intervals are smaller (see Figure 10.9).

Four bit quantisation is quite coarse, and leads to a large quantising error, since only sixteen steps are available. A system using 4 bit quantisation would sound very distorted because the difference between the quantised signal and the true shape of the signal would be considerable. What, though, is the effect of the error which arises in quantisation? It is really a form of distortion, since it changes the shape of the audio signal very slightly, depending on the magnitude of the error. The error

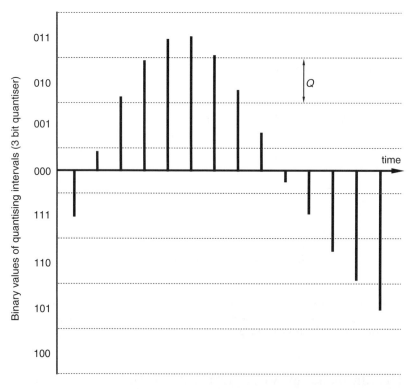

Figure 10.8 Each sample is mapped to the closest quantising interval Q, and given the binary value assigned to that interval. (Example of a 3 bit quantiser shown, with binary zero at 0 volts). On D/A conversion each binary value is assumed to represent the voltage at the mid point of the quantising interval

┌───┐

FACT FILE 10.4 — Representing negative values

Negative sample values (those that lie under the 0 volts line) must be designated as such, since a digital audio system may need to perform mathematical operations on the data in order to perform such functions as adding two signals together. A number system known as two's complement is normally used for this purpose, which represents negative binary numbers by making the leftmost bit a one. The leftmost bit is called the most significant bit or MSB, since it represents the greatest power of two, and for positive values the MSB is a zero.

Negative values are worked out by taking the positive equivalent, inverting all the bits and adding a one. Thus to get the 4 bit binary equivalent of decimal minus five (-5_{10}) in binary two's complement form:

$$5_{10} = 0101_2$$
$$-5_{10} = 1010 + 0001 = 1011_2$$

Two's complement numbers have the advantage that the MSB represents the sign (1 = negative, 0 = positive), and that arithmetic may be performed between positive and negative numbers giving the correct result:

e.g. (in decimal):

$$5$$
$$+ (-3)$$
$$= 2$$

or (in binary):

$$0101$$
$$+ 1101$$
$$= 0010$$

The carry bit which may result from adding the two MSBs is ignored.

An example was shown in Figure 10.8 of a 3 bit quantising scale using two's complement numbering, and it will be seen that the binary value changes from all zeros to all ones as it crosses the 0 volts line. Further, the maximum positive value is 011, and the maximum negative value is 100, so the values wrap around from max. positive to max. negative.

(a)

Figure 10.9 In (a) a 3 bit scale is used and only a small number of quantising intervals covers the analogue voltage range, making the maximum quantising error quite large. The second sample in this picture will be assigned the value 010, for example, the corresponding voltage of which is somewhat higher than that of the sample.

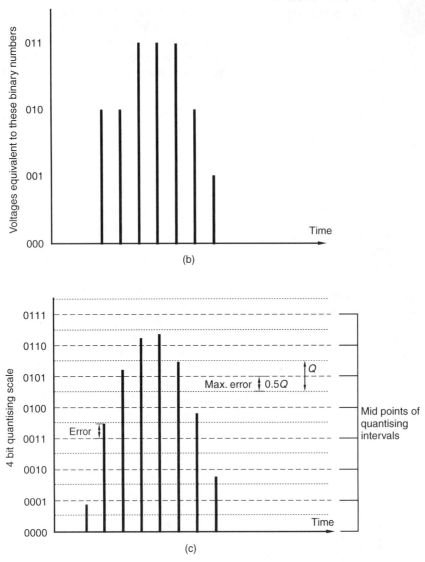

Figure 10.9 (contd.) During D/A conversion the binary sample values from (a) would be turned into pulses with the amplitudes shown in (b), where many samples have been forced to the same level owing to quantising. In (c) the 4 bit scale means that a larger number of intervals is used to cover the same range and the quantising error is reduced. (Expanded positive range only shown for clarity)

signal which results is called quantising noise or quantising distortion, and provided that the quantised audio signal is of suitably high level and has a semi-random nature (as most real signals do), the error will manifest itself as low level noise. The addition of an intentional low level noise signal to the audio signal prior to conversion, known as 'dither', helps to randomise the effect of the error and make it more noise like, particularly at low signal levels.

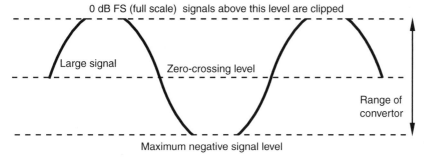

0 dB FS (full scale) signals above this level are clipped

Large signal   Zero-crossing level

Range of convertor

Maximum negative signal level

Figure 10.10 Signals exceeding peak level in a digital system are hard-clipped, since no more digits are available to represent the sample value

The dynamic range of a digital audio system is limited at high signal levels by the point at which the range of the convertor has been 'used up' (in other words, when there are no more bits available to represent a higher level signal). At this point the waveform will be hard clipped (see Figure 10.10) and will become very distorted. This point will normally be set to occur at a certain electrical input voltage, such as +24 dBu in some professional systems. (The effect is very different from that encountered in analogue tape recorders which tend to produce gradually more distortion as the recording level increases. Digital recorders remain relatively undistorted as the recording level rises until the overload point is reached, at which point very bad distortion occurs.) At low signal levels the dynamic range is limited by quantisation noise. (Fact File 10.5 addresses the choice of quantising resolution in digital audio systems.)

The quantised output of an A/D convertor is sometimes carried on a collection of wires, each carrying 1 bit of the binary word which represents the audio signal, so a 16 bit convertor would have sixteen single-bit outputs. When each bit of the audio sample is carried on a separate wire, the signal is said to be in a *parallel* format. If the data is transmitted down a single channel, one bit after the other, the data is said to be in *serial* format. The process of converting an analogue signal into serial binary data is called pulse-code modulation (PCM).

10.3.4 Oversampling in A/D conversion

Oversampling involves sampling audio at a higher frequency than strictly necessary to satisfy the Nyquist criterion (in other words, much higher than twice the highest audio frequency). The sampling rate required just to satisfy the Nyquist criterion is often known as the Nyquist rate. Using a sampling rate well above the audio range (often tens or hundreds of times the Nyquist rate) means that the spectral repetitions resulting from PAM are a long way from the upper end of the audio band (see Figure 10.11). Steep analogue filters are therefore not required, resulting in a better sound quality. The analogue anti-aliasing filter used in conventional convertors is replaced by a digital decimation filter which is capable of a more precise frequency and phase response.

FACT FILE **10.5**	Quantising resolution

The number of bits per sample dictates the signal-to-noise ratio or dynamic range of a digital audio system. The theoretical dynamic range of a linear PCM system is approximately 6 dB per bit. Thus a 16 bit system is theoretically capable of around 96 dB dynamic range. Unfortunately the situation in practice is nothing like so straightforward, since many systems are capable of reproducing signals below the theoretical limit due partly to modulation of the dither signal, and oversampling convertors with noise shaping exhibit noise characteristics which may appear subjectively to be lower. Additionally, many so-called 16 bit convertors do not have full 16 bit linearity, and thus the resulting noise floor may be higher than predicted in theory. The table is a summary of the applications for different sample resolutions.

For many years now 16 bit linear PCM has been the norm for high quality audio applications. This is the CD standard and is capable of offering good dynamic range of over 90 dB. For most purposes this is adequate, but it is often the case that for professional recording purposes one needs a certain amount of 'headroom' – in other words some unused dynamic range above the normal peak recording level which can be used in unforeseen circumstances such as when a signal overshoots its expected level. This can be particularly necessary in live recording situations where one is never quite sure what is going to happen with recording levels. 20 and 24 bit recording formats are therefore becoming increasingly popular because of their extra dynamic range.

At the lower quality end, some PC sound cards and internal sound generators operate at resolutions as low as 4 bits. 8 bit resolution is quite common in desktop computers, and this proves adequate for moderate quality sound through the PC's internal loudspeakers. It gives a dynamic range of nearly 50 dB undithered.

Linear quantising resolution

Bits per sample	Approx dyn. range with dither	Application
8	44 dB	Low–moderate quality for older PC internal sound generation. Some multimedia applications. Usually in the form of unsigned binary numbers
12	68 dB	Older Akai samplers, e.g. S900
14	80 dB	Original EIAJ format PCM adaptors, such as Sony PCM-100
16	92 dB	CD standard. DAT standard. Most widely used high quality resolution for consumer media and many professional recorders. Many multimedia PCs. Two's complement (signed) binary numbers
20	116 dB	High quality professional audio recording and mastering applications. Good convertors available
24	140 dB	Maximum resolution of most new prof. recording systems and DVD-Audio, also of AES/EBU digital interface. Dynamic range would exceed psychoacoustic requirements. Hard to convert at this resolution

Figure 10.11 (a) Oversampling in A/D conversion initially creates spectral repetitions that lie a long way from the top of the audio baseband. The dotted line shows the theoretical extension of the baseband and the potential for aliasing, but the audio signal only occupies the bottom part of this band. (b) Decimation and digital low pass filtering limits the baseband to half the sampling frequency, thereby eliminating any aliasing effects, and creates a conventional collection of spectral repetitions at multiples of the sampling frequency

Oversampling also makes it possible to introduce 'noise shaping' into the conversion process. Noise shaping is a means by which noise within the most audible parts of the audio frequency range (see Chapter 2) is reduced at the expense of increased noise at other frequencies, using a process which 'shapes' the spectral energy of the quantising noise so that most of it lies above the most sensitive regions of the human hearing range.

Although oversampling A/D convertors often use sampling rates of up to 128 times the standard Nyquist rates of 44.1 or 48 kHz, the actual rate at the digital output of the convertor may be no more than the Nyquist rate. Samples acquired at the high rate are quantised to only a few bits resolution and then digitally filtered to reduce the sampling rate back to the Nyquist rate, as shown in Figure 10.12. The digital low-pass filter limits the bandwidth of the signal to the Nyquist frequency in order to avoid aliasing, and this is coupled with 'decimation'. Decimation reduces the sampling rate by dropping samples from the oversampled stream. A result of the low-pass filtering operation is to increase the word length of the samples very considerably. This is not simply an arbitrary extension of the word length, but an accurate calculation of the correct value of each sample, based on the values of surrounding samples. The sample resolution can then be shortened as necessary to produce the desired word length.

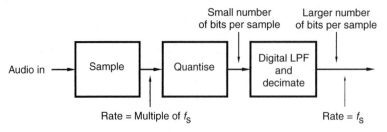

Figure 10.12 Block diagram of oversampling A/D conversion process

10.4 Digital-to-analogue (D/A) conversion

10.4.1 A basic D/A convertor

In order to reproduce digital recordings it is necessary to convert the binary data back into analogue sound signals. The basic D/A conversion process is shown in Figure 10.13. Audio sample words are converted back into a staircase-like chain of voltage levels corresponding to the quantised sample values. This is achieved in simple convertors by using the states of bits to turn current sources on or off, making up the required pulse amplitude by the combination of outputs of each of these sources. This staircase is then 'resampled' to reduce the width of the pulses before they are passed through a low-pass reconstruction filter, whose cut-off frequency is half the sampling frequency. The effect of the reconstruction filter is to join up the sample points to make a smooth waveform. Resampling is necessary because otherwise the averaging effect of the filter would result in a reduction in the amplitude of high-frequency audio signals (the so-called 'aperture effect'). Aperture effect may be reduced by limiting the width of the sample pulses to perhaps one-eighth of the sample period. Equalisation may be required to correct for aperture effect.

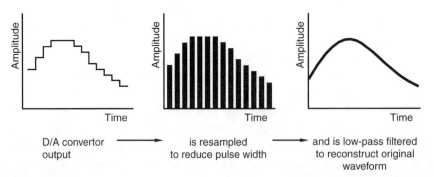

Figure 10.13 Processes involved in D/A conversion (positive sample values only shown)

10.4.2 Oversampling in D/A conversion

Oversampling may be used in D/A conversion, as well as in A/D conversion. In the D/A case new samples must be calculated and inserted at appropriate intervals in between the Nyquist rate samples, in order that conversion can be performed at a higher sampling rate. These samples are then converted back to analogue at the higher rate, thereby avoiding the need for steep analogue reconstruction filters. Noise shaping may also be introduced at the D/A stage, depending on the design of the convertor, to reduce the subjective level of the noise.

A number of advanced D/A convertor designs exist which involve oversampling at a high rate, creating samples with only a few bits of resolution. The extreme version of this approach involves very high rate conversion of single bit samples (so-called 'bit stream conversion'), with noise shaping to optimise the noise spectrum of the signal.

10.5 Direct Stream Digital

Bit stream conversion with delta–sigma analogue-to-digital converters forms the basis for the Direct Stream Digital (DSD) system developed by Philips and Sony for the Super Audio Compact Disc (SACD). DSD is likely to become increasingly important as a very high-quality audio coding system, and computer-based digital audio workstations using DSD have recently been developed to achieve levels of signal processing and editing comparable with those offered by conventional multi-bit PCM systems. Fact File 10.6 describes DSD.

10.6 Introduction to digital audio signal processing

Just as processing operations like equalisation, fading and compression can be performed on analogue sound signals, so they can on digital sound signals. Indeed it is often possible to achieve certain operations in the digital domain with fewer side-effects such as phase distortion. It is also possible to perform operations in the digital domain which are either very difficult or impossible in the analogue domain. High quality, authentic-sounding artificial reverberation is one such example, in which the reflection characteristics of different halls and rooms can be accurately simulated. Digital signal processing (DSP) involves the high speed manipulation of the binary data representing audio samples. This is carried out by computing devices capable of performing fast mathematical operations on data, often many millions per second. It may involve changing the values and timing order of samples, and it may involve the combining of two or more streams of audio data. DSP can affect the sound quality of digital audio in that it can add noise or distortion, although one must assume that the aim of good design is to minimise any such degradation in quality.

In the sections which follow an introduction will be given to some of the main applications of DSP in audio products such as mixers, effects devices and workstations without delving into the mathematical principles involved. In some cases the description is an over-simplification of the process, but the aim has been to illustrate concepts not to tackle the detailed design considerations involved.

10.6.1 Gain changing (level control)

It is relatively easy to change the level of an audio signal in the digital domain. It is most easy to shift its gain by 6 dB since this involves shifting the whole sample word either one step to the left or right (see Figure 10.14). Effectively the original value has been multiplied or divided by a factor of two. More precise gain control is obtained by multiplying the audio sample value by some other factor representing the increase or decrease in gain. The number of bits in the multiplication factor determines the accuracy of gain adjustment. The result of multiplying two binary numbers together is to create a new sample word which may have many more bits than the original, and it is common to find that digital mixers have internal structures capable of handling 32 bit words, even though their inputs and outputs may handle only 20.

The values used for multiplication in a digital gain control may be derived from any user control such as a fader, rotary knob or on-screen representation, or they may be derived from stored values in an automation system. A simple 'old-fashioned' way of deriving a digital value from an 'analogue' fader is to connect the

---

FACT FILE 10.6   Direct Stream Digital (DSD)

A very high sampling rate of 2.8224 MHz is used (i.e. 64× the CD standard of 44.1 kHz) to feed a delta–sigma A–D converter with a negative feedback loop. Unlike multi-bit PCM which uses, say, 16 bits to encode one of a possible 65 536 values (2^{16}) for the magnitude of the waveform at each sample point, the 1-bit DSD system can be said to encode the *difference* between a present sample and those which preceded it. A much simpler 1-bit stream of data is thereby generated. In order that amplitude as well as shape of the waveform is encoded, the negative feedback loop enables the level of the incoming signal to be compared against an accumulation of levels of previous samples. If it is higher, a 1 is generated. If it is lower, a 0 is generated.

During a period when a waveform has a high positive level, the data stream is therefore rich in 1s. Where the waveform has a high negative level, the stream is rich in 0s. A more even distribution of values obtains during periods between these extremes. Because there is a correlation between the amplitude of the waveform and the density of pulses, the system has been called Pulse Density Modulation (PDM).

Whereas earlier 1-bit systems then used decimation digital filtering to convert the 64×/1-bit output of the converter to a multi-bit, lower sample rate stream before recording, in DSD the 1-bit pulses are recorded directly. This eliminates both the decimation process and the reciprocal playback interpolation filtering from the chain.

In the decoding stage relatively simple analogue low-pass filtering and fifth-order noise-shaping filters are used to give a signal-to-noise ratio of 120 dB. The system has a bandwidth of 100 kHz.

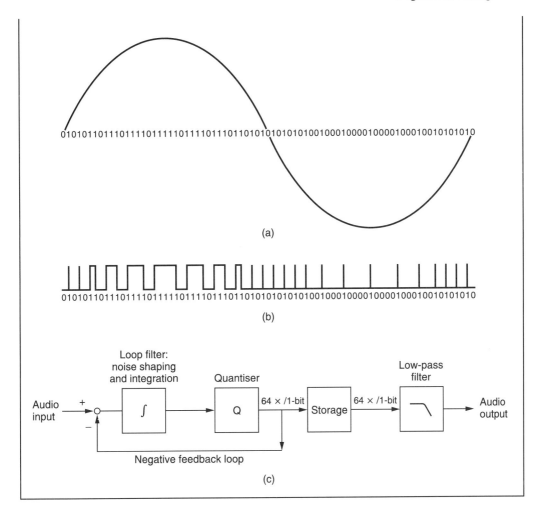

(a)

(b)

(c)

fader to a fixed voltage supply and connect the fader wiper to an A/D convertor, although it is quite common now to find controls capable of providing a direct binary output relating to their position. The 'law' of the fader (the way in which its gain is related to its physical position) can be determined by creating a suitable look-up table of values in memory which are then used as multiplication factors corresponding to each physical fader position.

Figure 10.14 The gain of a sample may be changed by 6 dB simply by shifting all the bits one step to the left or right

10.6.2 Cross-fading

Cross-fading is employed widely in digital audio workstations at points where one section of sound is to be joined to another (edit points). It avoids the abrupt change of waveform which might otherwise result in an audible click, and allows one sound to take over smoothly from the other. (This is discussed further later.)

The process is illustrated conceptually in Figure 10.15. It involves two signals each undergoing an automated fade (binary multiplication), one downwards and the other upwards, followed by an addition of time-coincident samples of the two signals. By controlling the rates and coefficients involved in the fades one can create different styles of cross-fade for different purposes.

10.6.3 Mixing

Mixing is the summation of independent data streams representing the different audio channels. Time coincident samples from each input channel are summed to produce a single output channel sample. Clearly it is possible to have many mix 'buses' by having a number of separate summing operations for different output channels. The result of summing a lot of signals may be to increase the overall level considerably, and the architecture of the mixer must allow enough headroom for this possibility. In the same way as an analogue mixer, the gain structure within a digital mixer must be such that there is an appropriate dynamic range window for the signals at each point in the chain, also allowing for operations such as equalisation which change the signal level.

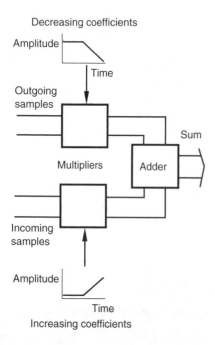

Figure 10.15 Conceptual block diagram of the cross-fading process, showing two audio signals multiplied by changing coefficients, after which they are added together

10.6.4 Digital filters and equalisation

Digital filtering is something of a 'catch-all' term, and is often used to describe DSP operations which do not at first sight appear to be filtering. A digital filter is essentially a process which involves the time delay, multiplication and recombination of audio samples in all sorts of configurations, from the simplest to the most complex. Using digital filters one can create low and high-pass filters, peaking and shelving filters, echo and reverberation effects, and even adaptive filters which adjust their characteristics to affect different parts of the signal.

To understand the basic principle of digital filters it helps to think about how one might emulate a certain analogue filtering process digitally. Filter responses can be modelled in two main ways – one by looking at their frequency domain response and the other by looking at their time domain response. (There is another approach involving the so-called z-plane transform, but this is not covered here.) The frequency domain response shows how the amplitude of the filter's output varies with frequency, whereas the time domain response is usually represented in terms of an impulse response (see Figure 10.16). An impulse response shows how the filter's output responds to stimulation at the input by a single short impulse. Every frequency response has a corresponding impulse response because the two are directly related. If you change the way a filter responds in time you also change the way it responds in frequency. A mathematical process known as the Fourier transform is often used as a means of transforming a time domain response into its equivalent frequency domain response (see Chapter 1). They are simply two ways of looking at the same thing.

Digital audio is time discrete because it is sampled. Each sample represents the amplitude of the sound wave at a certain point in time. It is therefore normal to create certain filtering characteristics digitally by operating on the audio samples in the time domain. In fact if it was desired to emulate a certain analogue filter characteristic digitally one would need only to measure its impulse response and model this in the digital domain. The digital version would then have the same frequency response as the analogue version, and one can even envisage the possibility for favourite analogue filters to be recreated for the digital workstation. The

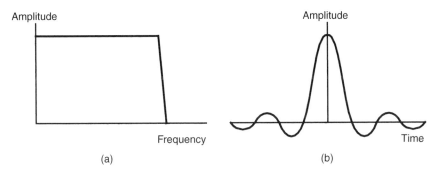

Figure 10.16 Examples of (a) the frequency response of a simple filter, and (b) the equivalent time domain impulse response

question, though, is how to create a particular impulse response characteristic digitally, and how to combine this with the audio data.

As mentioned earlier, all digital filters involve delay, multiplication and recombination of audio samples, and it is the arrangement of these elements that gives a filter its impulse response. A simple filter model is the finite impulse response (FIR) filter, or transversal filter, shown in Figure 10.17. As can be seen, this filter consists of a tapped delay line with each tap being multiplied by a certain coefficient before being summed with the outputs of the other taps. Each delay stage is normally a one sample period delay. An impulse arriving at the input would result in a number of separate versions of the impulse being summed at the output, each with a different amplitude. It is called a finite impulse response filter because a single impulse at the input results in a finite output sequence determined by the number of taps. The more taps there are the more intricate the filter's response can be made, although a simple low pass filter only requires a few taps.

The other main type is the infinite impulse response (IIR) filter which is also known as a recursive filter, because there is a degree of feedback between the output and the input (see Figure 10.18). The response of such a filter to a single impulse is an infinite output sequence, because of the feedback. IIR filters are often used in audio equipment because they involve fewer elements for most variable equalisers than equivalent FIR filters, and they are useful in effects devices. They are unfortunately not phase linear, though, whereas FIR filters can be made phase linear.

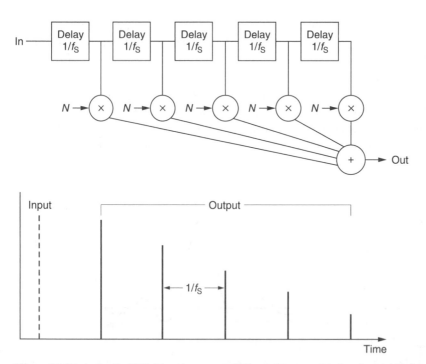

Figure 10.17 A simple FIR filter (transversal filter). N = multiplication coefficient for each tap. Response shown below indicates successive outputs samples multipled by decreasing coefficients

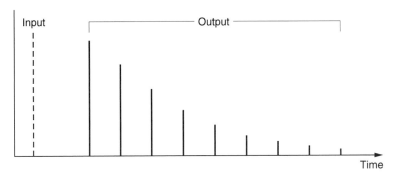

Figure 10.18 A simple IIR filter (recursive filter). The output impulses continue indefinitely but become very small. N in this case is about 0.8. A similar response to the previous FIR filter is achieved but with fewer stages

10.6.5 Digital reverberation and other effects

It can probably be seen that the IIR filter described in the previous section forms the basis for certain digital effects, such as reverberation. The impulse response of a typical room looks something like Figure 10.19, that is an initial direct arrival of sound from the source, followed by a series of early reflections, followed by a diffuse 'tail' of densely packed reflections decaying gradually to almost nothing. Using a number of IIR filters, perhaps together with a few FIR filters, one could create a suitable pattern of delayed and attenuated versions of the original impulse to simulate the decay pattern of a room. By modifying the delays and amplitudes of the early reflections and the nature of the diffuse tail one could simulate different rooms.

Figure 10.19 The impulse response of a typical reflective room

The design of convincing reverberation algorithms is a skilled task, and the difference between crude approaches and good ones is very noticeable. Some audio workstations offer limited reverberation effects built into the basic software package, but these often sound rather poor because of the limited DSP power available and the crude algorithms involved. More convincing reverberation processors are available which exist either as stand-alone devices or as optional plug-in cards to the workstation, having more DSP capacity and tailor-made software.

Other simple effects can be introduced without much DSP capacity, such as double-tracking and phasing/flanging effects. These often only involve very simple delaying and recombination processes. Pitch shifting can also be implemented digitally, and this involves processes similar to sample rate conversion, as described below. High quality pitch shifting requires quite considerable horsepower because of the number of calculations required.

10.6.6 Dynamics processing

Digital dynamics processing involves gain control which depends on the instantaneous level of the audio signal. A simple block diagram of such a device is shown in Figure 10.20. A side chain produces coefficients corresponding to the instantaneous gain change required, which are then used to multiply the delayed audio samples. First the r.m.s. level of the signal must be determined, after which it needs to be converted to a logarithmic value in order to determine the level change in decibels. Only samples above a certain threshold level will be affected, so a constant factor must be added to the values obtained, after which they are multiplied by a factor to represent the compression slope. The coefficient values are then anti-logged to produce linear coefficients by which the audio samples can be multiplied.

10.7 Low bit rate audio coding

Considerable advances have been made in recent years in the field of audio bit rate reduction or low bit rate coding. Conventional PCM audio has a high data rate, and

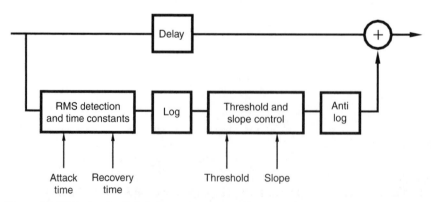

Figure 10.20 A simple digital dynamics processing operation

there are many applications for which it would be an advantage to have a lower data rate without much (or any) loss of sound quality. 16 bit linear PCM at a sampling rate of 44.1 kHz ('CD quality digital audio') results in a data rate of:

$$16 \times 44\,100 = 705\,600 \text{ bits per second per channel (about 700 kbit s}^{-1})$$

For multimedia applications, broadcasting, communications and some consumer purposes the data rate may be reduced to a fraction of that shown above with almost no effect on the perceived sound quality. Simple techniques for reducing the data rate, such as reducing the sampling rate or number of bits per sample would have a very noticeable effect on sound quality, so most modern low bit rate coding works by exploiting the phenomenon of auditory masking (see Fact File 2.3) to 'hide' the increased noise resulting from bit rate reduction in parts of the audio spectrum where it will hopefully be inaudible. The following is a very brief overview of how one approach works, based on the technology involved in the MPEG (Moving Pictures Expert Group) standards.

As shown in Figure 10.21, the incoming digital audio signal is filtered into a number of narrow frequency bands. Parallel to this a computer model of the human hearing process analyses a short portion of the audio signal (a few milliseconds). This analysis is used to determine what parts of the audio spectrum will be masked, and to what degree, during that short time period. In bands where there is a strong signal, quantising noise can be allowed to rise considerably without it being heard, because one signal is very efficient at masking another lower level signal in the same band as itself (see Figure 10.22). Provided that the noise is kept below the masking threshold in each band it should be inaudible.

Blocks of audio samples in each narrow band are scaled (low level signals are amplified so that they use more of the most significant bits of the range), and the scaled samples are then reduced in resolution (requantised) by reducing the number of bits available to represent each sample – a process which results in increased quantising noise. The output of the hearing model is used to control the requantising process so that the sound quality remains as high as possible for a given bit rate. The greatest number of bits is allocated to frequency bands where noise would be most audible, and the fewest to those bands where the noise would be effectively masked by the signal. Control information is sent along with the blocks of bit-rate-

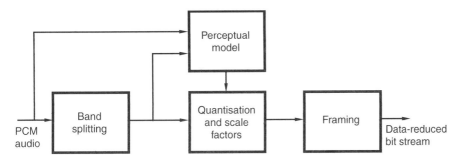

Figure 10.21 Generalised block diagram of a psychoacoustic low bit rate coder

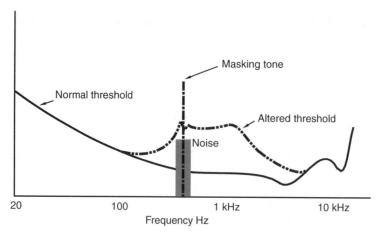

Figure 10.22 Quantising noise lying under the masking threshold will normally be inaudible

reduced samples to allow them to be reconstructed at the correct level and resolution upon decoding.

The above process is repeated every few milliseconds, so that the masking model is constantly being updated to take account of changes in the audio signal. Carefully implemented, such a process can result in a reduction of the data rate to anything from about one quarter to less than one tenth of the original data rate. A decoder uses the control information transmitted with the bit-rate-reduced samples to restore the samples to their correct level, and can determine how many bits were allocated to each frequency band by the encoder, reconstructing linear PCM samples and then recombining the frequency bands to form a single output (see Figure 10.23). A decoder can be much less complex, and therefore cheaper, than an encoder, because it does not need to contain the auditory model processing.

A standard known as MPEG, published by the International Standards Organisation (ISO 11172-3), defines a number of 'layers' of complexity for low bit rate audio coders as shown in Table 10.1.

Each of the layers can be operated at any of the bit rates within the ranges shown (although some of the higher rates are intended for stereo modes), and the user must make appropriate decisions about what sound quality is appropriate for each application. The lower the data rate, the lower the sound quality that will be obtained. At high data rates the encoding–decoding process has been judged by many to be audibly 'transparent' – in other words they cannot detect that the coded and decoded signal is different from the original input.

Figure 10.23 Generalised block diagram of an MPEG audio decoder

Table 10.1 MPEG-Audio layers

Layer	Complexity	Min. delay	Bit rate range	Target
1	Low	19 ms	32–448 kbit/s	192 kbit/s
2	Moderate	35 ms	32–384 kbit/s*	128 kbit/s
3	High	59 ms	32–320 kbit/s	64 kbit/s

* In Layer 2, bit rates of 224 kbit/s and above are for stereo modes only

Layer 2 has been chosen as the standard for Digital Audio Broadcasting (DAB) in Europe and some other parts of the world. It allows a large number of digital stereo radio channels to be transmitted in less radio frequency spectrum space than conventional VHF FM stereo channels, with higher sound quality and lower interference.

MP3 will be for many people the name associated with the personal stereo record/replay machines used for downloading music from the Internet, the recording medium being an internal flash memory card with no moving parts. The term MP3 has caused some confusion; it is short for MPEG-1 Layer 3, a standard described above, but MP3 has virtually become a generic term for the system used for receiving compressed audio from the Internet. The term MPEG ('empeg') is generally used as a reference to compressed video signal decoding. There is also MPEG-2 which can handle multichannel surround, and further developments in this and later systems will be briefly touched upon.

MPEG-2BC (Backwards Compatible with MPEG-1) additionally supports sampling frequencies from 16 kHz to 22.05 kHz and 24 kHz at bit rates from 32 to 256 kBit/s for Layer 1. For Layers 2 and 3, bit rates are from 8 to 160 kBit/s. Developments, intended to supersede MPEG-2BC, have included MPEG-2AAC (Advanced Audio Coding). This defines a standard for multichannel coding of up to 48 channels, with sampling rates from 8 kHz to 96 kHz. It also incorporates a Modified Discrete Cosine transform system as used in the Minidisc format which is described in section 10.10.1. MPEG-2AAC was not, however, designed to be backwards compatible with MPEG-1. These are covered further in Chapter 17. MPEG-4 is based on the standards outlined for MPEG-2AAC; it includes further coding techniques for reducing transmission bandwidth, and it can scale the bit rate according to the complexity of the decoder. Audio signals can be described in the form of 'objects' and 'scenes'.

There are a number of types of low bit rate coding used in audio systems, working on similar principles, and used for applications such as consumer disk and tape systems (DCC, MiniDisc, DVD) and digital cinema sound (Dolby SR-D, Sony SDDS, DTS).

10.8 Digital tape recording

10.8.1 The future for dedicated tape formats

In the sections that follow, the basic features of dedicated tape recording formats are outlined. Before the explosion in the use of computer workstations and mass storage for digital audio, tape recorders were the main means of recording and editing. Today there are still a considerable number of dedicated digital tape recording formats in existence, although they are slowly being superseded by computer-based products using removable disks or other mass storage media. Even though tape storage will continue to be used in the future (tape is much cheaper than most disks), it is more likely to be formatted in a computer file system manner than as a dedicated audio recording format, using tape drives connected by standard data interfaces to a host processor.

Tape's main advantages are its cheapness and familiarity. One can easily remove a tape from the recorder at the end of a session and walk away with it. Also, a dedicated tape format can easily be interchanged between recorders, provided that another machine operating to the same standard can be found. Disks, on the other hand, come in a very wide variety of sizes and formats, and even if the disk fits a particular drive it may not be possible to access the audio files there-on, owing to the multiplicity of levels at which compatibility must exist between systems before interchange can take place. This issue is discussed in much greater detail in *The Audio Workstation Handbook* by Francis Rumsey, as detailed at the end of the chapter.

Tape is relatively slow to access, because it is a linear storage medium. One must spool through the tape to reach the desired point. That said, the absence of a formal filing structure on dedicated tape formats makes it much more straightforward to use for people used to conventional recording, since an operator can drop directly in to record mode anywhere along the tape's length without having to worry about file names and so forth. It is clear to the operator exactly where one recording stops and the next one starts, whereas there is a greater degree of abstraction with computer filing structures.

10.8.2 Background to digital tape recording

When commercial digital audio recording systems were first introduced in the 1970s and early 1980s it was necessary to employ recorders with sufficient bandwidth for the high data rates involved (a machine capable of handling bandwidths of a few megahertz was required). Conventional audio tape recorders were out of the question because their bandwidths extended only up to around 35 kHz at best. Video recording, however, required high bandwidths and recording densities for a high-quality picture to be adequately represented, and so existing video tape recorders (VTRs) were drafted in as suitable storage devices. PCM adaptors converted digital audio data into a waveform which resembled a television waveform, suitable for recording on to a VTR. The Denon company of Japan developed such a system in partnership with the NHK broadcasting organisation (Japan's equivalent to the UK's BBC) and they released the world's first PCM recording on to LP in 1971. In the early 1980s, devices such as Sony's PCMF1 (pictured in Figure 10.24) became available at modest prices, allowing 16 bit, 44.1 kHz digital audio to be recorded on to a consumer VTR, resulting in widespread proliferation of stereo digital recording.

Figure 10.24 An early PCM adaptor: the Sony PCM-F1. (Courtesy of Sony Broadcast and Professional Europe)

Dedicated open-reel digital recorders using stationary heads were also developed (see Fact File 10.7). High-density tape formulations were then manufactured for digital use, and this, combined with new channel codes (see below), improvements in error correction and better head design, led to the use of a relatively low number of tracks per channel, or even single-track recording of a given digital signal, combined with playing speeds of 15 or 30 ips.

Dedicated rotary-head systems not using video tape recorders, have also been developed recently – the RDAT format being the most well known.

Digital recording tape is thinner (27.5 microns) than that used for analogue recordings; long playing times can be accommodated on a reel, but also thin tape contacts the machine's heads more intimately than does standard 50 micron thickness tape which tends to be stiffer. Intimate contact is essential for reliable recording and replay of such a densely packed and high-bandwidth signal.

10.8.3 Channel coding for dedicated tape formats

Since 'raw' binary data is normally unsuitable for recording directly on to tape, a 'channel code' is used which matches the data to the characteristics of the recording system, uses storage space efficiently, and makes the data easy to recover on replay. A wide range of channel codes exists, each with characteristics designed for a specific purpose. The channel code converts a pattern of binary data into a different pattern of transitions in the recording or transmission medium. It is another stage of modulation, in effect. Thus the pattern of bumps in the optical surface of a CD bears little resemblance to the original audio data, and the pattern of magnetic flux transitions on a DAT cassette would be similarly different. Given the correct code book, one could work out what audio data was represented by a given pattern from either of these systems.

Many channel codes are designed for a low DC content (in other words, the data is coded so as to spend, on average, half of the time in one state and half in the other) in cases where signals must be coupled by transformers (see section 12.1), and

FACT FILE 10.7 Rotary and stationary heads

There are two fundamental mechanisms for the recording of digital audio on tape, one which uses a relatively low linear tape speed and a quickly-rotating head, and one which uses a fast linear tape speed and a stationary head. In the rotary-head system the head either describes tracks almost perpendicular to the direction of tape travel, or it describes tracks which are almost in the same plane as the tape travel. The former is known as transverse scanning and the latter is known as helical scanning. Transverse scanning uses more tape when compared with helical scanning. It is not common for digital tape recording to use the transverse scanning method. The reason for using a rotary head is to achieve a high head-to-tape speed, since it is this which governs the available bandwidth. Rotary-head recordings cannot easily be splice edited because of the track pattern, but they can be electronically edited using at least two machines.

Stationary heads allow the design of tape machines which are very similar in many respects to analogue transports. With stationary-head recording it is possible to record a number of narrow tracks in parallel across the width of the tape. Tape speed can be traded off against the number of parallel tracks used for each audio channel, since the required data rate can be made up by a combination of recordings made on separate tracks. This approach is used in the DASH format, where the tape speed may be 30 ips (76 cm s^{-1}) using one track per channel, 15 ips using two tracks per channel, or 7.5 ips using four tracks per channel.

Typical track pattern

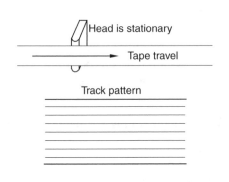

Head is stationary

Tape travel

Track pattern

others may be designed for narrow bandwidth or a limited high-frequency content. Certain codes are designed specifically for very high-density recording, and may have a low clock content with the possibility for long runs in one binary state or the other without a transition. Channel coding involves the incorporation of the data to be recorded with a clock signal, such that there is a sufficient clock content to allow the data and clock to be recovered on replay (see Fact File 10.8). Channel codes vary as to their robustness in the face of distortion, noise and timing errors in the recording channel.

FACT FILE
10.8

Data recovery

Channel-coded data must be decoded on replay, but first the audio data must be separated from the clock information which was combined with it before recording. This process is known as data and sync separation (see diagram).

oscillator will 'flywheel' over short losses or irregularities of the off-tape clock.

Recorded data is usually interspersed with synchronising patterns in order to give the PLL in the data separator a regular reference in the absence of regular dock data from the encoded audio signal, since many channel codes have long runs without a transition.

Even if the off-tape data and clock have timing irregularities, such as might manifest themselves as 'wow' and 'flutter' in analogue reproducers (see section A1.6), these can be

It is normal to use a phase-locked loop for the purpose of regenerating the clock signal from the replayed data (see diagram), this being based around a voltage-controlled oscillator (VCO) which runs at some multiple of the off-tape clock frequency. A phase comparator compares the relative phases of the divided VCO output and the clock data off tape, producing a voltage proportional to the error which controls the frequency of the VCO. With suitable damping, the phase-locked

removed in digital systems. The erratic data (from tape or disk, for example) is written into a short-term solid-state memory (RAM) and read out again a fraction of a second later under control of a crystal clock (which has an exceptionally stable frequency), as shown in the diagram. Provided that the average rate of input to the buffer is the same as the average rate of output, and the buffer is of sufficient size to soak up short-term irregularities in timing, the buffer will not overflow or become empty.

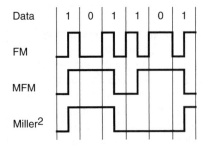

Figure 10.25 Examples of three channel codes used in digital recording. Miller[2] is the most efficient of those shown since it involves the least number of transitions for the given data sequence.

Some examples of channel codes used in audio systems are shown in Figure 10.25. FM is the simplest, being an example of binary frequency modulation. It is otherwise known as 'bi-phase mark', one of the Manchester codes, and is the channel code used by SMPTE/EBU timecode (see Chapter 15). MFM and Miller-squared are more efficient in terms of recording density. MFM is more efficient than FM because it eliminates the transitions between successive ones, only leaving them between successive zeros. Miller[2] eliminates the DC content present in MFM by removing the transition for the last one in an even number of successive ones.

Group codes, such as that used in the Compact Disc and R-DAT, involve the coding of patterns of bits from the original audio data into new codes with more suitable characteristics, using a look-up table or 'code book' to keep track of the relationship between recorded and original codes. This has clear parallels with coding as used in intelligence operations, in which the recipient of a message requires the code book to be able to understand the message. CD uses a method known as 8-to-14 modulation, in which 16 bit audio sample words are each split into two 8 bit words, after which a code book is used to generate a new 14 bit word for each of the 256 possible combinations of 8 bits. Since there are many more words possible with 14 bits than with 8, it is possible to choose those which have appropriate characteristics for the CD recording channel. In this case, it is those words which have no more than eleven consecutive bits in the same state, and no less than three. This limits the bandwidth of the recorded data, and makes it suitable for the optical pick-up process, whilst retaining the necessary clock content.

10.8.4 Error correction

There are two stages to the error correction process used in digital tape recording systems. Firstly the error must be detected, and then it must be corrected. If it cannot be corrected then it must be concealed. In order for the error to be detected it is necessary to build in certain protection mechanisms.

Two principal types of error exist: the burst error and the random error. Burst errors result in the loss of many successive samples and may be due to major momentary signal loss, such as might occur at a tape drop-out or at an instant of impulsive interference such as an electrical spike induced in a cable or piece of dirt on the surface of a CD. Burst error correction capability is usually quoted as the number of consecutive samples which may be corrected perfectly. Random errors result in the loss of single samples in randomly located positions, and are more likely to be the result of noise or poor signal quality. Random error rates are normally quoted as an average rate, for example: 1 in 10^6. Error correction systems must be able to cope with the occurrence of both burst and random errors in close proximity.

Audio data is normally interleaved before recording, which means that the order of samples is shuffled (as shown conceptually in Figure 10.26). Samples which had been adjacent in real time are now separated from each other on the tape. The benefit of this is that a burst error which destroys consecutive samples on tape will result in a collection of single-sample errors in between good samples when the data is de-interleaved, allowing for the error to be concealed. A common process, associated with interleaving, is the separation of odd and even samples by a delay. The greater the interleave delay, the longer the burst error that can be handled. A common example of this is found in the DASH tape format (an open-reel digital recording format), and involves delaying odd samples so that they are separated from even samples by 2448 samples, as well as reordering groups of odd and even samples within themselves.

Redundant data is also added before recording. Redundancy, in simple terms, involves the recording of data in more than one form or place. A simple example of the use of redundancy is found in the twin-DASH format, in which all audio data is recorded twice. On a second pair of tracks (handling the duplicated data), the odd–even sequence of data is reversed to become even–odd. Firstly, this results in double protection against errors, and secondly it allows for perfect correction at a splice, since two burst errors will be produced by the splice, one in each set of tracks. Because of the reversed odd–even order in the second set of tracks, uncorrupted odd data can be used from one set of tracks, and uncorrupted even data from the other set, obviating the need for interpolation (see Fact File 10.9).

Cyclic redundancy check (CRC) codes, calculated from the original data and recorded along with that data, are used in many systems to detect the presence and position of errors on replay. Complex mathematical procedures are also used to form codewords from audio data which allow for both burst and random errors to

Figure 10.26 Interleaving is used in digital recording and broadcasting systems to rearrange the original order of samples. This can have the effect of converting burst errors into random errors when the samples are de-interleaved

FACT FILE 10.9   Error handling

True correction

Up to a certain random error rate or burst error duration an error correction system will be able to reconstitute erroneous samples perfectly. Such corrected samples are indistinguishable from the originals, and sound quality will not be affected. Such errors are often signalled by green lights showing 'CRC' failure or 'Parity' failure.

Original sample amplitudes

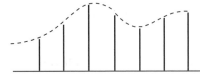

Sample missing due to error

Mathematically-interpolated value inserted

Interpolation

When the error rate exceeds the limits for perfect correction, an error correction system may move to a process involving interpolation between good samples to arrive at a value for a missing sample (as shown in the diagram). The interpolated value is the mathematical average of the foregoing and succeeding samples, which may or may not be correct. This process is also known as concealment or averaging, and the audible effect is not unpleasant, although it will result in a temporary reduction in audio bandwidth. Interpolation is usually signalled by an orange indicator to show that the error condition is fairly serious. In most cases the duration of such concealment is very short, but prolonged bouts of concealment should be viewed warily, since sound quality will be affected. This will usually point to a problem such as dirty heads or a misaligned transport, and action should be taken (see text).

Hold

In extreme cases, where even interpolation is impossible (when there are not two good samples either side of the bad one), a system may 'hold'. In other words, it will repeat the last correct sample value. The audible effect of this will not be marked in isolated cases, but is still a severe condition. Most systems will not hold for more than a few samples before muting. Hold is normally indicated by a red light.

Mute

When an error correction system is completely overwhelmed it will usually effect a mute on the audio output of the system. The duration of this mute may be varied by the user in some systems. The alternative to muting is to hear the output, regardless of the error. Depending on the severity of the error, it may sound like a small 'spit', click, or even a more severe break-up of the sound. In some cases this may be preferable to muting.

be corrected perfectly up to a given limit. Reed–Solomon encoding is another powerful system which is used to protect digital recordings against errors, but it is beyond the scope of this book to cover these codes in detail.

10.8.5 Overview of digital tape formats

There have been a considerable number of commercial recording formats over the last 20 years, and only a brief summary will be given here of the most common at the time of writing. The list is not intended to be exhaustive. Open reel two track formats are not covered here because they are used very rarely today. Although they were designed because of a perceived need for tape-cut editing on digital machines, they were not adopted widely by the market.

Sony's PCM-1610 and PCM-1630 adaptors dominated the CD-mastering market for a number of years, although by today's standards they use a fairly basic recording format and rely on 60 Hz/525 line Umatic cassette VTRs (as shown in Figure 10.27). The system operates at a sampling rate of 44.1 kHz and uses 16 bit quantisation, being designed specifically for the making of tapes to be turned into CDs. Recordings made in this format may be electronically edited using the Sony DAE-3000 editing system, and the playing time of tapes runs up to 75 minutes using a tape specially developed for digital audio use.

The R-DAT or DAT format is a small stereo, rotary-head, cassette-based format offering a range of sampling rates and recording times, including the professional rates of 44.1 and 48 kHz. Originally, consumer machines operated at 48 kHz to avoid the possibility for digital copying of CDs, but professional versions are available which will record at either 44.1 or 48 kHz. Modern consumer machines will record at 44.1 kHz, but usually only via the digital inputs. DAT is a 16 bit format. There is a fairly large number of machines available, including truly professional

Figure 10.27 Sony DMR-4000 digital master recorder. (Courtesy of Sony Broadcast and Professional Europe)

designs offering editing facilities, external sync and IEC-standard timecode. Various non-standard modifications have been introduced in recent years, including a 96 kHz sampling rate DAT machine for audiophile use, and adaptors which will store 20 bit audio on such a high sampling rate machine (but sacrificing the high sampling rate for more bits).

The IEC timecode standard for R-DAT was devised in 1990 as a common format for the insertion of timecode (see Chapter 15) into the subcode area of the helically scanned track. In practice it will allow for SMPTE/EBU timecode of any frame rate to be converted into the internal DAT 'running-time' code, and then converted back into any SMPTE/EBU frame rate on replay. This will involve the installation of an intelligent timecode reader/generator into the DAT machine.

A typical professional R-DAT machine is pictured in Figure 10.28. The format has become exceptionally popular with professionals owing to its low cost, high performance, portability and convenience.

The Nagra-D recorder (see Figure 10.29) was designed as a digital replacement for the world-famous Nagra analogue recorders, and as such is intended for professional use in field recording and studios. The format was designed to have considerable commonality with the audio format used in D1 and D2-format digital VTRs, having rotary heads, although it uses open reels for operational convenience. Allowing for 20–24 bits of audio resolution, the Nagra-D format is appropriate for use with high resolution convertors. The error correction and recording density used in this format are designed to make recordings exceptionally robust, and recording time may be up to 6 hours on a 7 inch (18 cm) reel, in two-track mode. The format is also designed for operation in a four-track mode at twice the stereo tape speed, such that in stereo the tape travels at 4.75 cm s^{-1}, and in four track at 9.525 cm s^{-1}.

The DASH (Digital Audio Stationary Head) format consists of a whole family of open-reel stationary-head recording formats from two tracks up to 48 tracks. DASH-format machines may operate at 44.1 kHz or 48 kHz rates (and sometimes optionally at 44.056 kHz), and they allow varispeed ±12.5 per cent. They are designed to allow gapless punch-in and punch-out, splice editing, electronic editing

Figure 10.28 Sony PCM7030 professional DAT machine. (Courtesy of Sony Broadcast and Professional Europe)

Figure 10.29 Nagra-D open reel digital tape recorder. (Courtesy of Sound PR)

and easy synchronisation. Multitrack DASH machines have gained wide acceptance in studios, but the stereo machines have not. An example of a machine is shown in Figure 10.30. Recent developments have resulted in DASH multitracks capable of storing 24 bit audio instead of the original 16 bits.

The ProDigi (PD) format was introduced by Mitsubishi in the mid 1980s, and a number of PD multitrack machines are to be found in studios around the world. The most common version of the format is in the 32 channel multitrack machines by Mitsubishi and Otari, but the format is now no longer marketed by these companies.

In recent years there has been a move towards the introduction of budget modular multitrack formats which are beginning to take over from the more expensive open reel formats. Most of these are based on 8 track cassettes using rotary head transports borrowed from consumer video technology. The most widely used at the time of writing are the DA-88 format (based on Hi-8 cassettes) and the ADAT format (based on VHS cassettes). These offer most of the features of open reel machines and a number of them may be synchronised to expand the channel capacity. An example is shown in Figure 10.31. (More recently, similar machines using Minidisc and M-O drives have been introduced, the latter capable of creating virtual tracks in addition to the eight simultaneous record and replay tracks.)

Figure 10.30 An open-reel digital multitrack recorder: the Sony PCM-3348. (Courtesy of Sony Broadcast and Professional Europe)

10.8.6 Performance and alignment

Crosstalk between the tracks of a digital recorder is virtually absent, and this frees the engineer from the need to allocate tracks carefully. For example, with an analogue multitrack machine one tends to record vocals on tracks which are physically far away on the tape from, say, drum tracks, since crosstalk from the latter can easily be audible. Unused tracks will often be used to form an extended guard band to separate, say, an electric guitar track or timecode track from a vocal. None of this is necessary with a digital multitrack recorder. Other recording artefacts commonly encountered with analogue recordings, such as wow and flutter, are also absent on digital machines.

Figure 10.31 A modular digital multitrack machine, Sony PCM-800. (Courtesy of Sony Broadcast and Professional Europe)

Digital tape machines require at least as much care and maintenance as do their analogue counterparts. In fact, the analogue machine can be monitored for slight deterioration in performance rather more easily than can the digital. Slight frequency response fall-off, for example, due to, say, head wear or azimuth misalignment, can quickly be checked for and spotted using a test tape on an analogue machine. Misalignment of a digital machine, however, will cause two types of problem, neither of which is easily spotted unless careful checks are regularly carried out. Firstly, a machine's transport may be misaligned such that although tapes recorded on it will replay satisfactorily on the same machine, another correctly aligned machine may well not be able to obtain sufficient data from the tape to reconstitute the signal adequately. Error correction systems will be working too hard and drop-outs will occur. Secondly, digital recording tends to work without noticeable performance deterioration until things get so bad that drop-outs occur. There is little warning of catastrophic drop-outs, although the regularity of errors is an excellent means of telling how close to the edge a system is in terms of random errors. If a recorder has error status indication, these can be used to tell the prevailing state of the tape or the machine's alignment. Different degrees of seriousness exist, as was discussed in Fact File 10.9. If the machine is generating an almost constant string of CRC errors or interpolations, then this is an indication that either the tape or the machine is badly out of alignment or worn. It is therefore extremely important to check alignment often and to clean the machine's heads using whatever means is recommended. Alignment may require specialised equipment which dealers will usually possess.

10.8.7 Editing digital tape recordings

Razor blade cut-and-splice editing is possible on open-reel digital formats, and the analogue cue tracks are monitored during these operations. This is necessary because the digital tracks are generally only capable of being replayed at speeds that are no more than about 10 per cent away from normal replay speed. The analogue tracks are of low quality – rather lower than a dedicated analogue machine – but usually adequate as a cue.

A 90° butt joint is used for the editing of digital tape. The discontinuity in the data stream caused by the splice would cause complete momentary drop-out of the digital signal if no further action were taken, so circuits are incorporated which sense the splice and perform an electronic cross-fade from one side of the splice to the other, with error concealment to minimise the audibility of the splice. It is normally advised that a 0.5 mm gap is left at the splice so that its presence will easily be detected by the cross-fade circuitry. The thin tape can easily be damaged during the cut-and-splice edit procedure and a safety copy would be most advisable. Electronic editing is far more desirable, and is the usual method.

Electronic editing normally requires the use of two machines plus a control unit, as shown in the example in Figure 10.32. A technique is employed whereby a finished master tape is assembled from source takes on player machines. This is a relatively slow process, as it involves real time copying of audio from one machine

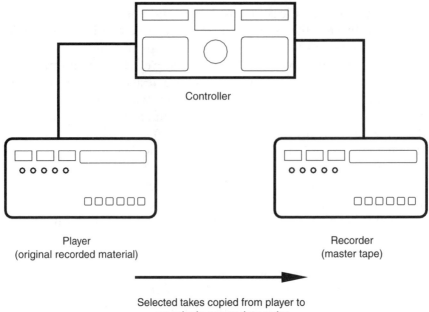

Controller

Player
(original recorded material)

Recorder
(master tape)

Selected takes copied from player to
recorder in appropriate order

Figure 10.32 In electronic tape-copy editing selected takes are copied in sequence from player to recorder with appropriate cross-fades at joins

to another, and modifications to the finished master are difficult. The digital editor can often store several seconds of programme in its memory and this can be replayed at normal speed or under the control of a search knob which enables very slow to-and-fro searches to be performed in the manner of rock and roll editing on an analogue machine. Edits may be rehearsed prior to execution. When satisfactory edit points have been determined the two machines are synchronised using timecode, and the record machine is switched to drop in the new section of the recording from the replay machine at the chosen moment. Here a cross-fade is introduced between old and new material to smooth the join. The original source tape is left unaltered.

10.9 Workstations and recording on mass storage media

Once audio is in a digital form it can be handled by a computer like any other data. The only real difference is that audio requires a high sustained data rate and large amounts of storage compared with more basic data such as text. The following is a very brief introduction to some of the technology associated with computer-based audio workstations. Much more detail will be found in *The Audio Workstation Handbook* as detailed in the Further Reading list. It should be noted that the terminology for such recording systems has not yet reached a point of universal agreement. It has become quite common to refer to 'non-linear recording', meaning recording on media where the data storage is not linear (as opposed to tape, which is a linear medium). This should not be confused with non-linear coding of audio, which is a technique used in A/D conversion and data reduction. Other terms used for recording on disk-based mass storage media include 'random access' and 'direct access' recording, also 'tapeless recording'.

10.9.1 Storage requirements of digital audio

Table 10.2 shows the data rates required to support a single channel of digital audio at various resolutions. Media to be used as primary storage would need to be able to sustain data transfer at a number of times these rates to be useful for multimedia workstations. The table also shows the number of megabytes of storage required per minute of audio, showing that the capacity needed for audio purposes is considerably greater than that required for text or simple graphics applications. Storage requirements increase pro rata with the number of audio channels to be handled.

Storage systems may use removable media but many have fixed media. It is advantageous to have removable media for audio purposes because it allows different jobs to be kept on different media and exchanged at will, but unfortunately the highest performance is still obtainable from storage systems with fixed media. Although the performance of removable media drives is improving all the time, fixed media drives have so far retained their advantage.

10.9.2 Disk drives

Disk drives are probably the most common form of mass storage. They have the advantage of being random-access systems – in other words any data can be accessed at random and with only a short delay. This may be contrasted with tape drives which only allow linear access – that is by winding through the tape until the desired data is reached, resulting in a considerable delay. Disk drives come in all shapes and sizes from the commonly encountered floppy disk at the bottom end to high performance hard drives at the top end. The means by which data are stored is usually either magnetic or optical, but some use a combination of the two, as described below. There exist both removable and fixed media disk drives, but in almost all cases the fixed media drives have a higher performance than removable media drives. This is because the design tolerances can be made much finer when the

Table 10.2 Data rates and capacities required for linear PCM

Sampling rate kHz	Resolution bits	Bit rate kbit/s	Capacity/min. Mbytes/min	Capacity/hour Mbytes/hour
96	16	1536	11.0	659
48	20	960	6.9	412
48	16	768	5.5	330
44.1	16	706	5.0	303
44.1	8	353	2.5	151

drive does not have to cope with removable media, allowing higher data storage densities to be achieved. Although removable disk media can appear to be expensive compared with tape media, the cost must be weighed against the benefits of random access and the possibility that some removable disks can be used for primary storage whereas a tape cannot.

The general structure of a disk drive is shown in Figure 10.33. It consists of a motor connected to a drive mechanism which causes one or more disk surfaces to rotate at anything from a few hundred to many thousands of revolutions per minute. This rotation may either remain constant or may stop and start, and it may either be at a constant rate or a variable rate, depending on the drive. One or more heads are mounted on a positioning mechanism which can move the head across the surface of the disk to access particular points, under the control of hardware and software called a disk controller. The heads read data from and write data to the disk surface by whatever means the drive employs. It should be noted that certain disk types are read-only, some are write-once–read-many (WORM), and some are fully erasable and rewritable.

The disk surface is normally divided up into tracks and sectors, not physically but by means of 'soft' formatting (see Figure 10.34). Formatting places logical markers which indicate block boundaries, amongst other processes. On most hard disks the tracks are arranged as a series of concentric rings, but with some optical disks there is a continuous spiral track.

Disk drives look after their own channel coding, error detection and correction so there is no need for system designers to devise dedicated audio processes for disk-based recording systems. The formatted capacity of a disk drive is all available for the storage of 'raw' audio data, with no additional overhead required for redundancy and error checking codes. 'Bad blocks' are mapped out during the formatting of a disk, and not used for data storage. If a disk drive detects an error when reading a block of data it will attempt to read it again. If this fails then an error

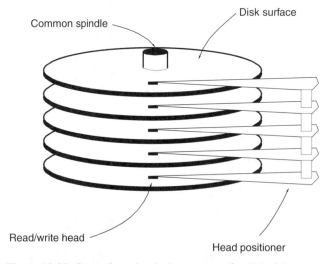

Figure 10.33 General mechanical structure of a disk drive

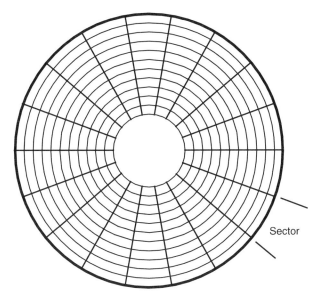

Figure 10.34 Disk formatting divides the storage area into tracks and sectors

is normally generated and the file cannot be accessed, requiring the user to resort to one of the many file recovery packages on the market. Disk-based audio systems do not resort to error interpolation or sample hold operations, unlike tape recorders. Replay is normally either correct or not possible.

10.9.3 Recording audio on to disks

The discontinuous 'bursty' nature of recording on to disk drives requires the use of a buffer RAM (Random Access Memory) during replay, which accepts this interrupted data stream and stores it for a short time before releasing it as a continuous stream. It performs the opposite function during recording, as shown in Figure 10.35. Several things cause a delay in the retrieval of information: the time it takes for the head positioner to move across a disk, the time it takes for the required data in a particular track to come around to the pickup head, and the transfer of the

Figure 10.35 Short-term memory buffering converts continuous audio data into bursts for recording onto disks, and vice versa on replay

data from the disk via the buffer RAM to the outside world, as shown in Figure 10.36. Total delay, or data access time, is in practice several milliseconds. The instantaneous rate at which the system can accept or give out data is called the transfer rate and varies with the storage device.

Sound is stored in named data files on the disk, the files consisting of a number of blocks of data stored either separately or together. A directory stored on the disk keeps track of where the blocks of each file are stored so that they can be retrieved in correct sequence. Each file normally corresponds to a single recording of a single channel of audio, although some stereo file formats exist.

Multiple channels are handled by accessing multiple files from the disk in a time-shared manner, with synchronisation between the tracks being performed subsequently in RAM. The storage capacity of a disk can be divided between channels in whatever proportion is appropriate, and it is not necessary to pre-allocate storage space to particular audio channels. For example, a 360 Mbyte disk will store about 60 minutes of mono audio at professional rates. This could be subdivided to give 30 minutes of stereo, 15 minutes of four track, etc., or the proportions could be shared unequally. A feature of the disk system is that unused storage capacity is not necessarily 'wasted' as can be the case with a tape system. During recording of a multi-track tape there will often be sections on each track with no information recorded, but that space cannot be allocated elsewhere. On a disk these gaps do not occupy storage space and can be used for additional space on other channels at other times.

The number of audio channels which can be recorded or replayed simultaneously depends on the performance of the storage device and the host computer. Slow

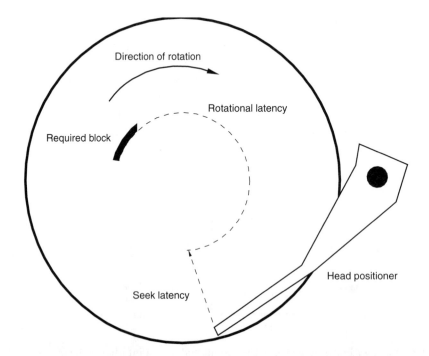

Figure 10.36 The delays involved in reading a block of data from a disk

systems may only be capable of handling a few channels whereas faster systems with multiple disk drives may be capable of expansion up to a virtually unlimited number of channels. Manufacturers are tending to make their systems modular, allowing for expansion of storage and other audio processing facilities as means allow, with all modules communicating over a high speed data bus, as shown in Figure 10.37.

10.9.4 Non-linear editing

Speed and flexibility of editing is probably one of the greatest benefits obtained from non-linear recording. With non-linear editing the editor may preview a number of possible masters in their entirety before deciding which should be the final one. Even after this, it is a simple matter to modify the edit list to update the master. Edits may also be previewed and experimented with in order to determine the most appropriate location and processing – an operation which is less easy with other forms of editing.

The majority of music editing is done today using digital audio workstations, indeed these are now taking over from dedicated audio editing systems because of the speed with which takes may be compared, cross-fades modified, and adjustments made to equalisation and levels, all in the digital domain. Non-linear editing

Figure 10.37 Arrangement of multiple disks in a typical modular system, showing how a number of disks can be attached to a single SCSI chain to increase storage capacity, and how additional disk I/O cards can be added to increase data throughput for additional audio channels

has also come to feature very widely in post-production for video and film, because it has a lot in common with film post-production techniques involving a number of independent mono sound reels.

Non-linear editing is truly non-destructive in that the edited master only exists as a series of instructions to replay certain parts of certain sound files at certain times, with certain signal processing overlaid, as shown in Figure 10.38. The original sound files remain intact at all times, and a single sound file can be used as many times as desired in different locations and on different tracks without the need for copying the actual audio data. Editing may involve the simple joining of sections, or it may involve more complex operations such as long cross-fades between one album track and the next, or gain offsets between one section and another. The beauty of non-linear editing is that all these things are possible without in any way affecting the original source material.

The manipulation of sound files on screen is normally under the control of a word processor-type of PC program, and it enables fast, intuitive editing. In the example illustrated here, the programme material is recorded onto the hard disk via a break-out box which carries analogue and digital inputs and outputs. Traditional-style transport controls are provided on-screen, these being operated with the mouse. Elsewhere on the screen a section will show mouse-operated mixer-type controls such as fader, pan, mute and solo for each channel, plus output sections (Figure 10.39). The main part of the screen is occupied by a horizontal display of empty recording tracks or 'streams', and these are analogous to the tracks of a multitrack tape recorder. A record icon associated with each stream is used to arm it ready for recording. As recording proceeds, the empty streams are filled from left to right across the screen in real time, led by a vertical moving cursor. These streams can

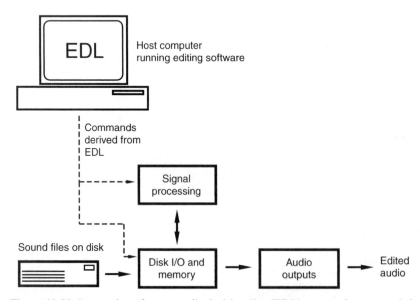

Figure 10.38 Instructions from an edit decision list (EDL) are used to control the replay of sound file segments from disk, which may be subjected to further processing (also under EDL control) before arriving at the audio outputs

Figure 10.39 The main screen display. (Courtesy SADiE UK)

be displayed either as solid continuous blocks or as waveforms, the latter being the usual mode when editing is undertaken. After recording, extra streams can be recorded if required simply by disarming the record icons of the streams already used and arming the record icons of empty streams below them, making it possible to build up a large number of 'virtual' tracks as required. The maximum number which can be replayed simultaneously depends upon the memory and DSP capacity of the system used. A basic two-input/four-output system will usually allow up to eight streams to be replayed (depending on the amount of DSP being used for other tasks), and a fully equipped system can allow up to at least 32 simultaneous streams of programme material to be recorded and replayed, i.e. it is a complete multitrack recording machine.

Replay involves either using the transport control display or clicking the mouse at a desired position on a time-bar towards the top of the screen, this positioning the moving cursor (which is analogous to the tape head) where one wishes replay to begin. Editing is performed by means of a razor-blade icon which will make the cut where the moving cursor is positioned. Alternatively, an edit icon can be loaded to the mouse's cursor for positioning anywhere on any individual stream to make a cut. Word processor-type operations such as delete, cut and paste, copy and paste and remove gap can then be performed on highlighted sections, streams or sections of streams can be dragged across to other positions and on to other streams, and there are many stages of undo and re-do so that nothing need be permanent at this stage. More precise edits can be achieved by loading a section into a special trim-editor which gives a greater degree of waveform magnification, and parameters such as cross-fade characteristics can be experimented with and auditioned (Figure 10.40). When a satisfactory edit is achieved, it can be written back to the main display where it will be incorporated. Scrub and jog actions for locating edit points are also possible. A useful 'lock to time' icon is provided which can be activated to prevent horizontal movement of the streams so that they cannot be accidentally moved out of sync with each other during editing.

The mixer section can be thought of in conventional terms, and indeed some systems offer physical plug-in interfaces with moving fader automation for those who prefer them. As well as mouse control of such things as fader, pan, solo and mute, processing such as eq, filters, aux send and compression can be selected from an effects 'rack', and each can be dragged across and dropped in above a fader where it will become incorporated into that channel. Third party 'plug-in' software is also available for many systems to enhance the signal processing features. Automation of faders and other processing is also possible. The recorded material itself resides on a (usually) removable hard disk drive, and the edit decision list (the information created during editing which tells the computer how to play the recorded material) resides on the computer's internal disk drive once the project has been 'saved'. When a project is complete, the latter can be loaded onto the removable disk so that the whole project is contained therein.

Similar MIDI-based systems are available for controlling and editing MIDI data streams, and they facilitate easy manipulation of sequencing and other editing processes. MIDI sequencing can also be integrated with the audio-based systems.

Figure 10.40 Editing in more detail. (Courtesy SADiE UK)

10.9.5 Digital audio workstations

There are fundamentally two types of audio workstation: dedicated systems or systems which use additional hardware and software installed in a standard desktop computer. Most systems conform to one or other of these models, with a tendency for dedicated systems to be more expensive than desktop computer-based systems (although this is by no means always the case).

It was most common in the early days of hard disk audio systems for manufacturers to develop dedicated systems with fairly high prices. This was mainly because mass produced desktop computers were insufficiently equipped for the purpose, and because large capacity mass storage media were less widely available than they are now, having a variety of different interfaces and requiring proprietary file storage strategies. It was also because the size of the market was relatively small to begin with, and considerable R&D investment had to be recouped. There are considerable advantages to dedicated systems, and they are very popular with professional facilities. Rather than a mouse and a QWERTY keyboard, the user controls the system using an interface designed specifically for the purpose, with perhaps more ergonomically appropriate devices. An example of such a system is pictured in Figure 10.41. It has a touch screen and dedicated controls for many functions, as well as rotary and slider controls for continuously variable functions. It is also becoming common for cheaper dedicated editing systems to be provided with an interface to a host computer so that more comprehensive display and control facilities can be provided.

In recent years desktop multimedia computers have been introduced with built-in basic AV (audio-video) facilities, providing limited capabilities for editing and sound manipulation. The quality of the built-in convertors in desktop computers is necessarily limited by price, but they are capable of 16 bit, 44.1 kHz audio operation in many cases. Better audio quality is achieved by using third party hardware.

Many desktop computers lack the processing power to handle digital audio and video directly, but by adding third-party hardware and software it is possible to turn a desktop computer into an AV workstation, capable of storing audio for an almost unlimited number of tracks with digital video alongside. An audio signal processing card is normally installed in an expansion slot of the computer, as shown in Figure 10.42. The card would be used to handle all sound editing and post-processing operations, using one or more DSP chips, with the host computer acting mainly as a user interface. The audio card would normally be connected to an audio interface, perhaps containing a number of A/D and D/A convertors, digital audio interfaces, probably a SMPTE/EBU timecode interface, and in some cases a MIDI interface. A SCSI interface (the most common high-speed peripheral interface for mass storage media) to one or more disk drives is often provided on the audio expansion card in order to optimise audio file transfer operations, although some basic systems use the computer's own SCSI bus for this purpose.

Figure 10.41 Dedicated digital audio workstations. (Courtesy of Digital Audio Research Ltd)

Figure 10.42 Typical system layout of a desktop computer-based audio workstation

10.10 Consumer digital formats

10.10.1 Minidisc

CD has been available for some years now as a 16-bit 44.1 kHz digital playback medium; it was joined by CD-ROM, CD-R (recordable) and CD-RW (recordable and rewritable). The minidisc (MD) is now an established consumer recording and playback format, and it is of the M-O (magneto-optical) type. Sampling frequency is fixed at 44.1 kHz, and resolution is nominally 16-bit. A coding system similar to those originally developed for digital audio broadcasting (DAB) known as Adaptive Transform Acoustic Coding (ATRAC) is used whereby the incoming signal is first split into three bands: below 5.5 kHz, 5.5–11 kHz, and above 11 kHz, and each band is individually analysed with respect to frequency content and level over successive short periods of time via Modified Discrete Cosine Transform (MDCT) filter blocks. Within the three blocks, non-uniform frequency splitting into 20, 16 and 16 further sub-bands takes place, and the circuit then discards material which it deems will be masked by other sounds which are present at higher signal levels and/or have a similar frequency content. A data rate of about one fifth that required for CD is adequate to encode the resulting signal (CD's data stream is 1.4 Mb/s, MD's is 292 Kb/s) and this allows usefully long playing times to be obtained from a disc which is somewhat smaller than a CD at 64 mm in diameter. Since the format involves considerable data compression (a slight misnomer for data reduction), it is not suitable for professional master recording or archiving, but is used quite widely in applications where the highest sound quality is not required such as broadcast journalism.

10.10.2 DVD-A and SACD

CD has now been with us for a couple of decades, and it was inevitable that new formats would be developed which offered larger storage capacities together with higher audio quality and the provision for multichannel surround sound. The two present contenders are DVD-Audio (DVD-A) and the Super Audio Compact Disc (SACD). There are a number of similarities between them, but also some differences which make them potentially incompatible. The similarities will be looked at first.

Physically, both are exactly the same size as the CD, and the greatly increased data storage capacity (up to 4.7 GBytes as against CD's 650 MBytes) has been achieved using a combination of reduced track pitch, reduced length of the pits on the surface of the disc, altered channel code and a higher linear reading speed. Lossless data compression has also been specified for both systems. Table 10.3 summarises the physical differences between the new formats and CD.

Both types of disc can be dual layer, the upper carrying the high-density information on a semi-transmissive layer which is read by a short wavelength laser, and on Super Audio CD the lower carrying normal CD information which can be read by a longer wavelength laser which shines through the top layer to come to a focus upon it. (The lower layer can also be another high-density layer.) The latter is provided

Table 10.3

Parameter	DVD/SACD	CD
Track pitch (microns)	0.74	1.6
Pit length (microns)	0.74	0.834
Speed (m/s)	3.49	1.2
Maximum data rate (Mbit/s)	11.08	1.536
Laser (nm)	635–650	780–790

to maintain backwards compatibility with CD players, but its provision is not a mandatory requirement.

DVD-A is one of a DVD family which also includes DVD-Video, DVD-ROM, DVD-R and DVD-RAM, in contrast to SACD which was conceived from the outset as a super-quality audio-only format, although the latter has an area on the disc reserved for other data which can include, for example, video clips, text and graphics.

DVD-A supports sampling rates of 44.1, 48, 88.2, 96, 176.4 and 192 kHz, split into two groups which are respectively multiples of 44.1 and 48 kHz. Resolution is 16, 20 or 24 bits, and combinations of these can be used to obtain the required playing times for up to six channels of audio. If more than two channels are used, these are split into two groups: group 1 contains the front channels, group 2 contains the others, and the latter could be coded to a lower standard in the interests of conserving data. This is called scalable audio. It is desirable that multichannel mixes give good results on stereo-only replay, and to this end a System Managed Audio Resource Technique (SMART), mandatory in players but optional at the production stage, is incorporated which makes possible a stereo mixdown of multichannel information in the playback machine which is under content producer control.

A data compression system called Meridian Lossless Packing (MLP) can be incorporated to reduce the data rate of audio signals without apparently rejecting any information. Unlike other data compression schemes which rely on psycho-acoustic masking to reject unessential data, MLP is more like the software programs that reduce the size of computer files without altering information. Repetitive patterns are looked for, and more efficient ways of packing the data are sought to make use of the available space in the most economical way.

DVD-Audio V is a similar disc which in addition holds video objects (video clips, Dolby AC-3 compressed audio and other information). Three types of player are envisaged: audio, video and universal, and there is a degree of cross-compatibility between them. The audio-only player will play just the audio part of a DVD-Audio V disc. The video-only player will play just the video part of a DVD-Audio V disc.

SACD uses a different coding system from DVD-A. Instead of multi-bit linear PCM, a 1-bit delta-sigma coding system called Direct Stream Digital is used. See Fact file 10.6.

A lossless data packing method called Direct Stream Transfer (DST) achieves a roughly 2:1 data compression to allow room for high-quality multichannel as well as stereo coding to be present on separate areas of the disc's playing surface.

Copy protection measures have been developed for both systems. DVD-A can include a 'watermark' in the bit stream. SACD uses Pit Signal Processing (PSP) which contains an invisible physical watermark which is read by the player. In addition, a visible watermark can be imprinted on the disc, this involving the injection of a modulation signal during the mastering stage which alters the actual size of the pits to produce an image which should be impossible to replicate.

10.11 Digital audio interfaces

It is often necessary to interconnect audio equipment so that digital audio data can be transferred between them without converting back to analogue form. This preserves sound quality, and is normally achieved using one of the standard point-to-point digital interfaces described below. These are different from computer data networks in that they are designed purely for the purpose of carrying audio data in real time, and cannot be used for general purpose file transfer applications.

10.11.1 Computer networks and digital audio interfaces compared

Real-time digital audio interfaces are the digital audio equivalent of signal cables, down which digital audio signals for one or more channels are carried in real time from one point to another, possibly with some auxiliary information attached. A real-time audio interface uses a data format dedicated specifically to audio purposes, unlike a computer data network which is not really concerned with what is carried in its data packets. A recording transferred over a digital interface to a second machine may be copied 'perfectly' or cloned, and this process takes place in real time, requiring the operator to put the receiving device into record mode such that it simply stores the incoming stream of audio data. The auxiliary information may or may not be recorded (usually most of it is not).

In contrast, a computer network typically operates asynchronously and data is transferred in the form of packets. One would expect a number of devices to be interconnected using a single network, and for an addressing structure to be used such that data might be transferred from a certain source to a certain destination. Bus arbitration is used to determine the existence of network traffic, and to avoid conflicts in bus usage. The 'direction' of data flow is determined simply by the source and destination addresses. The format of data on the network is not necessarily audio specific (although protocols optimised for audio transfer may be used), and one network may carry text data, graphics data and E-mail, all in separate packets, for example.

10.11.2 Interface standards

Professional digital audio systems, and some consumer systems, have digital interfaces conforming to one of the standard protocols and allow for a number of channels of digital audio data to be transferred between devices with no loss of sound quality. Any number of generations of digital copies may be made without affecting the sound quality of the latest generation, provided that errors have been fully corrected. The digital outputs of a recording device are taken from a point in the signal chain after error correction, which results in the copy being error cor-

---

FACT FILE

10.10

AES/EBU interface

The AES/EBU interface allows for two channels of digital audio to be transferred serially over one balanced interface. The interface allows audio to be transferred over distances up to 100 m, but longer distances may be covered using combinations of appropriate cabling, equalisation and termination. Standard XLR-3 connectors are used, often labelled DI (for digital in) and DO (for digital out).

One frame of data is made up of two subframes (see the diagram), and each subframe begins with one of three synchronising patterns to identify the sample as either A or B channel, or to mark the start of a new channel status block. Additional data is carried within the subframe in the form of 4 bits of auxiliary data (which may either be used for additional audio resolution or for other purposes such as low-quality speech),

a validity bit (V), a user bit (U), a channel status bit (C) and a parity bit (P), making 32 bits per subframe and 64 bits per frame.

One frame is transmitted in the time period of one audio sample, and thus the data rate varies with the sampling rate. Channel status bits are aggregated at the receiver to form a 24 byte word every 192 frames, and each bit of this word has a specific function relating to interface operation. Examples of bit usage in this word are the signalling of sampling rate and pre-emphasis, as well as the carrying of a sample address 'timecode' and labelling of source and destination. Bit 1 of the first byte signifies whether the interface is operating according to the professional (set to 1) or consumer (set to 0) specification.

Bi-phase mark coding, the same channel code as used for SMPTE/EBU timecode, is used in order to ensure that the data is self-clocking, of limited bandwidth, DC free, and polarity independent. The interface has to accommodate a wide range of cable types, and a nominal 110 ohms characteristic impedance is recommended rected. Thus the copy does not suffer from any errors which existed in the master, provided that those errors were correctable.

There are a number of types of digital interface, some of which are international standards and others of which are manufacturer-specific. They all carry digital audio with at least 16 bit resolution, and will operate at the standard sampling rates of 44.1 and 48 kHz, as well as at 32 kHz if necessary, with a degree of latitude for varispeed.

Manufacturer-specific interfaces may also be found on equipment from other manufacturers if they have been considered necessary for the purposes of signal interchange, especially on devices manufactured prior to the standardisation and proliferation of the AES/EBU interface (see Fact File 10.10). The interfaces vary as to how many physical interconnections are required, since some require one link per channel plus a synchronisation signal, whilst others carry all the audio information plus synchronisation information over one cable. The reader is referred to *The Digital Interface Handbook* by Rumsey and Watkinson, as well as to the standards themselves, if a greater understanding of the intricacies of digital audio interfaces is required.

The consumer interface (historically related to SPDIF – the Sony/Philips digital interface) is very similar to the professional AES/EBU interface, but uses unbalanced electrical interconnection over a coaxial cable having a characteristic impedance of 75 ohms. It can be found on many items of semi-professional or consumer digital audio equipment, such as CD players and DAT machines. It usually terminates in an RCA phono connector, although some hi-fi equipment makes use of optical fibre interconnects carrying the same data. Format convertors are available for converting consumer format signals to the professional format, and vice versa, and for converting between electrical and optical formats.

Two common manufacturer-specific multichannel interfaces are TDIF, devised by Tascam, and the ADAT 'light pipe', devised by Alesis. TDIF is most commonly found on eight-track tape machines conforming to the DTRS format and on some digital mixers. It is quite a cumbersome interface using 25-way D-type connectors. Owing to its small size and ease of use, the optical ADAT interface is increasingly used between computer sound cards and other digital audio equipment to carry up to eight channels of audio data.

Recommended further reading

Bailey, A.R. (2000) *Network Technology for Digital Audio*. Focal Press
Pohlmann, K. (1995) *Principles of Digital Audio*. McGraw Hill
Pohlmann, K. (1989) *The Compact Disc*. Oxford University Press
Rumsey, F. (1996) *The Audio Workstation Handbook*. Focal Press
Rumsey, F. and Watkinson, J. (1995) *The Digital Interface Handbook*. Focal Press
Watkinson, J. (1994) *An Introduction to Digital Audio*. Focal Press
Watkinson, J. (2000) *The Art of Digital Audio*. 3rd Edition. Focal Press

Chapter 11

Power amplifiers

Power amplifiers are uneventful devices. They are usually big and heavy, take up a lot of rack space, and feature very little (or sometimes nothing) beyond input and output sockets. Because one tends to ignore them, it is all the more important that they are chosen and used with due care. Coming in a variety of shapes, sizes and 'generations', they are all required to do the ostensibly simple job of providing voltage amplification – converting line levels of up to a volt or so into several tens of volts, with output currents in the ampere range to develop the necessary power across the loudspeaker terminals. Given these few requirements, it is perhaps surprising how many designs there are on the market.

11.1 Domestic power amplifiers

The domestic power amplifier, at its best, is designed for maximum fidelity in the true sense of that word, and this will usually mean that other considerations such as long-term overload protection and complete stability into any type of speaker load are not always given the type of priority which is essential in the professional field. A professional power amp may well be asked to drive a pair of 6 ohm speakers in parallel on the other end of 30 metres of cable, at near to maximum output level for hours on end if used in a rock PA rig. This demands large power supplies and heavy transformers, with plenty of heat sink area (the black fins usually found on the outer casing) to keep it from overheating. Cooling fans are frequently employed which will often run at different speeds depending on the temperature of the amplifier.

The domestic amplifier is unlikely to be operated at high output levels for a significant length of time, and the power supplies are often therefore designed to deliver high currents for short periods to take care of short, loud passages. A power supply big enough to supply high currents for lengthy periods is probably wasted in a domestic amplifier. Also, the thermal inertia of the transformer and the heat sinks means that unacceptable rises in temperature are unlikely. Although there are one or two domestic speakers which are notoriously difficult to drive due to various combinations of low impedance, low efficiency (leading to high power demand), and wide phase swings (current and voltage being out of step with each other due to crossover components and driver behaviour in a particular speaker enclosure), the majority of domestic hi-fi speakers are a comfortable load for an amplifier, and usually the speaker leads will be less than 10 metres in length.

It is unlikely that the amplifier will be driven into a short-circuit due to faulty speaker lines for any length of time (silence gives an immediate warning), which is not the case with a professional amplifier which may well be one of many, driving a whole array of speakers. A short-circuit developing soon after a show has begun may cause the amplifier to be driven hard into this condition for the whole evening. Protection circuitry needs to be incorporated into the design to allow the professional amplifier to cope with this without overheating or catastrophically failing which can affect other amplifiers in the same part of the rig.

Several 'classes' of amplifier design have appeared over the years, these being labels identifying the type of output stage topology employed to drive the speaker. These are outlined in Fact File 11.1.

11.2 Professional amplifier facilities

The most straightforward power amplifiers have input sockets and output terminals, and nothing else. Single-channel models are frequently encountered, and in the professional field these are often desirable because if one channel of a stereo power amplifier develops a fault then the other channel also has to be shut down, thus losing a perfectly good circuit. The single-channel power amplifier is thus a good idea when multi-speaker arrays are in use such as in rock PA systems and theatre sound.

Other facilities found on power amplifiers include input level controls, output level meters, overload indicators, thermal shutdown (the mains feed is automatically disconnected if the amplifier rises above a certain temperature), earth-lift facility to circumvent earth loops, and 'bridging' switch. This last facility, applicable to a stereo power amplifier, is a facility sometimes provided whereby the two channels of the amp can be bridged together to form a single-channel higher-powered one, the speaker(s) now being connected across the two positive output terminals with the negative terminals left unused. Only one of the input sockets is now used to drive it.

Cooling fans are often incorporated into an amplifier design. Such a force-cooled design can be physically smaller than its convection-cooled counterpart, but fans tend to be noisy. Anything other than a genuinely silent fan is unacceptable in a studio or broadcast control room, or indeed in theatre work, and such models will need to be housed in a separate well-ventilated room. Ventilation of course needs to be a consideration with all power amplifiers.

11.3 Specifications

Power amplifier specifications include sensitivity, maximum output power into a given load, power bandwidth, frequency response, slew rate, distortion, crosstalk

Amplifier Classes

Class A

The output stage draws a constant high current from the power supply regardless of whether there is an audio signal present or not. Low-current class A stages are used widely in audio circuits. The steady bias current as it is known is employed because transistors are non-linear devices, particularly when operated at very low currents. A steady current is therefore passed through them which biases them into the area of their working range at which they are most linear.

The constant bias current makes class A amplification inefficient due to heat generation, but there is the advantage that the output transistors are at a constant steady temperature. Class A is capable of very high sound quality, and several highly specified up-market domestic class A power amplifiers exist.

Class B

No current flows through the output transistors when no audio signal is present. The driving signal itself biases the transistors into conduction to drive the speakers. The technique is therefore extremely efficient because the current drawn from the power supply is entirely dependent upon the level of drive signal. Class B is therefore particularly attractive in battery-operated equipment. The disadvantage is that at low signal levels the output transistors operate in a non-linear region. It is usual for pairs (or multiples) of transistors to provide the output current of a power amplifier. Each of the pair handles opposite halves of the output waveform (positive and negative with respect to zero) and therefore as the output swings through zero from positive to negative and vice versa the signal suffers so-called 'crossover distortion'. The result is relatively low sound quality, but class B can be used in

applications which do not require high sound quality such as telephone systems, hand-held security transceivers, paging systems and the like.

Class A–B

In this design a relatively low constant bias current flows through the output transistors to give a low-power class A amplifier. As the input drive signal is increased, the output transistors are biased into appropriately higher-current conduction in order to deliver higher power to the speakers. This part of the operation is the class B part, i.e.: it depends on input drive signal level. But the low-level class A component keeps the transistors biased into a linear part of their operating range so that crossover distortion is largely avoided. The majority of high-quality amplifiers operate on this principle.

Other classes

Class C drives a narrow band of frequencies into a resonant load, and is appropriate to radio-frequency (RF) work where an amplifier is required to drive a single frequency into an appropriately tuned aerial. Class D is 'pulse width modulation' in which an ultrasonic frequency, modulated by the audio signal, is used to drive the output transistors. A low-pass filter is employed after the output stage. This technique has been revived in one or two designs in the late 1980s. A variation on class D called class T (from the Tripath company) has recently been seen. Here, the ultrasonic frequency is continuously varied in accordance with the amplitude of the audio signal. The frequency is about 1.2 MHz at low signal levels, falling to around 200 KHz for very high signal levels; a greater overall efficiency is claimed as a result. Classes E and F were concerned with increasing efficiency, and currently no commercial models conform to these particular categories. Class G incorporates several different voltage rails which progressively come into action as the drive signal voltage is increased. This technique can give very good efficiency because for much of the time only the lower-voltage, low-current supplies are in operation. Such designs can be rather smaller than their conventional class A–B counterparts of comparable output power rating. Class H is a variation on class G in that the power supply voltage rails are made to track the input signal continuously, maintaining just enough headroom to accommodate the amplifier's requirements for the necessary output voltage swing.

Since the early 1980s the MOSFET (Metal Oxide Semiconductor Field-Effect Transistor) has been widely employed for the output stages of power amplifiers. MOSFET techniques claim lower distortion, better thermal tracking (i.e.: good linearity over a wide range of operating temperatures), simpler output stage design, and greater tolerance of adverse loudspeaker loads without the need for elaborate protection circuitry.

between channels, signal-to-noise ratio, input impedance, output impedance, damping factor, phase response, and DC offset. Quite surprising differences in sound quality can be heard between certain models, and steady-state measurements do not, unfortunately, always tell a user what he or she can expect to hear.

11.3.1 Sensitivity

Sensitivity is a measurement of how much voltage input is required to produce the amplifier's maximum rated output. For example, a model may be specified '150 watts into 8 ohms, input sensitivity 775 mV = 0 dBu'. This means that an input voltage of 775 mV will cause the amplifier to deliver 150 watts into an 8 ohm load. Speakers exhibit impedances which vary considerably with frequency, so this is always a nominal specification when real speakers are being driven. Consideration of sensitivity is important because the equipment which is to drive the amp must not be allowed to deliver a greater voltage to the amplifier than its specification states, otherwise the amplifier will be overloaded causing 'clipping' of the output waveform (a squaring-off of the tops and bottoms of the waveform resulting in severe distortion). This manifests itself as a 'breaking-up' of the sound on musical peaks, and will often quickly damage tweeters and high-frequency horns.

Many amplifiers have input level controls so that if, for instance, the peak output level of the mixer which drives the amplifier is normally say 'PPM 6' – about 2 volts – then the amp's input levels can be turned down to prevent overload. In the given example, 2 volts is 8 dB higher than 775 mV (PPM 4 = 0 dBu) and so the input level control should be reduced by 8 dB to allow for this. If a dB calibration is not provided on the level control, and many are not particularly accurate anyway, a reasonable guide is that, compared with its maximum position of about '5 o'clock', reducing the level to about 2 o'clock will reduce the sensitivity by about 10 dB, or by a factor of three. In this position, the power amplifier with an input sensitivity of 775 mV will now require 0.775×3, or about 2 volts, to develop its full output.

a)

b)

Figure 11.1 (a) An unbalanced resistive attenuator. (b) A balanced resistive attenuator

If input level controls are not provided, one can build a simple resistive attenuator which reduces the voltage being fed to the amplifier's input. Two examples are shown in Figure 11.1. It is best to place such attenuators close to the power amp input in order to keep signal levels high while they are travelling down the connecting leads. In both cases the 3k3 resistor which is in parallel with the amplifier's input can be increased in value for less attenuation, and decreased in value for greater attenuation. With care, the resistors can be built into connecting plugs, the latter then needing to be clearly labelled.

11.3.2 Power output

A manufacturer will state the maximum power a particular model can provide into a given load, e.g.: '200 watts into 8 ohms', often with 'both channels driven' written after it. This last means that both channels of a stereo amplifier can deliver this simultaneously. When one channel only is being driven, the maximum output is often a bit higher, say 225 watts, because the power supply is less heavily taxed. Thus 200 watts into 8 ohms means that the amplifier is capable of delivering 40 volts into this load, with a current of 5 amps. If the load is now reduced to 4 ohms then the same amplifier should produce 400 watts. A theoretically perfect amplifier should then double its output when the impedance it drives is halved. In practice, this is beyond the great majority of power amplifiers and the 4 ohm specification of the above example may be more like 320 watts, but this only around 1 dB below the theoretically perfect value. A 2 ohm load is very punishing for an amplifier, and should be avoided even though a manufacturer sometimes claims a model is capable of, say, 800 watts of short-term peaks into 2 ohms. This at least tells us that the amp should be able to drive 4 ohm loads without any trouble.

Because 200 watts is only 3 dB higher than 100 watts, then, other things being equal, the exact wattage of an amplifier is less important than factors such as its ability to drive difficult reactive loads for long periods. Often, 'RMS' will be seen after the wattage rating. This stands for root-mean-square, and defines the raw 'heating' power of an amplifier, rather than its peak output. All amplifiers should be specified RMS so that they can easily be compared. The RMS value is 0.707 times the instantaneous peak capability, and it is unlikely that one would encounter a professional amplifier with just a peak power rating.

Power bandwidth is not the same as power rating, as discussed in Fact File 11.2.

11.3.3 Frequency response

Frequency response, unlike power bandwidth, is simply a measure of the limits within which an amplifier responds equally to all frequencies when delivering a very low power. The frequency response is usually measured with the amplifier delivering 1 watt into 8 ohms. A specification such as '20 Hz–20 kHz ±0.5 dB' should be looked for, meaning that the response is virtually flat across the whole of the audible band. Additionally, the −3 dB points are usually also stated, e.g.: '−3 dB at 12 Hz and 40 kHz', indicating that the response falls away smoothly below and above the audio range. This is desirable as it gives a degree of protection for the amp and speakers against subsonic disturbances and RF interference.

FACT FILE 11.2 Power bandwidth

Power bandwidth is a definition of the frequency response limits within which an amplifier can sustain its specified output. Specifically, a 3 dB drop of output power is allowed in defining a particular amplifier's power bandwidth. For example, a 200 watt amplifier may have a power bandwidth of 10 Hz to 30 kHz, meaning that it can supply 200 watts − 3 dB (= 100 watts) at 10 Hz and 30 kHz, compared with the full 200 watts at mid frequencies. Such an amplifier would be expected to deliver the full 200 watts at all frequencies between about 30 Hz and 20 kHz, and this should also be looked for in the specification. Often, though, the power rating of an amplifier is much more impressive when measured using single sine-wave tones than with broad-band signals, since the amplifier may be more efficient at a single frequency.

Power bandwidth can indicate whether a given amplifier is capable of driving a subwoofer at high levels in a PA rig, as it will be called upon to deliver much of its power at frequencies below 100 Hz or so. The driving of high-frequency horns also needs good high-frequency power bandwidth so that the amplifier never clips the high frequencies, which easily damages horns as has been said.

11.3.4 Distortion

Distortion should be 0.1 per cent THD (see section A1.3) or less across the audio band, even close to maximum-rated output. It often rises slightly at very high frequencies, but this is of no consequence. Transient distortion, or transient intermodulation distortion (TID), is also a useful specification. It is usually assessed by feeding both a 19 kHz and a 20 kHz sine wave into the amplifier and measuring the relative level of 1 kHz difference tone. The 1 kHz level should be at least 70 dB down, indicating a well-behaved amplifier in this respect. The test should be carried out with the amplifier delivering at least two-thirds of its rated power into 8 ohms. Slew rate distortion is also important (see Fact File 11.3).

11.3.5 Crosstalk

Crosstalk figures of around −70 dB at mid frequencies should be a reasonable minimum, degrading to around −50 dB at 20 kHz, and by perhaps the same amount at 25 Hz or so. 'Dynamic crosstalk' is sometimes specified, this manifesting itself mainly at low frequencies because the power supply works hardest when it is called upon to deliver high currents during high-level, low-frequency drive. Current demand by one channel can modulate the power supply voltage rails, which gets into the other channel. A number of amplifiers have completely separate power supplies for each channel, which eliminates such crosstalk, or at least separate secondary windings on the mains transformer plus two sets of rectifiers and reservoir capacitors which is almost as good.

11.3.6 Signal-to-noise ratio

Signal-to-noise ratio is a measure of the output residual noise voltage expressed as a decibel ratio between that and the maximum output voltage, when the input is short-circuited. Noise should never be a problem with a modern power amplifier and signal-to-noise ratios of at least 100 dB are common. High-powered models (200 watts upwards) should have signal-to-noise ratios correspondingly greater (e.g.: 110 dB or so) in order that the output residual noise remains below audibility.

11.3.7 Impedance

The input impedance of an amplifier ought to be at least 10 kΩ, so that if a mixer is required to drive, say, ten amplifiers in parallel, as is often the case with PA rigs, the total load will be 10 k ÷ 10, or 1 k, which is still a comfortable load for the mixer. Because speakers are of very low impedance, and because their impedance varies greatly with frequency, the amplifier's output impedance must not be greater than a fraction of an ohm, and a value of 0.1 ohms or less is needed. A power amplifier needs to be a virtually perfect 'voltage source', its output voltage remaining substantially constant with different load impedances.

The output impedance does, however, rise a little at frequency extremes. At LF, the output impedance of the power supply rises and therefore so does the amplifier's. It is common practice to place a low-valued inductor of a couple of micro-

FACT FILE 11.3   Slew rate

Slew rate is a measure of the ability of an amplifier to respond accurately to high-level transients. For instance, the leading edge of a transient may demand that the output of an amplifier swings from 0 to 120 watts in a fraction of a millisecond. The slew rate is defined in V μs^{-1} (volts per microsecond) and a power amplifier which is capable of 200 watts output will usually have a slew rate of at least 30 V μs^{-1}. Higher-powered models require a greater slew rate simply because their maximum output voltage swing is greater. A 400 watt model might be required to swing 57 volts into 8 ohms as compared with the 200 watt model's 40, so its slew rate needs to be at least:

$$30 \times (57 \div 40) = 43 \text{ V } \mu s^{-1}$$

In practice, modern power amplifiers achieve slew rates comfortably above these figures.

An absolute minimum can be estimated by considering the highest frequency of interest, 20 kHz, then doubling it for safety, 40 kHz, and considering how fast a given amplifier must respond to reproduce this accurately at full output. A sine wave of 40 kHz reaches its positive-going peak in 6.25 μs, as shown in the diagram. A 200 watt model delivers a peak voltage swing of 56.56 volts peak to peak (1.414 times the RMS voltage). It may seem then that it could therefore be required to swing from 0 V to +28.28 V in 6.25 μs, thus requiring a slew rate of 28.28 $\div$ 6.25, or 4.35 V μs^{-1}. But the actual slew rate requirement is rather higher because the initial portion of the sine wave rises steeply, tailing off towards its maximum level.

Musical waveforms come in all shapes and sizes of course, including near-square waves with their almost vertical leading edges, so a minimum slew rate of around eight times this (i.e.: 30 V μs^{-1}) might be considered as necessary. It should be remembered, though, that the harmonics of an HF square wave are well outside the audible spectrum, and thus slew rate distortion of such waves at HF is unlikely to be audible. Extremely high slew rates of several hundred volts per microsecond are sometimes encountered. These are achieved in part by a wide frequency response and 'fast' output transistors, which are not always as stable into difficult speaker loads as are their 'ordinary' counterparts. Excessive slew rates are therefore to be viewed with scepticism.

henrys in series with a power amp's output which raises its output impedance a little at HF, this being to protect the amp against particularly reactive speakers or excessively capacitive cables, which can provoke HF oscillation.

11.3.8 Damping factor

Damping factor is a numerical indication of how well an amplifier can 'control' a speaker. There is a tendency for speaker cones and diaphragms to go on vibrating a little after the driving signal has stopped, and a very low output impedance virtually short-circuits the speaker terminals which 'damps' this. Damping factor is the ratio between the amplifier's output impedance and the speaker's rated impedance, so a damping factor of '100 into 8 ohms' means that the output impedance of the amplifier is $8 \div 100$ ohms, or 0.08 ohms. One hundred is quite a good figure (the higher the better, but a number greater than 200 *could* imply that the amplifier is insufficiently well protected from reactive loads and the like), but it is better if a frequency is given. Damping factor is most useful at low frequencies because it is the bass cones which vibrate with greatest excursion, requiring the tightest control. A damping factor of '100 at 40 Hz' is therefore a more useful specification than '100 at 1 kHz'.

11.3.9 Phase response

Phase response is a measurement of how well the frequency extremes keep in step with mid frequencies. At very low and very high frequencies, 15° phase leads or phase lags are common, meaning that in the case of phase lag, there is a small delay of the signal compared with mid frequencies, and phase lead means the opposite. At 20 Hz and 20 kHz, the phase lag or phase lead should not be greater than 15°, otherwise this may imply a degree of instability when difficult loads are being driven, particularly if HF phase errors are present.

The absolute phase of a power amplifier is simply a statement of whether the output is in phase with the input. The amplifier should be non-phase-inverting overall. One or two models do phase invert, and this causes difficulties when such models are mixed with non-inverting ones in multi-speaker arrays when phase cancellations between adjacent speakers, and incorrect phase relationships between stereo pairs and the like, crop up. The cause of these problems is not usually apparent and can waste much time.

11.4 Coupling

The vast majority of power amplifier output stages are 'direct coupled', that is the output power transistors are connected to the speakers with nothing in between beyond perhaps a very low-valued resistor and a small inductor. The DC voltage operating points of the circuit must therefore be chosen such that no DC voltage appears across the output terminals of the amplifier. In practice this is achieved by using 'split' voltage rails of opposite polarity (e.g.: ±46 volts DC) between which the symmetrical output stage 'hangs', the output being the mid-point of the voltage rails (i.e.: 0 V). Small errors are always present, and so 'DC offsets' are produced which means that several millivolts of DC voltage will always be present across the output terminals. This DC flows through the speaker, causing its cone to deflect either forwards or backwards a little from its rest position. As low a DC offset as possible must therefore be achieved, and a value of ±40 mV is an acceptable maximum. Values of 15 mV or less are quite common.

Chapter 12

Lines and interconnection

This chapter is concerned with the interconnection of analogue audio signals, and the solving of problems concerned with analogue interfacing. It is not intended to cover digital interfacing systems here, since this subject is adequately covered elsewhere (see the Further reading list at the end of Chapter 10). The proper interconnection of analogue audio signals, and an understanding of the principles of balanced and unbalanced lines, is vital to the maintenance of high quality in an audio system, and will remain important for many years notwithstanding the growing usage of digital systems.

12.1 Transformers

Mains transformers are widely used throughout the electrical and electronics industries, usually to convert the 240 V AC mains voltage to a rather lower voltage. Audio transformers are widely used in audio equipment for balancing and isolating purposes, and whereas mains transformers are required only to work at 50 Hz, audio transformers must give a satisfactory performance over the complete audio spectrum. Fortunately most audio transformers are only called upon to handle a few volts at negligible power, so they are generally much smaller than their mains counterparts. The principles of transformer operation are outlined in Fact File 12.1.

12.1.1 Transformers and impedances

Consider Figure 12.1(a). The turns ratio is 1:2, so the square of the turns ratio (used to calculate the impedance across the secondary) is 1:4, and therefore the impedance across the secondary will be found to be $10 \times 4 = 40$ k. Another example is shown in Figure 12.1(b). The turns ratio is 1:4. 0.7 volts is applied across the primary and gives 2.8 volts across the secondary. The square of the turns ratio is 1:16, so the impedance across the secondary is $2 \text{ k} \times 16 = 32$ k. The transformer also works backwards, as shown in Figure 12.1(c). A 20 k resistor is now placed across the secondary. The square of the turns ratio is 1:16, and therefore the impedance across the primary is $20 \text{ k} \div 16 = 1\text{k}25$.

FACT FILE 12.1   The transformer

From the diagrams it can be seen that the transformer consists of a laminated core (i.e.: a number of thin sheets of metal 'laminated' together to form a single thick core) around which is wound a 'primary' winding and a 'secondary' winding. If an alternating current is passed through the primary winding, magnetic flux flows in the core (in a similar fashion to the principle of the tape head; see Fact File 8.1), and thus through the secondary winding. Flux changes in the secondary winding cause a current to be induced in it. The voltage across the secondary winding compared with that across the primary is proportional to the

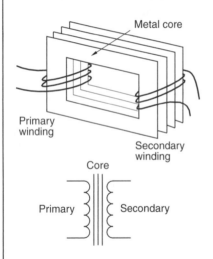

Primary winding
Metal core
Secondary winding
Core
Primary   Secondary

ratio between the number of turns on each coil. For example, if the primary and secondary windings each have the same number of turns, then 1 volt across the primary will also appear as 1 volt across the secondary. If the secondary has twice the number of turns as the primary then twice the voltage will appear across it. The transformer also works in reverse – voltage applied to the secondary will be induced into the primary in proportion to the turns ratio.

The current flowing through the secondary is in *inverse* proportion to the turns ratio, such that equal *power* exists on the primary and secondary sides of the transformer (it is not magic – the increased voltage across the secondary of a step-up transformer is traded off against reduced current!).

It is important to remember that the principle of operation of the transformer depends on AC in the windings inducing an alternating field into the core (i.e.: it is the *change* in direction of magnetic flux which induces a current in the secondary, not simply the presence of constant flux). A DC signal, therefore, is not passed by a transformer.

Impedances are proportional to the *square* of the turns ratio, as discussed in the main text. A transformer will 'reflect' the impedances between which it works. In the case of a 1:1 transformer the impedance across the secondary is equal to the impedance across the primary, but in the case of a 1:2 transformer the impedance seen across the secondary would be four times that across the primary.

a)   1:2   10k   ?
b)   1:4   0.7 V   2k   ?
c)   1:4   ?   20k

Figure 12.1 Examples of transformer circuits. (a) What is the impedance across the secondary? (b) What are the impedance and voltage across the secondary? (c) What is the impedance across the primary?

Mic output   200   Mixer input
impedance   ohms   2k impedance

Figure 12.2 The input impedance of the mixer is seen by the microphone, modified by the turns ratio of the transformer, and vice versa

Consider now a microphone transformer that is loaded with an impedance on both sides, as shown in Figure 12.2. The transformer presents the 2 k impedance of the mixer to the microphone, and the 200 ohm impedance of the microphone to the mixer. With a step-up ratio of 1:4 the square of the turns ratio would be 1:16. The microphone would be presented with an impedance of 2 k ÷ 16 = 125 ohms, whereas the mixer would be presented with an impedance of 200 × 16 = 3200 ohms. In this particular case a 1:4 step-up transformer is unsuitable because microphones like to work into an impedance five times or more their own impedance, so 125 ohms is far too low. Similarly, electronic inputs work best when driven by an impedance considerably lower than their own, so 3200 ohms is far too high.

12.1.2 Limitations of transformers

Earlier it was mentioned that an audio transformer must be able to handle the complete audio range. At very high and very low frequencies this is not easy to achieve, and it is usual to find that distortion rises at low frequencies, and also to a lesser extent at very high frequencies. The frequency response falls away at the frequency extremes, and an average transformer may well be 3 dB down at 20 Hz and 20 kHz compared with mid frequencies. Good (= expensive) transformers have a much better performance than this. All transformers are designed to work within certain limits of voltage and current, and if too high a voltage is applied a rapid increase in distortion results.

The frequency response and distortion performance is affected by the impedances between which the transformer works, and any particular model will be designed to give its optimum performance when used for its intended application. For example, a microphone transformer is designed to handle voltages in the millivolt range up to around 800 mV or so. The primary winding will be terminated with about 200 ohms, and the secondary will be terminated with around 1–2 k (or rather more if a step-up ratio is present). A line level transformer on the other hand must handle voltages up to 8 volts or so, and will probably be driven by a source impedance of below 100 ohms, and will feed an impedance of 10 k or more. Such differing parameters as these require specialised designs. There is no 'universal' transformer.

Transformers are sensitive to electromagnetic fields, and so their siting must be given consideration. Place an audio transformer next to a mains transformer and hum will be induced into it, and thus into the rest of the audio circuit. Most audio transformers are built into metal screening cans which considerably reduce their susceptibility to radio-frequency interference and the like.

12.2 Unbalanced lines

'Unbalanced' in this context does not mean unstable or faulty. The unbalanced audio line is to be found in virtually all domestic audio equipment, much semi-professional and some professional audio equipment as well. It consists of a 'send' and 'return' path for the audio signal, the return path being an outer screening braid which encloses the send wire and screens it from electromagnetic interference, shown in Figure 12.3. The screening effect considerably reduces interference such as hum, RF, and other induction, without eliminating it entirely. If the unbalanced line is used to carry an audio signal over tens of metres, the cumulative effect of interference may be unacceptable. Earth loops can also be formed (see Fact File 12.2). Unbalanced lines are normally terminated in connectors such as phono plugs, DIN plugs and quarter-inch 'A-gauge' jack plugs.

An improved means of unbalanced interconnection is shown in Figure 12.4. The connecting lead now has *two* wires inside the outer screen. One is used as the signal

---

FACT FILE 12.2   **Earth loops**

It is possible to wire cables such that the screening braid of a line is connected to earth at both ends. In many pieces of audio equipment the earth side of the audio circuit is connected to the mains earth. When two or more pieces of equipment are connected together this creates multiple paths to the mains earth, and low-level mains currents can circulate around the screening braids of the connecting leads if the earths are at even slightly different potentials. This induces 50 Hz mains hum into the inner conductor. A common remedy for this problem is to disconnect the earth wires in the mains plugs on all the interconnected pieces of equipment except one, the remaining connection providing the earth for all the other pieces of equipment via the audio screening braids. This, though, is potentially dangerous, since if a piece of equipment develops a fault and the mains plug with the earth connection is unplugged, then the rest of the system is now unearthed and the fault could in serious cases place a mains voltage on the metal parts of the equipment. A lot of units are now 'double insulated', so that internal mains wiring cannot place mains voltage on the metal chassis. The mains lead is just two core, live and neutral.

Figure 12.3 Simple unbalanced interconnection

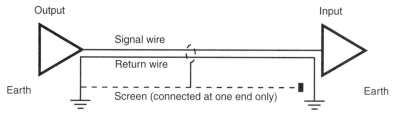

Figure 12.4 Alternative unbalanced interconnection

wire, and instead of the return being provided by the outer screen, it is provided by the second inner wire. The screening braid is connected to earth *at one end only*, and so it merely provides an interference screen without affecting the audio signal.

12.3 Cable effects with unbalanced lines

12.3.1 Cable resistance

'Loop' resistance is the total resistance of both the send and return paths for the signal, and generally, as long as the loop resistance of a cable is a couple of orders of magnitude (i.e.: a factor of 100) lower than the input impedance of the equipment it is feeding, it can be ignored. For example, the output impedance of a tape recorder might be 200 ohms. The input impedance of the amplifier it would be connected to would normally be 10 k or more. The DC resistance of a few metres of connecting cable would only be a fraction of an ohm and so would not need to be considered. But what about 100 metres of microphone cable? The input impedance of a micro-phone amplifier would normally be at least 1000 ohms. Two orders of magnitude lower than this is 10 ohms. Even 100 metres of mic lead will have a lower resistance than this unless very thin cheap wire is used, and so again the DC resistance of microphone cables can be ignored.

Speaker cables *do* need to be watched, because the input impedance of loudspea-kers is of the order of 8 ohms. Wiring manufacturers quote the value of DC resistance per unit length (usually 1 metre) of cable, and a typical cable suitable for speaker use would be of 6 amp rating and about 12 milliohms (0.012 ohms) resistance per metre. Consider a 5 metre length of speaker cable. Its total loop resistance then would be 10 metres multiplied by 0.012 ohms = 0.12 ohms. This is a bit too high to meet the criterion stated above, an 8 ohm speaker requiring a cable of around 0.08 ohm loop resistance. In practice, though, this would probably be perfectly adequate, since there are many other factors which will affect sound quality. Nevertheless, it does illustrate that quite heavy cables are required to feed speakers, otherwise too much power will be wasted in the cable itself before the signal reaches the speaker.

If the same cable as above were used for a 40 m feed to a remote 8 ohm loud-speaker, the loop resistance would be nearly 1 ohm and nearly one-eighth of the amplifier power would be dissipated in heat in the cable. The moral here is to use the shortest length of cable as is practicable, or if long runs are required use the 100 volt line system (see section 12.8).

12.3.2 Cable and transformer inductance

The effect of cable inductance (see section 1.8) becomes more serious at high frequencies, but at audio frequencies it is insignificant even over long runs of cable. Conversely, inductance is extremely important in transformers. The coils on the transformer cores consist of a large number of turns of wire, and the electromagnetic field of each turn works against the fields of the other turns. The metallic core greatly enhances this effect. Therefore, the inductance of each transformer coil is very high and presents a high impedance to an audio signal. For a given frequency, the higher the inductance the higher the impedance in ohms.

12.3.3 Cable capacitance

The closer the conductors in a cable are together, the greater the capacitance (see section 1.8). The surface area of the conductors is also important. Capacitance is the opposite of inductance in that, for a given frequency, the greater the capacitance the *lower* is the impedance in ohms. In a screened cable the screening braid entirely encloses the inner conductor and so the surface area of the braid, as seen by this inner conductor, is quite large. Since large surface area implies high-capacitance, screened cable has a much higher capacitance than ordinary mains wiring, for example. So when an audio signal looks into a connecting cable it sees the capacitance between the conductors and therefore a rather less-than-infinite impedance between them, especially at high frequencies. A small amount of the signal can therefore be conducted to earth via the screen.

In the diagram in Figure 12.5 there are two resistors of equal value. A voltage V_1 is applied across the two. Because the value of the resistors is the same, V_1 is divided exactly in half, and V_2 will be found to be exactly half the value of V_1. If the lower resistor were to be increased in value to 400 ohms then twice the voltage would appear across it than across the upper resistor. The ratio of the resistors equals the ratio of the voltages across them.

200 ohms

V_1

200 ohms V_2

Figure 12.5 The voltage V_2 across the output is half the input voltage (V_1)

Consider a 200 ohm microphone looking into a mic lead, as shown in Figure 12.6(a). C is the capacitance between the screening braid and the inner core of the cable. The equivalent of this circuit is shown in Figure 12.6(b). Manufacturers quote the capacitance of cables in picofarads (pF) per unit length. A typical value for screened cable is 200 pF (0.0002 µF) per metre. A simple formula exists for determining the frequency at which 3 dB of signal is lost for a given capacitance and source resistance:

$$f = 159\,155/RC$$

where f = frequency in hertz (Hz), R = resistance in ohms, and C = capacitance in microfarads (µF).

To calculate the capacitance which will cause a 3 dB loss at 40 kHz, putting it safely out of the way of the audio band, the formula must be rearranged to give the maximum value of acceptable capacitance:

$$C = 159\,155/R\,f$$

Thus, if $R = 200$ (mic impedance), $f = 40\,000$:

$$C = 159\,155 \div (200 \times 40\,000) \approx 0.02 \ \mu F$$

So a maximum value of 0.02 µF of lead capacitance is acceptable for a mic lead. Typical lead capacitance was quoted as 0.0002 µF per metre, so 100 metres will give 0.02 µF, which is the calculated acceptable value. Therefore one could safely use up to 100 metres of typical mic cable with a standard 200 ohm microphone without incurring significant signal loss at high frequencies.

Figure 12.6 (a) A microphone with a 200 ohm output impedance is connected to an amplifier. Lead capacitance conducts high frequencies to ground more than low frequencies, and thus the cable introduces HF roll-off. V_2 is lower at HF than at LF

The principle applies equally to other audio circuits, and one more example will be worked out. A certain tape recorder has an output impedance of 1 k. How long a cable can it safely drive? From the above formula:

$$C = 159\,155 \div (1000 \times 40\,000) \approx 0.004 \ \mu F$$

In this case, assuming the same cable capacitance, the maximum safe cable length is $0.004 \div 0.0002 = 20$ metres. In practice, modern audio equipment generally has a low enough source impedance to drive long leads, but it is always wise to check up on this in the manufacturer's specification. Probably of greater concern will be the need to avoid long runs of unbalanced cable due to interference problems.

12.4 Balanced lines

The balanced line is better at rejecting interference than the unbalanced line, and improvements upon the performance of the unbalanced line in this respect can be 80 dB or more for high-quality microphone lines.

As shown in Figure 12.7, the connecting cable consists of a pair of inner conductors enclosed by a screening braid. At each end of the line is a 'balancing' transformer. The output amplifier feeds the primary of the output transformer and its voltage appears across the secondary. The send and return paths for the audio signal are provided by the two inner conductors, and the screen does not form part of the audio circuit. If an interference signal breaks through the screen it is induced equally into both signal lines. At the secondary transformer's primary the induced interference current, flowing in the same direction in both legs of the balanced line, cancels out, thus rejecting the interference signal. Two identical signals, flowing in opposite directions, cancel out where they collide.

Such an interfering signal is called a 'common mode' signal because it is equal and common to both audio lines. The rejection of this in the transformer is termed 'common mode rejection' (CMR). A common mode rejection ratio (CMRR) of at least 80 dB may be feasible. Meanwhile, the legitimate audio signal flows through the primary of the transformer as before, because the signal appears at each end of the coil with equal strength but *opposite* phase. Such a signal is called a 'differential signal', and the balanced input is also termed a 'differential input' because it accepts differential mode signals but rejects common mode signals.

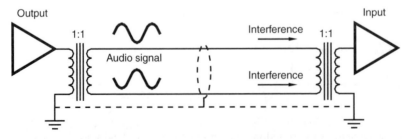

Figure 12.7 A balanced interconnection using transformers

So balanced lines are used for professional audio connections because of their greatly superior rejection of interference, and this is particularly useful when sending just a few precious millivolts from a microphone down many metres of cable to an amplifier.

12.5 Working with balanced lines

In order to avoid earth loops (see Fact File 12.2) with the balanced line, the earth screen is often connected at one end only, as shown in Figure 12.8(a), and still acts as a screen for the balanced audio lines. There is now no earth link between the two pieces of equipment, and so both can be safely earthed at the mains without causing an earth loop. The transformers have 'isolated' the two pieces of equipment from each other. The one potential danger with this is that the connecting lead with its earth disconnected in the plug at one end may later be used as a microphone cable. The lack of earth continuity between microphone and amplifier will cause inadequate screening of the microphone, and will also prevent a phantom power circuit being made (see section 4.9), so such cables and tie-lines should be marked 'earth off' at the plug without the earth connection.

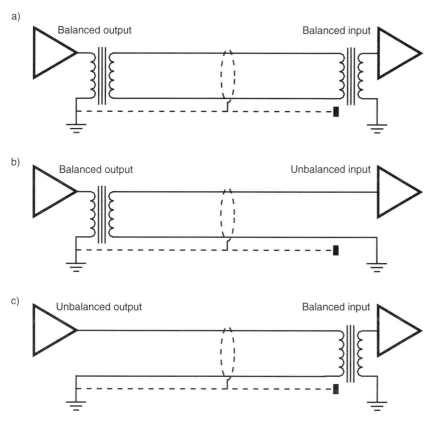

Figure 12.8 (a) Balanced output to balanced input with screen connected to earth only at output. (b) Balanced output to unbalanced input. (c) Unbalanced output to balanced input

FACT FILE **12.3** **XLR-3 connectors**

Pin 1 Screen
Pin 2 Signal (Live or 'hot')
Pin 3 Signal (Return or 'cold')

The most common balanced connector in professional audio is the XLR-3. This connector has three pins (as shown in the diagram), carrying respectively:

Viewed from end of
male pins

It is easy to remember this configuration, since X–L–R stands for Xternal, Live, Return. Unfortunately, an American convention still hangs on in some equipment which reverses the roles of pins 2 and 3, making pin 2 return and pin 3 live (or 'hot'). The result of this is an apparent absolute phase reversal in signals from devices using this convention when compared with an identical signal leaving a standard device. Modem American equipment mostly uses the European convention, and American manufacturers have now agreed to standardise on this approach.

Unfortunately, not all pieces of audio equipment have balanced inputs and outputs, and one may be faced with the problem of interfacing a balanced output with an unbalanced input, and an unbalanced output with a balanced input. A solution is shown in Figure 12.8(b), were the output transformer is connected to the signal and earth of the unbalanced input to give signal continuity. Because the input is unbalanced, there is no common mode rejection and the line is as susceptible to interference as an ordinary unbalanced line is. But notice that the screen is connected at one end only, so at least one can avoid an earth loop.

Figure 12.8(c) illustrates an unbalanced output feeding a balanced input. The signal and earth from the output feed the primary of the input transformer. Again the screen is not connected at one end so earth loops are avoided. Common mode rejection of interference at the input is again lost, because one side of the transformer primary is connected to earth. A better solution is to use a balancing transformer as close to the unbalanced output as possible, preferably before sending the signal over any length of cable. In the longer term it would be a good idea to fit balancing transformers inside unbalanced equipment with associated three-pin XLR-type sockets (see Fact File 12.3) if space inside the casing will allow. (Wait until the guarantee has elapsed first!)

12.6 Star-quad cable

Two audio lines can never occupy exactly the same physical space, and any interference induced into a balanced line may be slightly stronger in one line than in the

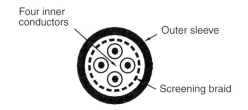

Four inner conductors

Outer sleeve

Screening braid

Figure 12.9 Four conductors are used in star-quad cable

other. This imbalance is seen by the transformer as a small differential signal which it will pass on, so a small amount of the unwanted signal will still get through. To help combat this, the two audio lines are twisted together during manufacture so as to present, on average, an equal face to the interference along both lines. A further step has been taken in the form of a cable called 'star-quad'. Here, four audio lines are incorporated inside the screen, as shown in Figure 12.9.

It is connected as follows. The screen is connected as usual. The four inner cores are connected in pairs such that two of the opposite wires (top and bottom in the figure) are connected together and used as one line, and the other two opposite wires are used as the other. All four are twisted together along the length of the cable during manufacture. This configuration ensures that for a given length of cable, both audio lines are exposed to an interference signal as equally as possible so that any interference is induced as equally as possible. So the input transformer sees the interference as a virtually perfect common mode signal, and efficiently rejects it. This may seem like taking things to extremes, but star-quad is in fact quite widely used for microphone cables. When multicore cables are used which contain many separate audio lines in a single thick cable, the balanced system gives good immunity from crosstalk, due to the fact that a signal in a particular wire will be induced equally into the audio pair of the adjacent line, and will therefore be a common mode signal. Star-quad multicores give even lower values of crosstalk.

12.7 Electronic balancing

Much audio equipment uses an electronically balanced arrangement instead of a transformer, and it is schematically represented as in Figure 12.10. The transformers have been replaced by a differential amplifier. The differential amplifier is designed to respond only to differential signals, as is the case with the transformer, and has one

Differential output

Balanced input
Differential amplifier

Figure 12.10 An electronically balanced interconnection using differential amplifiers positive and one negative input. Electronically balanced and transformer-balanced equipment can of course be freely intermixed. Reasons for dispensing with transformers include lower cost (a good transformer is rather more expensive than the electronic components which replace it), smaller size (transformers take up at least a few cubic centimetres, the alternative electronics rather less), less susceptibility to electromagnetic interference, and rather less sensitivity to the impedances between which they work with respect to distortion, frequency response and the like.

Good electronic balancing circuitry is, however, tricky to design, and the use of high-quality transformers in expensive audio equipment may well be a safer bet than electronic balancing of unknown performance. The best electronic balancing is usually capable of equal CMR performance to the transformer. Critics of transformer balancing cite factors such as the low-frequency distortion performance of a transformer, and its inability to pass extremely low frequencies, whereas critics of electronic balancing cite the better CMR available from a transformer when compared with a differential amplifier, and the fact that only the transformer provides true isolation between devices. Broadcasters often prefer to use transformer balancing because signals are transferred over very long distances and isolation is required, whereas recording studios often prefer electronic balancing, claiming that the sound quality is better.

12.8 100 volt lines

12.8.1 Principles

In section 12.3.1 it was suggested that the resistance of even quite thick cables was still sufficient to cause signal loss in loudspeaker interconnection unless short runs were employed. But long speaker lines are frequently unavoidable, examples being: back-stage paging and show relay speakers in theatres; wall-mounted speakers in lecture theatres and halls; paging speakers in supermarkets and factories; and open-air 'tannoy' horns at fairgrounds and fêtes. All these require long speaker runs, or alternatively a separate power amplifier sited close to each speaker, each amplifier being driven from the line output of a mixer or microphone amplifier. The latter solution will in most cases be considered an unnecessarily expensive and complicated solution. So the '100 volt line' was developed so that long speaker cable runs could be employed without too much signal loss along them.

The problem in normal speaker connection is that the speaker cable has a resistance comparable with, or even greater than, the speaker's impedance over longer runs. It was shown in section 12.1.2 that a transformer reflects impedance according to the square of the turns ratio. Suppose a transformer with a turns ratio of 5:1 is connected to the input of an 8 ohm speaker, as shown in Figure 12.11. The square of the turns ratio is 25:1 so the impedance across the primary of the transformer is $25 \times 8 = 200$ ohms. Now, the effective impedance of the speaker is much greater than the cable resistance, so most of the voltage will now reach the primary of the transformer and thence to the secondary and the speaker itself. But the transformer

Figure 12.11 Transformer coupling to a loudspeaker as used in 100 volt line systems

also transforms voltage, and the voltage across the secondary will only be a fifth of that across the primary. To produce 20 volts across the speaker then, one must apply 100 volts to the primary.

In a 100 V line system, as shown in Figure 12.12, a 50 W power amplifier drives a transformer with a step-up ratio of 1:5. Because the output impedance of a power amplifier is designed to be extremely low, the impedance across the secondary is also low enough to be ignored. The 20 volts, 2.5 amp output of the 50 watt amplifier is stepped up to 100 volts. The current is correspondingly stepped down to 0.5 amps (see Fact File 12.1), so that the total power remains the same. Along the speaker line there is a much higher voltage than before, and a much lower current. The voltage drop across the cable resistance is proportional to the current flowing through it, so this reduction in current means that there is a much smaller voltage drop due to the line. At the speaker end, a transformer restores the voltage to 20 volts and the current to 2.5 amps, and so the original 50 watts is delivered to the speaker.

A 50 watt amplifier has been used in the discussion. Any wattage of amplifier can be used, the transformer being chosen so that the step-up ratio gives the standard 100 volts output when the amplifier is delivering its maximum power output. For example, an amplifier rated at 100 watts into 8 ohms produces about 28 V. The step-up ratio of the line transformer would then have to be 28:100, or 1:3.6, to give the standard 100 volt output when the amplifier is being fully driven.

Returning to the loudspeaker end of the circuit. What if the speaker is only rated at 10 watts? The full 100 watts of the above amplifier would burn it out very quickly, and so a step-down ratio of the speaker transformer is chosen so that it receives only 10 watts. As 10 watts across 8 ohms is equivalent to around 9 volts, the required speaker transformer would have a step-down ratio of 100:9, or approximately 11:1.

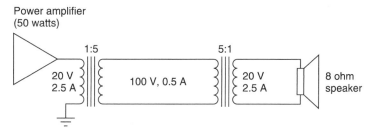

Figure 12.12 Voltage/current relationships in an example of 100 volt line operation

12.8.2 Working with 100 V lines

Speaker line transformers usually have a range of terminals labelled such that the primary side has a choice of wattage settings (e.g.: 30 W, 20 W, 10 W, 2 W) and the secondary gives a choice of speaker impedance, usually 15 ohms, 8 ohms and 4 ohms. This choice means that a number of speaker systems can be connected along the line (a transformer being required for each speaker enclosure), the wattage setting being appropriate to the speaker's coverage. For example, a paging system in the back-stage area of a theatre could be required to feed paging to six dressing rooms, a large toilet, and a fairly noisy green room. The dressing rooms are small and quiet, and so small speakers rated at 10 watts are employed with line transformers wired for 2 watts output. The large toilet requires greater power from the speaker, so one could use a 10 W speaker with a line transformer wired for 10 W output. The noisy green room could have a rather larger 20 W speaker with a line transformer wired for 20 W output. In this way, each speaker is supplied only with the wattage required to make it loud enough to be clearly heard in that particular room. A 20 watt speaker in a small dressing room would be far too loud, and a 2 watt speaker in a larger, noisier room would be inadequate.

As a string of loudspeakers is added to the system, one must be careful that the *total* wattage of the speakers does not exceed the output wattage of the power amplifier, or the latter will be overloaded. In the example, the six dressing rooms were allocated 2 W each, total 12 W. The toilet was allocated 10 W, the green room 20 W. The total is therefore 42 W, and a 50 W amplifier and line transformer would be adequate. In practice, a 100 W amplifier would be chosen to allow for both a good safety margin and plenty of power in hand if extra speakers need to be connected at a later date.

From the foregoing, it might well be asked why the 100 volt line system is not automatically used in all speaker systems. One reason is that 100 volts is high enough to give an electric shock and so is potentially dangerous in the domestic environment and other places where members of the public could interfere with an inadequately installed system. Secondly, the ultimate sound quality is compromised by the presence of transformers in the speaker lines – they are harder to design than the microphone and line level transformers already discussed, because they have to handle high voltages as well as several amps – and whereas they still give a perfectly adequate and extremely useful performance in paging and background music applications, they are not therefore used in high-quality PA systems or hi-fi and studio monitor speakers.

12.9 600 ohms

One frequently sees 600 ohms mentioned in the specifications of mixers, microphone amplifiers, and other equipment with line level outputs. Why is 600 ohms so special? The short answer is: it is not.

12.9.1 Principles

As has been shown, the output impedances of audio devices are low, typically 200 ohms for microphones and the same value or rather less for line level outputs. The input impedances of devices are much higher, at least 1–2 k for microphone inputs and 10 k or more for line inputs. This is to ensure that virtually the whole of the output voltage of a piece of equipment appears across the input it is feeding. Also, lower input impedances draw a higher current for a given voltage, the obvious example of this being an 8 ohm loudspeaker which draws several amps from a power amplifier. So a high input impedance means that only very small currents need to be supplied by the outputs, and one can look upon microphone and line level signals as being purely voltage without considering current.

This works fine, unless you are a telephone company that needs to send its signals along miles of cable. Over these distances, a hitherto unmentioned parameter comes into play which would cause signal loss if not dealt with, namely the wavelength of the signal in the cable. Now the audio signal is transmitted along a line at close to the speed of light (186 000 miles per second or 3×10^8 metres per second). The shortest signal wavelength will occur at the upper limit of the audio spectrum, and will be around 9.3 miles (14.9 km) at 20 kHz.

When a cable is long enough to accommodate a whole wavelength or more, the signal can be reflected back along the line and cause some cancellation of the primary signal. Even when the cable run is somewhat less than a wavelength some reflection and cancellation still occurs. To stop this from happening the cable must be terminated correctly, to form a so-called 'transmission line', and input and output impedances are chosen to be equal. The value of 600 ohms was chosen many decades ago as the standard value for telecommunications, and therefore the '600 ohm balanced line' is used to send audio signals along lines which need to be longer than a mile or so. It was chosen because comparatively little current needs to be supplied to drive this impedance, but on the other hand it is not high enough to allow much interference, as it is much easier for interference to affect a high-impedance circuit than a low one. Thus, professional equipment began to appear which boasted '600 ohms' to make it compatible with these lines. Unfortunately, many people did not bother to find out, or never understood, why 600 ohms was sometimes needed, and assumed that this was a professional audio standard *per se*, rather than a telecommunications standard. It was used widely in broadcasting, which has parallels with telecommunications, and may still be found in many cases involving older equipment.

The 600 ohm standard also gave rise to the standard reference level unit of 0 dBm, which corresponds to 1 mW of power dissipated in a resistance of 600 ohms. The corresponding voltage across the 600 ohm resistance at 0 dBm is 0.775 volts, and this leads some people still to confuse dBm with dBu, but 0 dBu refers simply to 0.775 volts with no reference to power or impedance: dBu is much more appropriate in modern equipment, since, as indicated above, the current flowing in most interfaces is negligible and impedances vary; dBm should only correctly be used in 600 ohm systems, unless an alternative impedance is quoted (e.g. dBm (75 ohms) is sometimes used in video equipment where 75 ohm termination is common).

12.9.2 Problems with 600 ohm equipment

A 600 ohm output impedance is too high for normal applications. With 600 ohms, 200 pF per metre cable, and an acceptable 3 dB loss at 40 kHz, the maximum cable length would be only around 33 m, which is inadequate for many installations. (This HF loss does not occur with the properly terminated 600 ohm system because the cable assumes the properties of a transmission line.) Furthermore, consider a situation where a mixer with a 600 ohm output impedance is required to drive five power amplifiers, each with an input impedance of 10 k. Five lots of 10 k in parallel produce an effective impedance of 2 k, as shown in Figure 12.13(a). Effectively then 600 ohms is driving 2 k, as shown in Figure 12.13(b). If V_1 is 1 volt then V_2 (from Ohm's law) is only 0.77 volts. Almost a quarter of the audio signal has been lost, and only a maximum of 33 metres of cable can be driven anyway.

Despite this, there are still one or two manufacturers who use 600 ohm impedances in order to appear 'professional'. It actually renders their equipment *less* suitable for professional use, as has been shown. One specification that is frequently

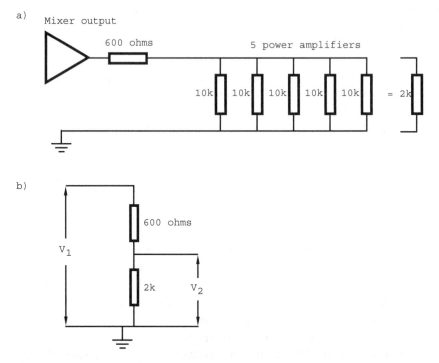

Figure 12.13 (a) Considerable signal loss can result if a 600 ohm output is connected to a number of 10 k inputs in parallel. (b) Electrical equivalent

encountered for a line output is something like: 'capable of delivering +20 dBu into 600 ohms'. Here +20 dBu is 7.75 volts, and 600 ohms is quite a low impedance, thus drawing more current from the source for a given voltage than, say, 10 kΩ does. The above specification is therefore useful, because it tells the user that the equipment can deliver 7.75 volts even into 600 ohms, and can therefore safely drive, say, a stack of power amplifiers, and/or a number of tape recorders, etc., without being overloaded. A domestic cassette recorder, for instance, despite having a low output impedance, could not be expected to do this. Its specification may well state '2 volts into 10 kΩ or greater'. This is fine for domestic applications, but one should do a quick calculation or two before asking it to drive all the power amplifiers in the building at once.

12.10 DI boxes

12.10.1 Overview

A frequent requirement is the need to interface equipment that has basically non-standard unbalanced outputs with the standard balanced inputs of mixers, either at line level or microphone level. An electric guitar, for example, has an unbalanced output of fairly high impedance – around 10 kΩ or so. The standard output socket is the 'mono' quarter-inch jack, and output voltage levels of around a volt or so (with the guitar's volume controls set to maximum) can be expected. Plugging the guitar directly into the mic or line level input of a mixer is unsatisfactory for several reasons: the input impedance of the mixer will be too low for the guitar, which likes to drive impedances of 500 kΩ or more; the guitar output is unbalanced so the interference-rejecting properties of the mixer's balanced input will be lost; the high output impedance of the guitar renders it incapable of driving long studio tie-lines; and the guitarist will frequently wish to plug the instrument into an amplifier as well as the mixer, and simply using the same guitar output to feed both via a splitter lead electrically connects the amplifier to the studio equipment which causes severe interference and low-frequency hum problems. Similar problems are encountered with other instruments such as synthesisers, electric pianos, and pickup systems for acoustic instruments.

To connect such an instrument with the mixer, a special interfacing unit known as a DI box (DI = direct injection) is therefore employed. This unit will convert the instrument's output to a low-impedance balanced signal, and also reduce its output level to the millivolt range suitable for feeding a microphone input. In addition to the input jack socket, it will also have an output jack socket so that the instrument's unprocessed signal can be passed to an amplifier as well. The low-impedance balanced output appears on a standard three-pin XLR panel-mounted plug which can now be looked upon as the output of a microphone. An earth-lift switch is also provided which isolates the earth of the input and output jack sockets from the XLR output, to trap earth loop problems.

12.10.2 Passive DI boxes

The simplest DI boxes contain just a transformer, and are termed 'passive' because they require no power supply. Figure 12.14 shows the circuit. The transformer in this case has a 20:1 step-down ratio, converting the fairly high output of the instrument to a lower output suitable for feeding microphone lines. Impedance is converted according to the square of the turns ratio (400:1), so a typical guitar output impedance of 15 kΩ will be stepped down to about 40 ohms which is comfortably low enough to drive long microphone lines. But the guitar itself likes to look into a high impedance. If the mixer's microphone input impedance is 2 kΩ, the transformer will step this up to 800 kΩ which is adequately high for the guitar. The 'link output jack socket' is used to connect the guitar to an amplifier if required. Note the configuration of the input jack socket: the make-and-break contact normally short-circuits the input which gives the box immunity from interference, and also very low noise when an instrument is not plugged in. Insertion of the jack plug opens this contact, removing the short-circuit. The transformer isolates the instrument from phantom power on the microphone line.

This type of DI box design has the advantages of being cheap, simple, and requiring no power source – there are no internal batteries to forget to change. On the other hand, its input and output impedances are entirely dependent on the reflected impedances each side of the transformer. Unusually low microphone input impedances will give insufficiently high impedances for many guitars. Also, instruments with passive volume controls can exhibit output impedances as high as 200 kΩ with the control turned down a few numbers from maximum, and this will cause too high an impedance at the output of the DI box for driving long lines. The fixed turns ratio of the transformer is not equally suited to the wide variety of instruments the DI box will encounter, although several units have additional switches which alter the transformer tapping giving different degrees of attenuation.

Figure 12.14 A simple passive direct-injection box

12.10.3 Active DI boxes

The active DI box replaces the transformer with an electronic circuit which presents a constant very high impedance to the instrument and provides a constant low-impedance output. Additionally, the presence of electronics provides scope for including other features such as several switched attenuation values (say −20 dB, −40 dB, −60 dB), high and low filters and the like. The box is powered either by internal batteries, or preferably by the phantom power on the microphone line. If batteries are used, the box should include an indication of battery status; a 'test' switch is often included which lights an LED when the battery is good. Alternatively, an LED comes on as a warning when the voltage of the battery drops below a certain level. The make-and-break contacts of the input jack socket are often configured so that insertion of the jack plug automatically switches the unit on. One should be mindful of this because if the jack plug is left plugged into the unit overnight, for instance, this will waste battery power. Usually the current consumption of the DI box is just a few milliamps, so the battery will last for perhaps a hundred hours. Some guitar and keyboard amplifiers offer a separate balanced output on an XLR socket labelled 'DI' or 'studio' which is intended to replace the DI box, and it is often convenient to use this instead.

DI boxes are generally small and light, and they spend much of their time on the floor being kicked around and trodden on by musicians and sound engineers. Therefore, rugged metal boxes should be used (not plastic) and any switches, LEDs, etc. should be mounted such that they are recessed or shrouded for protection. Switches should not be easily moved by trailing guitar leads and feet. The DI box can also be used for interfacing domestic hi-fi equipment such as cassette recorders and radio tuners with balanced microphone inputs.

12.11 Splitter boxes

The recording or broadcasting of live events calls for the outputs of microphones and instruments to be fed to at least two destinations, namely the PA mixer and the mixer in the mobile recording or outside broadcast van. The PA engineer can then balance the sound for the live audience, and the recording/broadcast balancer can independently control the mix for these differing requirements. It is possible of course to use two completely separate sets of microphones for this, but when one considers that there may be as many as ten microphones on a drum kit alone, and a vocalist would find the handling of two mics strapped together with two trailing leads rather unacceptable, a single set of microphones plugged into splitter boxes is the obvious way to do it. Ten or fifteen years ago recording studios and broadcasting companies would have frowned on this because the quality of some of the microphones then being used for PA was insufficient for their needs; but today PA mics tend to be every bit as good as those found in studios, and indeed many of the same models are common to both environments.

A microphone cannot be plugged into two microphone inputs of two separate mixers directly because on the one hand this will electrically connect one mixer with the other causing ground loop and interference problems, not to mention the fact that one phantom power circuit will be driving directly into the other phantom power circuit; and on the other hand the impedance seen by the microphone will now be the parallel result of the two mixers resulting in impedances of as low as 500 ohms which is too low for many microphones. A splitter box is therefore used which isolates the two mixers from each other and maintains a suitable impedance for the microphone. A splitter box will often contain a transformer with one primary winding for the microphone and two separate secondary windings giving the two outputs, as shown in Figure 12.15.

The diagram requires a bit of explanation. Firstly, phantom power must be conveyed to the microphone. In this case, output 2 provides it via the centre tap of its winding which conveys the power to the centre tap of the primary. The earth screen, on pin 1 of the input and output XLR sockets, is connected between the input and output 2 only, to provide screening of the microphone and its lead and also the phantom power return path. Note that pin 1 of output 1 is left unconnected so that earth loops cannot be created between the two outputs.

The turns ratio of the transformer must be considered next. The 1:0.7:0.7 indicates that each secondary coil has only 0.7 times the windings of the primary, and therefore the output voltage of the secondaries will each be 0.7 times the microphone output, which is about 3 dB down. There is therefore a 3 dB insertion loss in the splitter transformer. The reason for this is that the impedance as seen by the microphone must not be allowed to go too low. If there were the same number of turns on each coil, the microphone would be driving the impedance across each output directly in parallel. Two mixers with input impedances of 1 kΩ would therefore together load the microphone with 500 ohms, which is too low. But the illustrated turns ratio means that each 1 kΩ impedance is stepped up by a factor of $1:(0.7)^2 \approx 1:0.5$, so each mixer input appears as 2 kΩ to the microphone, giving a resultant parallel impedance of 1 kΩ, equal to the value of each mixer on its own which is fine. The 3 dB loss of signal is accompanied by an effective halving of

Figure 12.15 A simple passive splitter box

the microphone impedance as seen by each mixer, again due to the transformer's impedance conversion according to the square of the turns ratio, so there need not be a signal-to-noise ratio penalty.

Because of the simple nature of the splitter box, a high-quality transformer and a metal case with the necessary input and output sockets are all that is really needed. Active electronic units are also available which eliminate the insertion loss and can even provide extra gain if required. The advantages of an active splitter box over its transformer counterpart are, however, of far less importance than, say, the advantages that an active DI box has over its passive counterpart.

12.12 Jackfields (patchbays)

12.12.1 Overview

A jackfield (or patchbay) provides a versatile and comprehensive means of inter-connecting equipment and tie-lines in a non-permanent manner such that various source and destination configurations can quickly and easily be set up to cater for any requirements that may arise.

For example, a large mixing console may have microphone inputs, line inputs, main outputs, group outputs, auxiliary outputs, and insert send and returns for all input channels and all outputs. A jackfield, which usually consists of banks of 19 inch (48 cm) wide rack-mounting modules filled with rows of quarter-inch 'GPO'- (Telecom) type balanced jack sockets, is used as the termination point for all of the above facilities so that any individual input or output can be separately accessed. There are usually 24 jack sockets to a row, but sometimes 20 or 28 are encountered. Multicore cables connect the mixer to the jackfield, multipin connectors normally being employed at the mixer end. At the jackfield end multipin connectors can again be used, but as often as not the multicores will be hard wired to the jack socket terminals themselves. The layout of a mixer jackfield was discussed in section 6.4.9.

In addition to the mixer's jackfield there will be other jackfields either elsewhere in the rack or in adjacent racks which provide connection points for the other equipment and tie-lines. In a recording studio control room there will be such things as multitrack inputs and outputs, mic and line tie-lines linking control room to studio, outboard processor inputs and outputs, and tie-lines to the other control rooms and studios within a studio complex. A broadcasting studio will have similar arrangements, and there may also be tie-lines linking the studio with nearby concert halls or transmitter distribution facilities. A theatre jackfield will in addition carry tie-lines leading to various destinations around the auditorium, back stage, in the wings, in the orchestra pit, and almost anywhere else. There is no such thing as too many tie-lines in a theatre.

12.12.2 Patch cords

Patch cords (screened leads of around 1 metre in length terminated at each end with a 'B'-gauge jack plug (small tip) giving tip/ring/ sleeve connections) are used to link two appropriate sockets. The tip is live (corresponding to pin 2 on an XLR), the ring is return (pin 3 on an XLR), and the sleeve is earth (pin 1 on an XLR), providing balanced interconnection. The wire or 'cord' of a normal patch cord is coloured red. Yellow cords should indicate that the patch cord reverses the phase of the signal (i.e.: tip at one end is connected to ring of the other end) but this convention is not followed rigidly, leading to potential confusion. Green cords indicate that the earth is left unconnected at one end, and such cords are employed if an earth loop occurs when two separately powered pieces of equipment are connected together via the jackfield.

A large-tipped 'A'-gauge jack plug, found on the end of most stereo headphone leads and often called a stereo jack (although it was originally used as a single-line balanced jack before stereo came into vogue), may sometimes be inserted into a GPO-type socket. This works, but the spring-loaded tip connector is positioned so as to make secure contact with the small 'B'-type tip and the large 'A' tip can sometimes bend the contact. When the correct type of jack plug is later inserted there will be intermittent or no contact between the tip and the socket's tip connector. The insertion of large tipped jack plugs should therefore be discouraged.

12.12.3 Normalling

Normally, jackfield insertion points will be unused. Therefore the insertion send socket must be connected to the insertion return socket so that signal continuity is achieved. When an outboard unit is to be patched in, the insertion send socket is used to feed the unit's input. The unit's output is fed to the insertion return socket on the jackfield. This means that the send signal must be disconnected from the return socket and replaced by the return from the processor. To effect this requirement, extra make-and-break 'normalling' contacts on the jack socket are employed. Figure 12.16 shows how this is done. The signal is taken from the top jack socket to the bottom jack socket via the black triangle make-and-break contactors on the bottom socket. There is signal continuity. If a jack plug is now inserted into the bottom socket the contacts will be moved away from the make-and-break triangles,

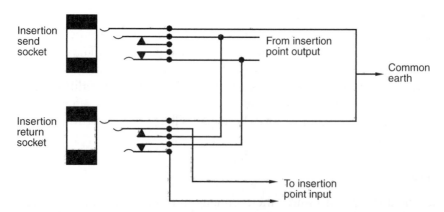

Figure 12.16 Normalling at jackfield insertion points

disconnecting the upper socket's signal from it. Signal from that jack plug now feeds the return socket. The make-and-break contacts on the upper jack socket are left unused, and so insertion of a jack plug into this socket alone has no effect on signal continuity. The send socket therefore simply provides an output signal to feed the processor.

Sometimes these make-and-break contacts are wired so that the signal is also interrupted if a jack plug is inserted into the send socket alone. This can be useful if, for instance, an alternative destination is required for, say, a group output. Insertion of a jack plug into the send socket will automatically mute the group's output and allow its signal to be patched in elsewhere, without disturbing the original patching arrangement. Such a wiring scheme is, however, rather less often encountered.

In addition to providing insert points, normalling can also be used to connect the group outputs of a mixer to the inputs of a multitrack recorder or a set of power amplifiers. The recorder or amplifier inputs will have associated sockets on the jackfield, and the mixer's group output sockets will be normalled to these as described. If need be, the input sockets can be overplugged in order to drive the inputs with alternative signals, automatically disconnecting the mixer's outputs in the process.

12.12.4 Other jackfield facilities

Other useful facilities in a jackfield include multiple rows of interconnected jacks, or 'mults'. These consist of, say, six or eight or however many adjacent sockets which are wired together in parallel so that a mixer output can be patched into one of the sockets, the signal now appearing on all of the sockets in the chain which can then be used to feed a number of power amplifier inputs in parallel, or several tape machines. The disadvantage of this poor man's distribution amplifier is that if there is a short-circuit or an interference signal on any of the inputs it is feeding this will affect all of the sockets on the mult. There is no isolation between them.

Outboard equipment will generally be equipped with XLR-type input and output connectors (or sometimes 'mono' jack sockets). It is useful therefore to have a panel of both male and female XLR sockets near to where such outboard gear is usually placed, these being wired to a row of sockets on the jackfield to facilitate connection between the mixer or tielines and these units.

Special additional make-and-break contacts can be included in the jack socket which are operated in the usual manner by jack plug insertion but have no contact with any part of the audio signal. These can be used to activate warning indicators to tell operators that a certain piece of equipment is in use, for example.

Since most if not all of the interconnections in a rig pass through the jackfield it is essential that the contacts are of good quality, giving reliable service. Palladium metal plating is employed which is tough, offering good resistance to wear and oxidation. This should always be looked for when jackfield is being ordered. Gold or silver plating is not used because it would quickly wear away in the face of professional use. The latter also tarnishes rather easily.

There is a miniature version known as the Bantam jack. This type is frequently employed in the control surface areas of mixers to give convenient access to the patching. Very high-density jackfields can be assembled, which has implications for the wiring arrangements on the back. Several earlier examples of Bantam-type jackfields were unreliable and unsuited to professional use. Later examples are rather better, and of course palladium contacts should always be specified.

Electronically controlled 'jackfields' dispense with patch cords altogether. Such systems consist of a digitally controlled 'stage box' which carries a number of input and output sockets into which the mixer inputs and outputs and any other tie-lines, processor and tape machine inputs and outputs are plugged. The unit is controlled by a keypad, and a VDU displays the state of the patch. Information can also be entered identifying each input and output by name according to the particular plug-up arrangement of the stage box. Any output can be routed to any input, and an output can be switched to drive any number of inputs as required. Various patches can be stored in the system's memory and recalled; rapid repatching is therefore possible, and this facility can be used in conjunction with timecode to effect automatic repatches at certain chosen points on a tape during mixdown for instance. MIDI control is also a possibility.

12.13 Distribution amplifiers

A distribution amplifier is an amplifier used for distributing one input to a number of outputs, with independent level control and isolation for each output. It is used widely in broadcast centres and other locations where signals must be split off and routed to a number of independent locations. This approach is preferable to simple parallel connections, since each output is unaffected by connections made to the others, preventing one from loading down or interfering with the others.

Chapter 13

Outboard equipment

Outboard equipment includes such things as parametric equalisers, graphic equalisers, delay lines, echo devices, dynamics units, multi-effects processors, gates, and room simulators. They provide extra facilities which will not normally be present in the mixing console itself, although several consoles incorporate such things as parametric equalisation and dynamics control.

13.1 The graphic equaliser

The graphic equaliser, pictured in Figure 13.1, consists of a row of faders (or sometimes rotary controls), each of which can cut and boost a relatively narrow band of frequencies. Simple four- or five-band devices exist which are aimed at the electronic music market, these being multiband tone controls. They perform the useful function of expanding existing simple tone controls on guitars and amplifiers, and several amplifiers actually incorporate them.

The professional rack-mounting graphic equaliser will have at least ten frequency bands, spaced at octave or one-third-octave intervals. The ISO (International Standards Organisation) centre frequencies for octave bands are 31 Hz, 63 Hz,

Figure 13.1 A typical two-channel graphic equaliser. (Courtesy of Klark-Teknik Research Ltd)

125 Hz, 250 Hz, 500 Hz, 1 kHz, 2 kHz, 4 kHz, 8 kHz, and 16 kHz. Each fader can cut or boost its band by typically 12 dB or more. Figure 13.2 shows two possible types of filter action. The 1 kHz fader is chosen, and three levels of cut and boost are illustrated. Maximum cut and boost of both types produces very similar Q (see Fact File 6.6). A high Q result is obtained by both types when maximum cut or boost is applied. The action of the first type is rather gentler when less cut or boost is applied, and the Q varies according to the degree of deviation from the fader's central position.

Many graphic equalisers conform to this type of action, and it has the disadvantage that a relatively broad band of frequencies is affected when moderate degrees of boost or cut are applied. The second type maintains a tight control of frequency bandwidth throughout the cut and boost range, and such filters are termed constant Q, the Q remaining virtually the same throughout the fader's travel. This is parti-

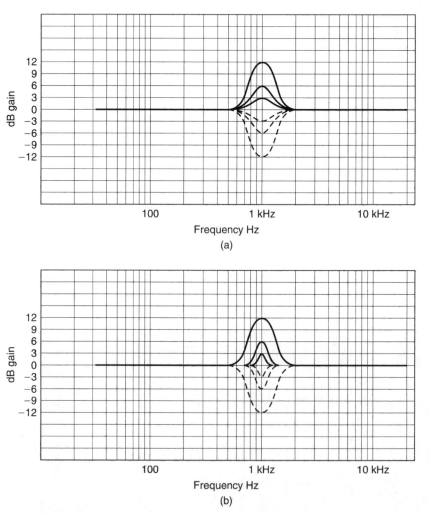

Figure 13.2 Two types of filter action shown with various degrees of boost and cut. (a) Typical graphic equaliser with Q dependent upon degree of boost/cut. (b) Constant Q filter action

cularly important in the closely spaced one-third-octave graphic equaliser which has 30 separate bands, so that adjacent bands do not interact with each other too much. The ISO centre frequencies for 30 bands are 25 Hz, 31 Hz, 40 Hz, 50 Hz, 63 Hz, 80 Hz, 100 Hz, 125 Hz, 180 Hz, 200 Hz, 250 Hz, 315 Hz, 400 Hz, 500 Hz, 630 Hz, 800 Hz, 1 kHz, 1k25 Hz, 1k8 Hz, 2 kHz, 2 k5 Hz, 3k15 Hz, 4 kHz, 5 kHz, 6k3 Hz, 8 kHz, 10 kHz, 12k5 Hz, 16 kHz, and 20 kHz. The value of using standard centre frequencies is that complementary equipment such as spectrum analysers which will often be used in conjunction with graphic equalisers have their scales centred on the same frequencies.

Even with tight constant Q filters, the conventional analogue graphic equaliser still suffers from adjacent filter interaction. If, say, 12 dB of boost is applied to one frequency and 12 dB of cut applied to the next, the result will be more like a 6 dB boost and cut, the response merging in between to produce an ill-defined Q value. Such extreme settings are, however, unlikely. The digital graphic equaliser applies cut and boost in the digital domain, and extreme settings of adjacent bands can be successfully accomplished without interaction if required.

Some graphic equalisers are single channel, some are stereo. All will have an overall level control, a bypass switch, and many also sport separate steep-cut LF filters. A useful facility is an overload indicator – usually an LED which flashes just before the signal is clipped – which indicates signal clipping anywhere along the circuit path within the unit. Large degrees of boost can sometimes provoke this. Some feature frequency cut only, these being useful as notch filters for getting rid of feedback frequencies in PA/microphone combinations. Some can be switched between cut/boost, or cut only. It is quite possible that the graphic equaliser will be asked to drive very long lines, if it is placed between mixer outputs and power amplifiers for example, and so it must be capable of doing this. The '+20 dBu into 600 ohms' specification should be looked for as is the case with mixers. It will be more usual though to patch the graphic equaliser into mixer output inserts, so that the mixer's output level meters display the effect on level the graphic equaliser is having. Signal-to-noise ratio should be at least 100 dB.

The graphic equaliser can be used purely as a creative tool, providing tone control to taste. It will frequently be used to provide overall frequency balance correction for PA rigs. It has formerly been used to equalise control room speakers, but poor results are frequently obtained due to the fact that a spectrum analyser's microphone samples the complete room frequency response whereas the perceived frequency balance is a complex combination of direct and reflected sound arriving at different times. The graphic equaliser can also change the phase response of signals, and there has been a trend away from their use in the control room for monitor EQ, adjustments being made to the control room acoustics instead.

The parametric equaliser was fully described in section 6.4.4.

13.2 The compressor/limiter

The compressor/limiter (see Fact File 13.1) is used in applications such as dynamics control and as a guard against signal clipping. Such a device is pictured in Figure

FACT FILE
13.1

Compression and limiting

A compressor is a device whose output level can be made to change at a different rate to input level. For example, a compressor with a ratio of 2:1 will give an output level that changes by only half as much as the input level above a certain threshold (see diagram). For example, if the input level were to change by 6 dB the output level would change by only 3 dB. Other compression ratios are available such as 3:1, 5:1, etc. At the higher ratios, the output level changes only a very small amount with changes in input level, which makes the device useful for reducing the dynamic range of a signal. The threshold of a compressor determines the signal level above which action occurs.

A limiter is a compressor with a very high compression ratio. A limiter is used to ensure that signal level does not rise above a given threshold. A 'soft' limiter has an action which comes in only gently above the threshold, rather than acting as a brick wall, whereas a 'hard' limiter has the effect almost of clipping anything which exceeds the threshold.

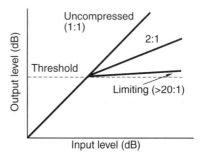

13.3. The three main variable parameters are attack, release and threshold. The attack time, in microseconds (μs) and milliseconds (ms), is the time taken for a limiter to react to a signal. A very fast attack time of 10 μs can be used to avoid signal clipping, any high-level transients being rapidly brought under control. A fast release time will rapidly restore the gain so that only very short-duration peaks will be truncated. A ducking effect can be produced by using rapid attack plus a release of around 200–300 ms. A threshold level is chosen which causes the limiting to come in at a moderate signal level so that peaks are pushed down before the gain is quickly reinstated. Such a ducking effect is ugly on speech, but is useful for overhead cymbal mics for example.

Figure 13.3 A typical compressor/limiter. (Courtesy of Drawmer Distribution Ltd)

A slow release time of several seconds, coupled with a moderate threshold, will compress the signal dynamics into a narrower window, allowing a higher mean signal level to be produced. Such a technique is often used in vocals to obtain consistent vocal level from a singer. AM radio is compressed in this way so as to squeeze wide dynamic range material into this narrow dynamic range medium. It is also used on FM radio to a lesser extent, although very bad examples of its application are frequently heard on pop stations. An oppressive, raspy sound is the result, and in the pauses in between items or speech one hears the system gain creeping back up, causing pumping noises. Background noise rapidly ducks back down when the presenter again speaks. The effect is like listening to a tape with noise reduction encoding, the reciprocal decoding being absent.

Many units offer separate limiting and compressing sections, their attack, release and threshold controls in each section having values appropriate to the two applications. Some also include gates (see section 9.5.2) with variable threshold and ratio, rather like an upside-down limiter, such that *below* a certain level the level drops faster than would be expected. 'Gain make-up' is often available to compensate for the overall level-reducing effect of compression. Meters may indicate the amount of level reduction occurring.

13.3 Echo and reverb devices

13.3.1 Echo chamber

Echo, or more properly reverb, used to be produced literally by setting aside a suitable reflective room in which was placed a loudspeaker and two microphones, the mics being placed a fair distance apart from each other, and from the speaker (see Figure 13.4). Signal was sent to the speaker at a fairly high level, which excited reflections in the room. The two microphones provided the 'stereo' echo return signals. The technique was undoubtedly successful, quality varying according to room size, shape and acoustic treatment. Pre-delay (see Fact File 13.2) could be used to simulate reflections from more distant surfaces.

13.3.2 Echo plate

The echo plate consists of a large, thin metal plate, several square metres in area. It is enclosed in an acoustically isolated housing. A driving transducer is attached close to one edge, and two or more receiving transducers are positioned at various other locations on the plate (see Figure 13.5). Signal is applied to the driving transducer which acoustically excites the plate. Its resonant properties ensure that vibrations persist for a usefully long period of time. The receiving transducers pick up the signal to provide the echo signal. The several receiving transducers can give various types of echo according to their positions, and more than one driving transducer can also be used to excite different resonant modes. A pre-delay signal (tape machine or solid-state delay) may be used to simulate a larger room. The plate is capable at its best of quite pleasing effects, and the twangy 'plate-echo'

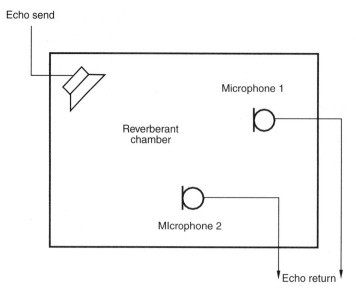

Figure 13.4 In a reverberation chamber the send is fed to a loudspeaker which in turn is used to excite the room. The returns are fed from two microphones placed in the reverberant field

character of some settings is deliberately simulated in digital reverb devices in some of the programs because it suits some instruments well. The characteristics of plate reverb can be altered using a graphic equaliser if necessary. The plate must be well isolated acoustically because it is sensitive to both airborne and

FACT FILE

13.2

Simulating reflections

Pre-delay in a reverb device is a means of delaying the first reflection to simulate the effect of a large room with distant surfaces. Early reflections may then be programmed to simulate the first few reflections from the surfaces as the reverberant field builds up, followed by the general decay of reverberant energy in the room as random reflections lose their energy (see diagram).

Pre-delay and early reflections have an important effect on one's perception of the size of a room, and it is these first few milliseconds which provide the brain with one of its main

clues as to room size. Reverberation time (RT) alone is not a good guide to room size, since the RT is affected both by room volume and absorption (see Fact Files 1.5 and 1.6); thus the same RT could be obtained from a certain large room and a smaller, more reflective room. Early reflections, though, are dictated only by the distance of the surfaces.

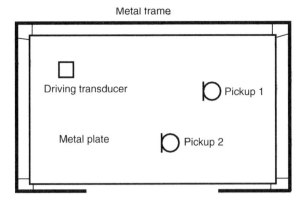

Metal frame

Driving transducer

Pickup 1

Metal plate

Pickup 2

Figure 13.5 A reverberation plate is a metal sheet suspended from a frame, with transducers mounted at appropriate locations. Often the damping of the plate may be varied to alter the decay time of vibrations

structure-borne vibrations. If it is mounted in a noisy area the echo return may contain echoey versions of the sound in the plate's surroundings! A separate floor from the studio is often chosen.

13.3.3 Spring reverb

The spring reverb, popularised by the American Hammond organ company many years ago, consists basically of a coil of springy wire several millimetres in diameter and up to a metre or so long. It is acoustically excited by a transducer at one end, and other transducers, both at the far end and sometimes along the length, pick up the vibrations. Some are fairly sophisticated, utilising several springs and a number of transducers. The modest quality of their performance does not, however, prevent them from having an application in guitar amplifiers, and quite a pleasing reverb quality can be obtained. The springs are acoustically sensitive, and one frequently hears loud twangs and crashes through the speakers when guitar amplifiers are being moved about.

13.3.4 Digital reverb

The present-day digital reverb, such as that pictured in Figure 13.6, can be quite a sophisticated device. Research into path lengths, boundary and atmospheric absorption, and the physical volume and dimensions of real halls, have been taken into account when algorithms have been designed. Typical front panel controls will include selection of internal pre-programmed effects, labelled as 'large hall', 'medium hall', 'cathedral', 'church', etc., and parameters such as degree of pre-delay, decay time, frequency balance of delay, dry-to-wet ratio (how much direct untreated sound appears with the effect signal on the output), stereo width, and relative phase between the stereo outputs can often be additionally altered by the user. A small display gives information about the various parameters.

Figure 13.6 The Lexicon 960-L multichannel effects system. (Courtesy of Stirling/Pure Distribution)

Memory stores generally contain a volatile and a non-volatile section. The non-volatile section contains factory pre-set effects, and although the parameters can be varied to taste the alterations cannot be stored in that memory. Settings can be stored in the volatile section, and it is usual to adjust an internal pre-set to taste and then transfer and store this in the volatile section. For example, a unit may contain 100 pre-sets. The first 50 are non-volatile, and cannot be permanently altered. The last 50 can store settings arrived at by the user by transferring existing settings to this section. The method of doing this varies between models, but is usually a simple two- or three-button procedure, for example by pressing 5, 7 and 'store'. This means that a setting in the non-volatile memory which has been adjusted to taste will be stored in memory 57. Additional adjustments can then be made later if required.

Several units provide a lock facility so that stored effects can be made safe against accidental overwriting. An internal battery backup protects the memory contents when the unit is switched off. Various unique settings can be stored in the memories, although it is surprising how a particular model will frequently stamp its own character on the sound however it is altered. This can be both good or bad of course, and operators may have a preference for a particular system. Certain processors sound a bit like clangy spring reverbs whatever the settings. Some sound dull due to limited bandwidth, the frequency response extending only up to around 12 kHz or so. This must be carefully looked for in the specification. Sometimes the bandwidth reduces with increasing reverberation or echo decay times. Such a shortcoming is sometimes hard to find in a unit's manual.

In all the above reverb devices it should be noted that the input is normally mono and the output stereo. In this way 'stereo space' can be added to a mono signal, there being a degree of decorrelation between the outputs. Occasionally, a reverb device may have stereo inputs, so that the source can be assumed to be other than a point.

13.4 Multi-effects processors

Digital multi-effects processors such as that shown in Figure 13.7 can offer a great variety of features. Parametric equalisation is available, offering variations in Q, frequency and degree of cut and boost. Short memory capacity can store a sample, the unit being able to process this and reproduce it according to the incoming signal's command. MIDI interfacing (see Chapter 14) has become popular for the selection of effects under remote control, as has the RS 232 computer interface, and a floppy-disk drive is sometimes encountered for loading information. Repeat echo,

Figure 13.7 The tc 2290 multi-effects processor. (Courtesy of tc Electronics)

autopan, phase, modulation, flange, high and low filters, straight signal delay, pitch change, gating, and added harmony may all be available in the pre-sets, various multifunction nudge buttons being provided for overall control. Many units are only capable of offering one type of effect at a time. Several have software update options so that a basic unit can be purchased and updates later incorporated internally to provide, say, longer delay times, higher sample storage capacity, and new types of effect as funds allow and as the manufacturer develops them. This helps to keep obsolescence at bay in an area which is always rapidly changing.

13.5 Frequency shifter

The frequency shifter shifts an incoming signal by a few hertz. It is used for acoustic feedback control in PA work, and operates as follows. Feedback is caused by sound from a speaker re-entering a microphone to be reamplified and reproduced again by the speaker, forming a positive feedback loop which builds up to a continuous loud howling noise at a particular frequency. The frequency shifter is placed in the signal path such that the frequencies reproduced by the speakers are displaced by several hertz compared with the sound entering the microphone, preventing additive effects when the sound is recycled, so the positive feedback loop is broken. The very small frequency shift has minimal effect on the perceived pitch of the primary sound.

13.6 Miscellaneous devices

The Aphex company of America introduced a unit called the *Aural Exciter* in the 1970s, and for a time the mechanisms by which it achieved its effect were shrouded in a certain amount of mystery. The unit made a signal 'sparkle', enhancing its overall presence and life, and it was usually applied to individual sounds in a mix such as solo instruments and voices, but sometimes also to the complete stereo signal. Such devices succeed entirely by their subjective effect, and several companies later produced similar units. They achieve their psycho-acoustic effect by techniques such as comb filtering, selective boosting of certain frequencies, and by introducing relatively narrow-band phase shifts between stereo channels.

The de-esser cleans up closely miced vocals. Sibilant sounds can produce a rasping quality, and the de-esser dynamically filters the high-level, high-frequency component of the sound to produce a more natural vocal quality.

13.7 Connection of outboard devices

A distinction needs to be made between processors which need to interrupt a signal for treatment, and those which basically add something to an existing signal. Graphic and parametric equalisers, compressors, de-essers and gates

Figure 13.8 Outboard processors such as compressors are normally patched in at an insertion point of the required mixer channel

Figure 13.9 Reverberation and echo devices are usually fed from a post-fader auxiliary send, and brought back to a dedicated echo return or another channel input

need to be placed in the signal path. One would not normally wish to mix, say, an uncompressed signal with its compressed version, or an ungated with the gated sound. Such processors will generally be patched in via the mixer's channel insertion send and returns (see Figure 13.8), or patched in ahead of the incoming

Figure 13.10 The tc Electronics System 6000 multi-effects processor. (Courtesy of tc Electronics)

signal or immediately after an output. Devices such as echo, reverb, chorus, flange are generally used to add something to an existing signal and usually a channel aux send will be used to drive them. Their outputs will be brought back to additional input channels and these signals mixed to taste with the existing dry signal (see Figure 13.9).

Sometimes just the effects signal will be required, in which case either the aux send will be switched to pre-fade and that channel's fader closed, or the channel will simply be de-routed from the outputs. The channel is then used merely to send the signal to the effects unit via the aux. The returns will often contain a degree of dry signal anyway, the ratio of dry to effect being adjusted on the processor.

Figure 13.10 shows a multi-effects processor.

Chapter 14

MIDI

MIDI, the Musical Instrument Digital Interface, now forms an integral part of many studio and multimedia operations. It can be used for controlling musical instruments such as samplers and synthesisers, internal sound cards within PCs (on which may be provided FM and wavetable synthesis functions, as well as audio sampling), and a range of other studio equipment such as mixers, effects devices and recording devices. MIDI software is often integrated with digital audio software, so as to produce a package capable of recording, editing and replaying both musical performance data and sampled digital audio data, combining the benefits of both approaches.

MIDI has become popular as a means of controlling and storing sound information in an economical way. As previously explained, sampled digital audio can occupy very large amounts of memory space (around 5 Mbytes per minute for high quality sound), whereas the MIDI data for a minute of music might only consume 5 kbytes. Whilst the two are not directly comparable, the widespread adoption of General MIDI and the availability of cheap sound cards for PCs has meant that many high quality computer sound applications can be handled using MIDI control instead of digital sound files.

This chapter will provide an introduction to MIDI and its applications. For more detailed information the reader is referred to the book *MIDI Systems and Control* by Francis Rumsey, as detailed at the end of this chapter.

14.1 What is MIDI?

MIDI is principally a means of remote control for electronic musical instruments. It integrates timing and system control functions with note triggering and other performance information. It is possible to control a number of musical instruments polyphonically from one controller using a single MIDI data stream. The interface is serial (see Fact File 14.1) which offers a number of advantages.

FACT FILE	A serial
14.1	interface

A serial interface is one in which the bits of data making up each byte are transferred one after another. It is thus different from a parallel interface in which a number of bits may be transferred together, using a number of lines in parallel. A serial link between two devices requires a single communications channel, whereas a parallel link requires many; so the serial link is useful in cases where economy in wiring is paramount, or where only a single channel exists (over a telephone link, for example). Most computers, and this includes those inside MIDI devices, use parallel data

internally, requiring some conversion to and from serial form for MIDI communication. Parallel data is converted to serial form using a device known as a shift register, illustrated below. The rate of the clock signal determines the rate at which the bits of the parallel input word are shifted out of the serial output. The reverse process can be used for converting serial inputs to parallel form.

The adoption of a simple serial standard for MIDI was dictated largely by economic and practical considerations, as it was intended that the interface should be installed on relatively cheap items of equipment, and that it should be available to as wide a range of users as possible. The simplicity and ease of installation of a MIDI system has been largely responsible for its rapid proliferation as an international standard.

Parallel input

0   1   1   1   0   0   1   0

Clock (bit rate)

Serial output

14.2 MIDI and digital audio contrasted

For many the distinction between MIDI and digital audio may be a clear one, but those new to the subject often confuse the two. Any confusion is often due to both MIDI and digital audio equipment appearing to perform the same task – that is the recording of multiple channels of music using digital equipment – and is not helped by the way in which some manufacturers refer to MIDI sequencing as digital recording.

Digital audio involves a process whereby an audio waveform (such as the line output of a musical instrument) is sampled regularly and then converted into a series of binary words which actually represent the sound itself (see Chapter 10). A digital audio recorder stores this sequence of data and can replay it by passing the original data through a digital-to-analogue convertor which turns the data back into a sound waveform, as shown in Figure 14.1. MIDI, on the other hand, handles digital information which controls the generation of sound. The data represents

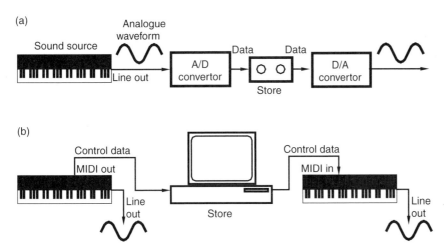

Figure 14.1 (a) Digital audio recording and (b) MIDI contrasted. In (a) the sound waveform is converted into digital data and stored, whereas in (b) only control information is stored, and a MIDI-controlled sound generator is required during replay

events such as notes pressed, wheels moved and switches operated. When a music recording is made using a MIDI sequencer this event data is stored, and can be replayed by transmitting it to a collection of MIDI-controlled musical instruments. It is the instruments which actually reproduce the recording.

A digital audio recorder, then, allows any sound to be stored and replayed without the need for additional hardware. It is useful for recording acoustic sounds such as voices, where MIDI is not much help. A MIDI sequencer is almost useless without a collection of sound generators, and the replayed sound quality and timbre are not fixed – they depend on the voices of the sound generators. An advantage of the MIDI recording is that the stored data represents the events in a piece of music, making it possible to change the music by changing the event data.

14.3 Basic MIDI principles

14.3.1 System specifications

The MIDI standard specifies a uni-directional asynchronous serial interface running at a data rate of 31.25 kbits ± 1 per cent (see Fact File 14.2). The rate had to be slow enough to be carried over simple cables and interface hardware, but fast enough to allow musical information to be transferred from one instrument to another without noticeable delays, giving the impression of real-time control.

There is a defined hardware interface which should be incorporated in all MIDI equipment, which ensures electrical compatibility within a system. This is shown in Fact File 14.3. Concerning any potential for delay between IN and THRU connectors, it should be stated at the outset that the delay in data passed through this

FACT FILE

14.2

Asynchronous data transmission

MIDI uses what is known as asynchronous serial communication. In asynchronous communication data is transmitted without an accompanying clock signal, and the sending device expects the receiver to lock on to the incoming data whenever it arrives. This makes it very easy to interlink two devices, as there is only really a need for a single wire to carry the data (although in practice there are two to complete the circuit, and a screen to prevent interference).

In an asynchronous interface the clocks of the transmitter and the receiver must be running at exactly the same rate, otherwise data may be lost due to timing errors. This close tolerance in clock rates is fundamental to the satisfactory operation of an asynchronous interface, as it is vital that the receiver can rely on a particular bit of a received word occurring at a particular time. The tolerance in clock frequency allowed for MIDI is ±1 per cent. Also, data may be

transmitted at any time, and there may be long gaps in between messages. Thus there is a need for a means of signalling the start and end of a data word. This is achieved using start and stop bits.

A start bit precedes every byte transmitted, and a stop bit follows every byte. On recognition of the leading edge of the start bit, the receiver adjusts the phase of its clock such that it thereafter accepts the bits of the following word in their correct time slots, and the stop bit terminates the word (see diagram). A device known as a UART (Universal Asynchronous Receiver/Transmitter), which is really just a more advanced form of shift register, automatically handles the timing of the serial interface and converts data to and from the parallel form.

The MIDI standard specifies a unidirectional serial interface running at 31.25 kbaud. The baud rate is the maximum number of bits per second that can be transferred over the interface. Data is transmitted unidirectionally, that is it has only one direction: from the transmitter to the receiver. Standard MIDI messages typically consist of 1, 2, or 3 bytes, although there are longer messages for special purposes.

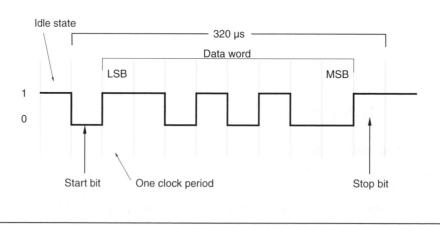

circuit is only a matter of microseconds, so this contributes little to any audible delays perceived in the musical outputs of some instruments in a large system. The bulk of any perceived delay will be due to other factors, which are covered in later sections.

FACT FILE

14.3

MIDI
hardware
interface

Most equipment using MIDI has three interface connectors: IN, OUT and THRU (sic), as shown below. The OUT connector carries data which the device itself has generated, relating to keys pressed, controls altered, etc. It is transmitted from the device's UART. The IN connector receives data from other devices and relays it to the device's UART. Data messages received will be acted upon by the instrument concerned if it is capable of handling the particular message, and if it is destined for that instrument (see text).

The THRU connector is a direct relay of the data that is present at the IN. It is simply a buffered feed of the input data,

and it has not been processed in any way. A few cheaper devices do not have THRU connectors, but it is possible to buy 'MIDI THRU boxes' which provide a number of THRUs from one input. Occasionally, devices without a THRU socket allow the OUT socket to be switched between OUT and THRU. The THRU socket can be used to 'daisy-chain' MIDI devices together, so that transmitted information from one controller can be sent to a number of receivers without the need for multiple outputs from the controller.

The opto-isolator (an encapsulated device containing a light-emitting diode and a photo-transistor) between the MIDI IN and the UART couples the data optically, and means that there is no direct electrical link between devices. It helps to reduce the effects of any problems which might occur if one instrument in a system were to develop an electrical fault.

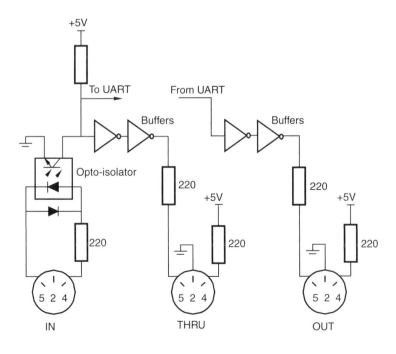

14.3.2 Simple interconnection

Before going further it is necessary to look at a simple practical application of MIDI, to see how various aspects of the standard message protocol have arisen. For the present it will be assumed that we are dealing with a music system, although, as will be seen, MIDI systems can easily incorporate non-musical devices in various ways.

In the simplest MIDI system, one instrument could be connected to another using a standard MIDI cable (see Fact File 14.4) as shown in Figure 14.2. Here, instrument 1 sends information relating to actions performed on its own controls (notes pressed, pedals pressed, etc.) to instrument 2, which imitates these actions as far as it is able. This type of arrangement can be used for 'doubling-up' sounds, 'layering' or 'stacking', such that a composite sound can be made up from two synthesisers' outputs. (It should be pointed out that the audio outputs of the two instruments would have to be mixed together for this effect to be heard.) Larger MIDI systems could be built up by further 'daisy-chaining' of instruments, such that instruments further down the chain all received information generated by the first (see Figure 14.3), although this is not a very satisfactory way of building a large MIDI system as described later in this chapter.

14.3.3 MIDI channels

MIDI messages are made up of a number of bytes. Each part of the message has a specific purpose, and one of these is to define the receiving channel to which the message refers. In this way, a controlling device can make a message device specific – in other words it can define which receiving instrument will act on the data. This is most important in large systems which use a computer sequencer as a master controller, when a large amount of information will be present on the MIDI data bus, not all of which is intended for every instrument. If a device is set to receive on a specific channel or on a number of channels it will act only on information which is 'tagged' with its own channel numbers. It will usually ignore everything else. There are sixteen basic MIDI channels and instruments can usually be set to receive on any specific channel or channels (omni off mode), or to receive on all channels (omni on mode).

Later it will be seen that the limit of sixteen MIDI channels can be exceeded easily by using multiport MIDI interfaces connected to a computer. In such cases it is important not to confuse the MIDI data channel with the physical port to which a

Figure 14.2 In the simplest form of MIDI interconnection, the OUT of one instrument is connected to the IN of another

FACT FILE 14.4   Connectors and cables

The connectors used for MIDI interfaces are like the five-pin DIN plugs used in some hi-fi systems. The specification also allows for the use of professional XLR connectors, although these are rarely encountered in practice. Only three of the pins of a five-pin DIN plug are actually used in most equipment (the three innermost pins). The cable should be a shielded twisted pair with the shield connected to pin 2 at both ends, although within the receiver itself the MIDI IN does not have pin 2 connected to earth. This is to avoid earth loops (see Fact File 12.2), but makes it possible to use a cable either way round. In the cable, pin 5 at one end should be connected to pin 5 at the other, and likewise pin 4 to pin 4, and pin 2 to pin 2. Unless DIN cables follow this convention they will not work.

A current loop is created between a MIDI OUT or THRU and a MIDI IN, when connected with the appropriate cable, as shown below, and data is represented by the sending device turning this current on and off – a binary zero is represented by a current flowing through the loop whereas a one is represented by a lack of current.

It is recommended that no more than 15 m of cable is used for a single cable run in a simple MIDI system, although deterioration of the signal is gradual and depends on the conditions, cable, and equipment in use. Longer distances may be accommodated with the use of buffer or 'booster' boxes. Professional microphone cable terminated in DIN connectors may be used as a higher quality solution, because domestic cables will not always be a shielded twisted pair and thus will be more susceptible to external interference, as well as radiating more themselves which could interfere with adjacent audio signals.

It is recommended that the correct cable is used in professional installations where MIDI cables are installed in the same trunking as audio cables, to avoid any problems with crosstalk.

Figure 14.3 Devices can be 'daisy-chained' using the THRU connector, in which case all subsequent devices receive commands from the first device in the chain

device may be connected, since each physical port will be capable of transmitting on all sixteen data channels.

14.4 MIDI communications

14.4.1 Message format

There are two basic types of MIDI message byte: the status byte and the data byte. Status bytes always begin with a binary one to distinguish them from data bytes, which always begin with a zero. As shown in Figure 14.4, the first half of the status byte denotes the message type and the second half denotes the channel number. Because the most significant bit (MSB) of each byte is reserved to denote the type (status or data) there are only seven active bits per byte which allows 2^7 (that is 128) possible values. (Some basic information about binary data was given in Chapter 10 and Fact File 14.5 gives further details of the method of binary shorthand known as hexadecimal.)

The first byte in a MIDI message is normally a status byte, which contains information about the channel number to which the message applies. It can be seen that four bits of the status byte are set aside to indicate the channel number, which allows for 2^4 (or 16) possible channels. The status byte is the label that denotes which receiver or part of a receiver the message is intended for, and it also denotes which type of message is to follow (e.g. a note on message). There are only three bits to denote the message type because the first bit must always be a one. This theoretically allows for eight message types, but there are some special cases in the form of system messages, all of which begin with binary 1111 (&F).

Standard MIDI messages can be one, two or three bytes long, but System Exclusive messages are special exceptions which often exceed this length for a

Figure 14.4 General format of a MIDI message. The 'sss' bits are used to define the message type, the 'nnnn' bits define the channel number, whilst the 'xxxxxxx' and 'yyyyyyy' bits carry the message data

FACT FILE

14.5

Hexadecimal

Bin	Hex	Dec
0000	0	0
0001	1	1
0010	2	2
0011	3	3
0100	4	4
0101	5	5
0110	6	6
0111	7	7
1000	8	8
1001	9	9
1010	A	10
1011	B	11
1100	C	12
1101	D	13
1110	E	14
1111	F	15

Long binary numbers can be fairly unwieldy to write down, so various forms of shorthand are used to make them more manageable. The most common of these is hexadecimal. The hexadecimal system (base 16) represents decimal values from 0 to 15 using the sixteen symbols 0–9 and A–F, as shown in the table. Each hex character corresponds to a group of four binary digits (bits), or one 'nibble' of a binary word. An example showing how a long binary word may be written in hex is shown in the diagram – it is simply a matter of breaking the word up into 4 bit nibbles and representing each nibble as the appropriate hex character. Hex can be converted back to binary using the reverse process.

```
0 0 1 0   1 1 1 1   1 0 1 1   1 1 1 0
    2         F         B         E
```

variety of reasons. Table 14.1 shows the format and content of the main MIDI message types. The letter 'n' is used to denote any channel number from 1–16, although these are actually encoded as 0–15 in the message. A short description of some of the most commonly encountered MIDI messages is given in the following sections, although there is not space here to go into detail on every type.

14.4.2 Channel and system messages contrasted

Two primary classes of message exist: those which relate to specific MIDI channels and those which relate to the system as a whole. Channel messages start with status bytes in the range &8n to &En (they start at hexadecimal eight because the MSB must be a one for a status byte). System messages all begin with &F, and do not contain a channel number. Instead the second part of the status byte (that which normally would contain the channel number) is used for further identification of the system message, allowing for sixteen possible system messages running from &FO to &FF.

System messages are themselves split into three groups: system common, system exclusive and system realtime. The common messages may apply to any device on the MIDI bus, depending only on the device's ability to handle the message. The exclusive messages apply to whichever manufacturer's devices are specified later in

Table 14.1 MIDI messages summarised

Message	Status	Data 1	Data 2
Note off	&8n	Note number	Velocity
Note on	&9n	Note number	Velocity
Polyphonic aftertouch	&An	Note number	Pressure
Control change	&Bn	Controller number	Data
Program change	&Cn	Program number	–
Channel aftertouch	&Dn	Pressure	–
Pitch wheel	&En	LSbyte	MSbyte
System exclusive			
System exclusive start	&F0	Manufacturer ID	Data, (Data), (Data)
End of SysEx	&F7	–	
System common			
Quarter frame	&F1	Data	–
Song pointer	&F2	LSbyte	MSbyte
Song select	&F3	Song number	–
Tune request	&F6	–	–
System realtime			
Timing clock	&F8	–	–
Start	&FA	–	–
Continue	&FB	–	–
Stop	&FC	–	–
Active sensing	&FE	–	–
Reset	&FF	–	–

the message (see below), and the realtime messages are intended for devices which are to be synchronised to the controller's musical tempo.

14.4.3 Note on and note off messages

Much of the musical information sent over a typical MIDI interface will consist of these two message types. As indicated by the titles, the note on message turns on a musical note, and the note off message turns it off. Note on takes the general format:

   [&8n] [Note number] [Velocity]

and note off takes the form:

   [&9n] [Note number] [Velocity]

A MIDI instrument will generate note on messages at its MIDI OUT corresponding to whatever notes are pressed on the keyboard, on whatever channel the instrument is set to transmit. Also, any note which has been turned on must subsequently be turned off in order for it to stop sounding, thus if one instrument receives a note on message from another and then loses the MIDI connection for any reason, the note will continue sounding ad infinitum. This situation can occur if a MIDI cable is pulled out during transmission.

Table 14.2 MIDI note numbers related to the musical scale

Musical note	MIDI note number	
C–2	0	
C–1	12	
C0	24	
C1	36	
C2	48	
C3 (middle C)	60	(Yamaha convention)
C4	72	
C5	84	
C6	96	
C7	108	
C8	120	
G8	127	

MIDI note numbers relate directly to the western musical chromatic scale, and the format of the message allows for 128 note numbers which cover a range of a little over ten octaves – adequate for the full range of most musical material. This quantisation of the pitch scale is geared very much towards keyboard instruments, and is perhaps less suitable for other instruments and cultures where the definition of pitches is not so black-and-white. None the less, means have been developed of adapting control to situations where unconventional tunings are required. Note numbers normally relate to the musical scale as shown in Table 14.2, although there is a lack of consistency here. Yamaha established the use of C3 for middle C, whereas others have used C4. Some software allows the user to decide which convention will be used for display purposes.

14.4.4 Velocity information

Note messages are associated with a velocity byte, and this is used to represent the speed with which a key was pressed or released. The former will correspond to the force exerted on the key as it is depressed: in other words, 'how hard you hit it' (called 'note on velocity'). It is applied to control parameters such as the volume or timbre of a note, and can be used to scale the effect of one or more of the envelope generators in a synthesiser. This velocity value has 128 possible states, but not all MIDI instruments are able to generate or interpret the velocity byte, in which case they will set it to a value half way between the limits, i.e.: 64_{10}. Some instruments may act on velocity information even if they are unable to generate it themselves. Note off velocity (or 'release velocity') is not widely used. None the less it is available for special effects if a manufacturer decides to implement it.

The note on, velocity zero value is reserved for the special purpose of turning a note off, for reasons which will become clear under 'Running status' below. If an instrument sees a note number with a velocity of zero, its software should interpret this as a note off message.

14.4.5 Running status

When a large amount of information is transmitted over a single MIDI interface, delays naturally arise due to the serial nature of transmission wherein data such as the concurrent notes of a chord must be sent one after the other. It will be advantageous, therefore, to reduce the amount of data transmitted as much as possible, in order to keep the delay as short as possible and to avoid overloading the devices on the bus with unnecessary data.

Running status is an accepted method of reducing the amount of data transmitted, and one which all MIDI software should understand. It involves the assumption that once a status byte has been asserted by a controller there is no need to reiterate this status for each subsequent message of that status, so long as the status has not changed in between. Thus a string of note on messages could be sent with the note on status only sent at the start of the series of note data, for example:

[&9n] [Data] [Velocity] [Data] [Velocity] [Data] [Velocity]

It will be appreciated that for a long string of note data this could reduce the amount of data sent by nearly one third. But as in most music each note on is almost always followed quickly by a note off for the same note number, this method would clearly break down, as the status would be changing from note on to note off very regularly, thus eliminating most of the advantage gained by running status. This is the reason for the adoption of note on, velocity zero as equivalent to a note off message, because it avoids a change of status during running status, allowing a string of what appear to be note on messages, but which represent both note on and note off.

Running status is not used at all times for a string of same-status messages, and will often only be called upon by an instrument's software when the rate of data exceeds a certain point. Indeed, an examination of the data from a typical synthesiser indicates that running status is not used during a large amount of ordinary playing.

14.4.6 Control change messages

As well as note information, a MIDI device may be capable of transmitting control information which corresponds to the various switches, control wheels and pedals associated with it. The controller messages have proliferated enormously since the early days of MIDI, and not all devices will implement all of them. It should be noted that there are two distinct kinds of controller: that is, the switch type, and the continuous type. The continuous controller is any variable wheel, lever, slider or pedal that might have one of a number of positions, whereas the switch normally only has two states.

Most of the continuous controllers in real instruments are only represented to 7 bit resolution, giving 2^7 or 128 possible steps over their range, but there is provision for a second message to be sent carrying a further seven bits to add finer control.

This is why there are two controller numbers for each 14 bit controller, as exemplified in the table, one for coarse control and one for fine control.

On/off switches can be represented easily in binary form (0 for OFF, 1 for ON), and it would be possible to use just a single bit for this purpose, but, in order to conform to the standard format of the message, switch states are normally represented by data values between &00 and &3F for OFF, and &40&7F for ON. In other words switches are now considered as 7 bit continuous controllers, and it may be possible on some instruments to define positions in between off and on in order to provide further degrees of control, such as used in some 'sustain' pedals (although this is not common in the majority of equipment). In older systems it may be found that only &00 = OFF and &7F = ON.

14.4.7 Channel modes

Although grouped with the controllers under the same status, the channel mode messages set the mode of operation of the instrument receiving on that particular channel.

'Local on/off' is used to make or break the link between an instrument's keyboard and its own sound generators. Effectively there is a switch between the output of the keyboard and the control input to the sound generators which allows the instrument to play its own sound generators in normal operation when the switch is closed (see Figure 14.5). If the switch is opened, the link is broken and the output from the keyboard feeds the MIDI OUT while the sound generators are controlled from the MIDI IN. In this mode the instrument acts as two separate devices: a keyboard without any sound, and a sound generator without a keyboard. This configuration can be useful when the instrument in use is the master keyboard for a large sequencer system, where it may not always be desired that everything played on the master keyboard results in sound from the instrument itself.

Figure 14.5 The Local Off switch disconnects a keyboard from its associated sound generators in order that the two parts may be treated independently in a MIDI system

'Omni off' ensures that the instrument will only act on data tagged with its own channel number(s), as set by the instrument's controls. 'Omni on' sets the instrument to receive on all of the MIDI channels. In other words, the instrument will ignore the channel number in the status byte and will attempt to act on any data that arrives, whatever its channel. Devices should power-up in this mode according to the original specification, but more recent devices will tend to power-up in the mode that they were left. Mono mode sets the instrument such that it will only reproduce one note at a time, as opposed to 'Poly' (phonic) in which a number of notes may be sounded together.

In poly mode the instrument will sound as many notes as it is able at the same time. Instruments differ as to the action taken when the number of simultaneous notes is exceeded: some will release the first note played in favour of the new note, whereas others will refuse to play the new note. The more intelligent of them may look to see if the same note already exists in the notes currently sounding, and only accept a new note if it is not already sounding. Even more intelligently, some devices may release the quietest note (that with the lowest velocity value), or the note furthest through its velocity envelope, to make way for a later arrival. It is also common to run a device in poly mode on more than one receive channel, provided that the software can handle the reception of multiple polyphonic channels. A multitimbral sound generator may well have this facility, commonly referred to as 'multi' mode, making it act as if it were a number of separate instruments each receiving on a separate channel. In multi mode a device may be able to dynamically assign its polyphony between the channels and voices in order that the user does not need to assign a fixed polyphony to each voice.

14.4.8 Program change

The program change message is used most commonly to change the 'patch' of an instrument or other device. A patch is a stored configuration of the device, describing the set-up of the tone generators in a synthesiser, and the way in which they are interconnected. It often corresponds to a preprogrammed voice on a synthesiser. Program change is channel-specific, and there is only a single data byte associated with it, specifying to which of 128 possible stored programs the receiving device should switch. On non-musical devices such as effects units, the program change message is often used to switch between different effects, and the different effects programs may be mapped to specific program change numbers.

14.4.9 System exclusive

A system exclusive message is one which is unique to a particular manufacturer, and often a particular instrument. The only thing that is defined about such messages is how they are to start and finish, with the exception of the use of system exclusive messages for standardised universal information, which makes use of the three highest numbered manufacturer IDs.

System exclusive messages generated by a device will naturally be produced at the MIDI OUT, not at the THRU, so a deliberate connection must be made between the transmitting device and the receiving device before data transfer may take place. Occasionally it is necessary to make a return link from the OUT of the receiver to the IN of the transmitter so that two-way communication is possible, and so that the receiver can control the flow of data to some extent by telling the transmitter when it is ready to receive and when it has received correctly (a form of handshaking).

The message takes the general form:

&[FO] [ident.] [data] [data] ...[F7]

where [ident.] identifies the relevant manufacturer ID, which is a number defining which manufacturer's message is to follow. Originally, manufacturer IDs were a single byte but the number of IDs has been extended by setting aside the [00] value of the ID to indicate that two further bytes of ID follow. Manufacturer IDs are therefore either one or three bytes long.

Data of virtually any sort can follow the ID. It can be used for a variety of miscellaneous purposes which have not been defined in the MIDI standard, and the message can have virtually any length that the manufacturer requires, although it is often split into packets of a manageable size in order not to cause receiver memory buffers to overflow. Exceptions are data bytes which look like other MIDI status bytes (except realtime messages), as they will naturally be interpreted as such by any receiver, which might terminate reception of the system exclusive message. The message should be terminated with &F7.

Examples of applications for such messages can be seen in the form of sample data dumps (from a sampler to a computer and back again for editing purposes), although this is painfully slow, and voice data dumps (from a synthesiser to a computer for storage and editing of user-programmed voices). There are now an enormous number of uses of system exclusive messages, both in the universal categories and in the manufacturer categories.

14.5 Synchronisation

MIDI information is also used to synchronise devices within a system. It depends on the type of device as to whether or not MIDI synchronisation data is recognised. Sequencers normally need a timing reference, either internal or external, to determine the rate at which music and other data are replayed. Drum machines also need timing information because a drum machine usually contains a sequencer which stores the patterns of rhythm which a player may lay down. If two or more sequencers or drum machines are connected together using MIDI cables it is possible to run them in synchronism with each other, stopping and starting in the same places, and autolocating to the same positions in the song. In such cases one device acts as the master and all the others as slaves. Wherever the master goes the slave follows.

A normal synthesiser, effects unit or sampler is not concerned with timing information, because it has no functions which would be affected by a timing clock. Such devices do not normally store rhythm patterns, although there are some keyboards with onboard sequencers which ought to handle timing data.

As MIDI has become more involved with the professional recording studio it has become possible for MIDI timing information to be locked to a pre-recorded time-code which may be derived from a video or audio recorder.

14.5.1 System Realtime

A group of system messages called the System Realtime messages (see Table 14.1)) control the execution of timed sequences in a MIDI system, and these are often used in conjunction with the Song Pointer (which is really a System Common message) for autolocation. The MIDI clock byte, a single status byte (&F8), is transmitted by the master device six times per MIDI beat. A MIDI beat is equivalent to a musical semi-quaver or sixteenth note. At any one musical tempo a MIDI beat could be said to represent a fixed amount of time, but this would change if the tempo changed. Any slaves lock to the tempo rate indicated by these clock bytes.

START (&FA), transmitted by the master, should result in the playback of the sequence from the very beginning. STOP (&FC) is used to halt the execution of a song which is running on a synchronised instrument. CONTINUE (&FB) is used to restart the execution from the point at which the sequence was stopped, not from the beginning.

14.5.2 Song position pointers (SPPs)

SPPs are used when the master device needs to tell slaves where it is in a song. For example, one might 'fast-forward' through a song and start again twenty bars later, in which case the other timed devices in the system would have to know where to restart. An SPP would be sent followed by 'continue' and then regular clocks.

The SPP represents the position in a stored song in terms of number of MIDI beats (not clocks) from the start of the song. Thus again it relates directly to musical time rather than real time, and does not necessarily represent the number of seconds into the song. It uses two data bytes, and the MSB of data bytes must be zero, so the resulting fourteen bits can specify up to 16 384 MIDI beats. SPP is often used in conjunction with &F3 (Song Select), which is used to define one of a collection of stored songs to be recalled.

SPPs are fine for directing the movements of an entirely musical system, in which every action is related to a particular beat or subdivision of a beat, but not so fine when actions must occur at a particular point in real time. If, for example, one was using a MIDI system to dub music and effects to a picture in which an effect was intended to be triggered at a particular visual occurrence, that effect would have to maintain its position in time no matter what happened to the music. If the effect was to be triggered by a sequencer at a particular number of beats from the beginning of the song, this point could change in real time if the tempo of the music was altered slightly to fit a particular visual scene. Clearly some means of real-time synchronisation is required either instead of, or as well as, the clock and song pointer arrangement, such that certain events in a MIDI-controlled system may be triggered at specific times in hours, minutes and seconds.

14.5.3 MIDI TimeCode (MTC)

MIDI Timecode (MTC) is used widely as a means of synchronising MIDI-controlled equipment to a real time reference. SMPTE/EBU longitudinal timecode (LTC) is described in Chapter 15, and MTC is a means of transferring LTC around a MIDI system.

In an LTC timecode frame two binary data groups are allocated to each of hours, minutes, seconds and frames, these groups representing the tens and units of each, so there are eight binary groups in total representing the time value of a frame. In order to transmit this information over MIDI, it has to be turned into a format which is compatible with other MIDI data (i.e. a status byte followed by relevant data bytes). There are two types of MTC synchronising message: one which updates a receiver regularly with running timecode, and another which transmits one-time updates of the timecode position for situations such as exist during the high-speed spooling of tape machines, where regular updating of each single frame would involve too great a rate of transmitted data. The former is known as a quarter-frame message (see Fact File 14.6), denoted by the status byte (&F1), whilst the latter is known as a full-frame message and is transmitted as a universal realtime system exclusive (SysEx) message, the details of which will not be given here.

14.6 Interfacing a computer to a MIDI system

In order to use a computer as a central controller for a MIDI system it must have at least one MIDI interface, consisting of at least an IN and an OUT port. (THRU is not strictly necessary in most cases.) Unless the computer has a built-in interface, as found on old Atari machines, some form of third-party hardware interface must be added, and there are many ranging from simple single ports to complex multiple port products.

14.6.1 Single port MIDI interfaces

A typical single port MIDI interface will be connected either to one of the spare I/O ports of the computer, or plugged into an expansion slot. On the Macintosh, for example, a MIDI interface is normally connected directly to one of the two serial ports as shown in Figure 14.6. Provided that electrical conditions were properly controlled it would be possible to connect a number of receiving devices to a single port, either in a chain or using a splitter box or router. Some examples of simple setups are shown in Figure 14.7. As shown in one of these figures, some sound generators also double as a serial MIDI interface for the host computer, avoiding the need for a separate box.

318   MIDI

Figure 14.6 A spare general purpose serial interface, such as the Macintosh modem port, can be used to connect an external MIDI interface

---

FACT FILE

14.6

Quarter-frame MTC messages

One timecode frame is represented by too much information to be sent in a standard three byte MIDI message, so it is broken down into eight separate messages called quarter-frame messages. Four of these quarter-frame messages are transmitted in the period of one timecode frame (in order to limit the message rate), so it takes two frames to transmit a complete frame value and the receiver is updated every two frames. Receivers therefore need to maintain a two frame offset between their displayed timecode and the last decoded value, because by the time a frame value has been completely transmitted two frames will have elapsed. Internally, the timing resolution of software can be made higher than that of the timecode messages, with the messages being used as a reference point every so often.

Each message of the group of eight represents a part of the timecode frame value, as shown in the diagram, and takes the general form:

&[F1] [DATA]

The data byte begins with zero (as always), and the next seven bits of the data word are made up of a 3 bit code defining whether the

message represents hours, minutes, seconds or frames, MSnibble or LSnibble, followed by the four bits representing the binary value of that nibble. In order to reassemble the correct timecode value from the eight quarter-frame messages, the LS and MS nibbles of hours, minutes, seconds and frames are each paired within the receiver to form 8 bit words as follows:

Frames: rrr qqqqq

where rrr is reserved for future use and qqqqq represents the frames value from 0 to 29;

Seconds: rr qqqqqq

where rr is reserved for future use and qqqqqq represents the seconds value from 0 to 59;

Minutes: rr qqqqqq

as for seconds;

Hours. r qq ppppp

where r is undefined, qq represents the timecode type (see below), and ppppp is the hours value from 0 to 23.
The timecode frame rate is denoted as follows in the hours count:

00 = 24 fps
01 = 25 fps
10 = 30 fps drop-frame
11 = 30 fps non-drop-frame

Unassigned bits should be set to zero.

Figure 14.7 (a) Simple single port system with MIDI IF integrated with computer. (b) Using sound module as a MIDI IF. (c) Rear panel of Yamaha TG100, which will act as a MIDI IF for a computer, showing 'TO HOST' port for serial connection to computer

The limitation of single port MIDI interfaces is that they can only address the basic sixteen channels of the MIDI protocol, and so one can run short of channels when controlling large systems with many voices and devices.

14.6.2 Multiport interfaces

Multiport interfaces have a number of independent MIDI OUT ports. Such an interface is connected to the host computer using either a parallel or serial I/O port, or using an expansion card, as shown in Figure 14.8. Most multiport systems allow up to 16 channels to be addressed by each of the MIDI OUTs, thus expanding

Figure 14.8 A multiport MIDI interface allows interconnection via a large number of independent MIDI ports

Figure 14.9 (a) Example of an intermediate-level computer-controlled MIDI system for music production, using a 4 port MIDI interface. All devices can both receive and send MIDI information from and to the computer

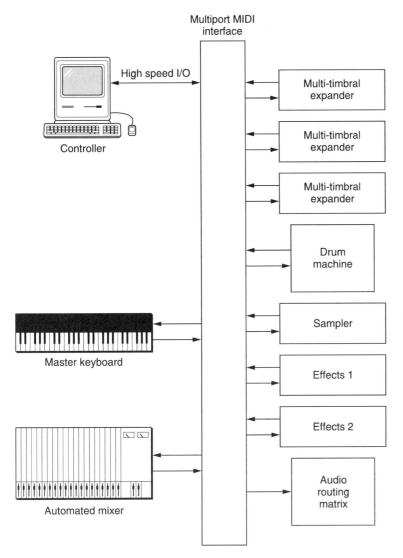

Figure 14.9 (b) Example of a large computer-controlled MIDI system, incorporating MIDI-controlled audio effects and an automated mixer. The audio routing is not shown here

the total number of channels addressable to 16 times the number of ports. In such systems it is common for each instrument to be connected to its own port, both IN and OUT, which is especially useful with multi-timbral sound modules capable of operating on all 16 channels simultaneously. A number of possibilities are illustrated in Figure 14.9. It also allows any instrument to be used as a controller, and makes it possible for instruments to send system exclusive information back to the computer without replugging. Multiport interfaces will often allow data received from more than one port to be merged, or allow recording from more than one source at a time.

The more advanced multiport interfaces may also incorporate a timecode reader and generator, allowing for the synchronisation of video and audio tape recorders with the MIDI system. The conversion of SMPTE/EBU timecode to MIDI time-code or an equivalent then takes place within the interface before being transferred to the host computer. MIDI Machine Control (MMC), as described below, can be used to control the tape machine from a MIDI port. Some examples of system configurations are shown in Figure 14.10.

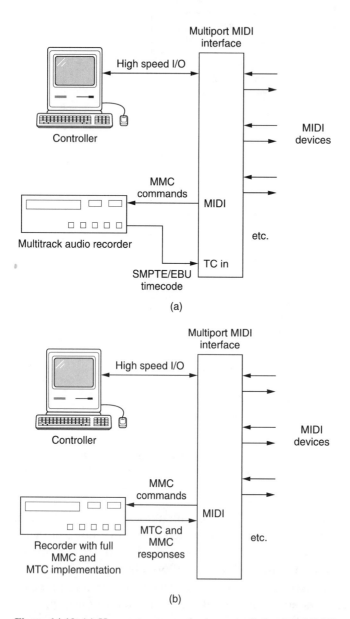

(a)

(b)

Figure 14.10 (a) Here a tape recorder is controlled using MMC commands from the sequencer, with positional information being fed back to the sequencer via the SMPTE/EBU timecode reader in the MIDI interface. (b) A tape recorder with a built-in timecode reader and a full MMC/MTC implementation does not require a timecode reader in the MIDI interface, because positional information is returned to the sequencer in the form of MTC messages

14.6.3 Interface driver software

Quite commonly, a software driver is provided for an external MIDI interface or expansion card. This driver is used by the computer's operating system to address the MIDI interface concerned, and it takes care of handling the various routines necessary to carry data to and from the physical I/O ports. The MIDI application therefore simply talks to the driver, and it follows that you must have the correct driver installed for the interface which you intend to use. High-end MIDI software usually comes complete with a number of drivers and configuration documents for the most popular MIDI interfaces.

14.7 An overview of software for MIDI

Sequencers are probably the most ubiquitous of the available MIDI software packages. A sequencer will be capable of storing a number of 'tracks' of MIDI information, editing it and otherwise manipulating it for musical composition purposes. It is also capable of storing MIDI events for non-musical purposes such as studio automation, and may be equipped with digital audio recording capabilities in some cases. Some of the more advanced packages are available in modular form (allowing the user to buy only the functional blocks required) and in cut-down or 'entry-level' versions for the new user.

The dividing line between sequencer and music notation software is a grey one, since there are features common to each. Music notation software is designed to allow the user control over the detailed appearance of the printed musical page, rather as desktop publishing packages work for typesetters, and such software often provides facilities for MIDI input and output. MIDI input is used for entering note pitches during setting, whilst output is used for playing the finished score in an audible form. Most major packages will read and write standard MIDI files (see Fact File 14.7), and can therefore exchange data with sequencers, allowing sequenced music to be exported to a notation package for fine tuning of printed appearance. It is also common for sequencer packages to offer varying degrees of music notation capability, although the scores which result are rarely as professional in appearance as those produced by dedicated notation software.

Librarian and editor software is used for managing large amounts of voice data for MIDI-controlled instruments. Such packages communicate with MIDI instruments using system exclusive messages in order to exchange parameters relating to voice programs. The software may then allow these voice programs or 'patches' to be modified using an editor, offering a rather better graphical interface than those usually found on the front panels of most sound modules. Banks of patches may be stored on disk by the librarian, in order that libraries of sounds can be managed easily, and this is often cheaper than storing patches in the various 'memory cards' offered by synth manufacturers. Banks of patch information may be accessed by sequencer software in order that the operator may choose voices for particular tracks by name, rather than by program change numbers. Sample editors are also available, offering similar facilities, although sample dumps using system exclusive

FACT FILE

14.7

Standard MIDI files

Sequencers and notation packages typically store data on disk in their own unique file formats. Occasionally it is possible for a file from one package to be read by others, especially if the packages are from the same manufacturer, but this is rare. The standard MIDI file was developed in an attempt to make interchange of information between packages more straightforward, and is now used widely in the industry in addition to manufacturers' own file formats.

Three types of standard MIDI file exist to encourage the interchange of sequencer data between software packages. The MIDI file contains data representing events on individual sequencer tracks, as well as containing labels such as track names, instrument names and time signatures. It is the intention that not only should software packages running on the same computer type be able to read these universal files, but that such files may be transferred to other computers (either over a network, or on a disk that can be read by the other computer), so that a package running under a completely different operating system may read the data.

File type 0 is the simplest, and is used for single-track data, whilst file type 1 supports multiple tracks which are 'vertically' synchronous with each other (such as the parts of a song), and file type 2 contains multiple tracks which have no direct timing relationship and may thus be asynchronous. Type 2 could be used for transferring song files which are made up of a number of discrete sequences, each with a multiple track structure.

are not really recommended, unless they are short, since the time taken can be excessive. Sample data can be transferred to a computer using a faster interface than MIDI (such as SCSI) and the sample waveforms can be edited graphically.

Amongst other miscellaneous software packages available for the computer are MIDI mixer automation systems, guitar sequencers, MIDI file players, multimedia authoring systems and alternative user interfaces. Development software is also available for MIDI programmers, providing a programming environment for the writing of new MIDI software applications. There are also a number of software applications designed principally for research purposes or for experimental music composition.

14.8 General MIDI

If MIDI files are to be exchanged between systems and replayed on different sound generating hardware, one needs a means of ensuring that the music will sound at least similar on the two systems. One of the problems with MIDI-controlled sound generators used to be that although voice programs could be selected using MIDI program change commands, there was no standard mapping of voices to program numbers. In other words, program change 3 might correspond to 'alto sax' on one instrument and 'grand piano' on another. Consequently a music sequence could sound completely different when replayed on two different multitimbral sound generators. General MIDI was introduced to standardise some basic elements of sound generator control, so that MIDI files could be exchanged more easily between systems, ensuring that an approximation to the sounds and instruments intended by the composer would be heard no matter what General MIDI sound generator was used for replay. Currently, General MIDI is specified at Level 1, although there are proposals to extend the concept to further levels.

General MIDI specifies a number of other things as well as standard sounds. For example, it specifies a minimum degree of polyphony, and requires that a sound generator should be able to receive MIDI data on all 16 channels simultaneously and polyphonically, with a different voice on each channel. There is also a requirement that the sound generator should support percussion sounds in the form of drum kits, so that a General MIDI sound module is capable of acting as a complete 'band in a box'. Some multimedia computers now incorporate General MIDI synthesisers as part of the sound hardware, and have operating systems capable of addressing MIDI instruments.

Dynamic voice allocation is the norm in GM sound modules, with a requirement either for at least 24 dynamically allocated voices in total, or 16 for melody and 8 for percussion. In order to ensure compatibility between sequences that are replayed on GM modules, percussion sounds are always allocated to MIDI channel 10. Program change numbers are mapped to specific voice names, with ranges of numbers allocated to certain types of sounds, as shown in Table 14.3. Precise voice names may be found in the GM documentation (see the end of the chapter). Channel 10, the percussion channel, has a defined set of note numbers on which particular sounds are to occur, so that the composer may know for example that key 39 will always be a 'hand clap'.

Table 14.3 General MIDI program number ranges (except channel 10)

Program change (decimal)	Sound type
0–7	Piano
8–15	Chromatic percussion
16–23	Organ
24–31	Guitar
32–39	Bass
40–47	Strings
48–55	Ensemble
56–63	Brass
64–71	Reed
72–79	Pipe
80–87	Synth lead
88–95	Synth pad
96–103	Synth effects
104–111	Ethnic
112–119	Percussive
121–128	Sound effects

14.9 MIDI Machine Control (MMC)

MIDI may be used for remotely controlling tape machines and other studio equipment. MMC uses universal realtime SysEx messages, and has a lot in common with a remote control protocol known as 'ESbus' which was devised by the EBU and SMPTE as a universal standard for the remote control of tape machines, VTRs and other studio equipment.

There are a number of levels of complexity at which MMC can be made to operate, and communication is possible in both closed and open loop modes (either with or without a return connection to the controlling computer). By allowing this flexibility, MMC makes it possible for designers to implement it at anything from a very simple level (i.e. cheaply) to a very complicated level involving all the finer points. MMC is gaining increasing popularity in semi-professional equipment, because it is somewhat cheaper to implement than ESbus and allows equipment to be integrated easily with a MIDI-based studio system. There are a number of tape machines and synchronisers now on the market with MIDI interfaces, and some sequencer packages handle the remote control of studio machines using the MMC protocol. Studio machines may be connected to the main studio computer by connecting them to one port on a multiport MIDI interface, as was shown in Figure 14.10.

MMC could be used simply to control the basic transport functions of an audio tape recorder. In such a case only a very limited set of commands would need to be implemented in the tape recorder, and very little would be needed in the way of responses. In fact it would be quite feasible to operate the transport of a tape recorder using an open-loop approach – simply sending 'play', 'stop', 'rewind', etc. as required by the controlling application.

Recommended further reading

MMA (1983) *MIDI 1.0 Detailed Specification*. MIDI Manufacturers Association
MMA (1991) *General MIDI System Level* 1. MIDI Manufacturers Association
MMA (1993) *4.2 Addendum to MIDI 1.0 Specification*. MIDI Manufacturers Association
Rumsey, F. (1994) *MIDI Systems and Control*, 2nd edn., Focal Press
Yavelow, C. (1992) *Macworld Music and Sound Bible*. IDG Books Worldwide, Inc., San Mateo, CA, USA

Chapter 15

Timecode and synchronisation

The boundaries between audio and video operations are less clear these days, and subjects such as timecode which used to be almost universally the domain of the video engineer are now as pertinent to the audio engineer. Timecode is used widely in the audio post-production industry for synchronising machines and providing a real-time positional reference on tapes. It is used in video editing and in the editing of digital audio recordings, and it is used in hard-disk recording systems for compiling edit lists and for synchronisation. Many modern analogue tape recorders have timecode facilities as do professional digital recorders, some even being equipped with 'chase' synchronisers.

In the following chapter the basics of timecode and machine synchronisation are discussed, but omitting discussion of the many systems which have been used in the past (and are still used in certain cases) for the synchronisation of film systems. MIDI Timecode (MTC) is discussed in Chapter 14.

15.1 SMPTE/EBU timecode

The American Society of Motion Picture and Television Engineers proposed a system to facilitate the accurate editing of video tape in 1967. This became known as SMPTE ('simpty') code, and it is basically a continuously running eight-digit clock registering time from an arbitrary start point (which may be the time of day) in hours, minutes, seconds and frames, against which the programme runs. The clock information is encoded into a signal which can be recorded on the audio track of a tape. Every single frame on a particular video tape has its own unique number called the timecode address and this can be used to pinpoint a precise editing position.

A number of frame rates are used, depending on the television standard to which they relate, the frame rate being the number of still frames per second used to give the impression of continuous motion: 30 frames per second (fps), or true SMPTE, was used for monochrome American television, and is now only used for CD mastering in the Sony 1630 format; 29.97 fps is used for colour NTSC television (mainly USA, Japan and parts of the Middle East), and is called 'SMPTE drop-

327

frame' (see Fact File 15.1); 25 fps is used for PAL and SECAM TV and is called 'EBU' (Europe, Australia, etc.); and 24 fps is used for some film work.

Each timecode frame is represented by an 80 bit binary 'word', split principally into groups of 4 bits, with each 4 bits representing a particular parameter such as tens of hours, units of hours, and so forth, in BCD (binary-coded decimal) form (see Figure 15.1). Sometimes, not all four bits per group are required – the hours only go up to '23', for example – and in these cases the remaining bits are either used for special control purposes or set to zero (unassigned): 26 bits in total are used for time address information to give each frame its unique hours, minutes, seconds, frame value; 32 are 'user bits' and can be used for encoding information such as reel number, scene number, day of the month and the like; bit 10 can denote drop-frame mode if a binary 1 is encoded there, and bit 11 can denote colour frame mode if a binary 1 is encoded. The end of each word consists of 16 bits in a unique sequence, called the 'sync word', and this is used to mark the boundary between one frame and the next. It also allows the reader to tell in which direction the code is being read, since the sync word begins with 11 in one direction and 10 in the other.

This binary information cannot be recorded to tape directly, since its bandwidth would be too wide, so it is modulated in a simple scheme known as 'bi-phase mark', or FM, such that a transition from one state to the other (low to high or high to low) occurs at the edge of each bit period, but an additional transition is forced within the period to denote a binary 1 (see Figure 15.2). The result looks like a square wave with two frequencies, depending on the presence of ones and zeros in the code. Depending on the frame rate, the maximum frequency of square wave contained within the timecode signal is either 2400 Hz (80 bits × 30 fps) or 2000 Hz (80 bits × 25 fps), and the lowest frequency is either 1200 Hz or 1000 Hz, and thus

FACT FILE 15.1   Drop-frame timecode

When colour TV (NTSC standard) was introduced in the USA it proved necessary to change the frame rate of TV broadcasts slightly in order to accommodate the colour information within the same spectrum. The 30 fps of monochrome TV, originally chosen so as to lock to the American mains frequency of 60 Hz, was thus changed to 29.97 fps, since there was no longer a need to maintain synchronism with the mains owing to improvements in oscillator stability. In order that 30 fps timecode could be made synchronous with the new frame rate it became necessary to drop two frames every minute, except for every tenth minute, which resulted in minimal long-term drift between timecode and picture (75 ms over 24 hours). The drift in the short term gradually increased towards the minute boundaries and was then reset.

A flag is set in the timecode word to denote NTSC drop-frame timecode. This type of code should be used for all applications where the recording might be expected to lock to an NTSC video programme.

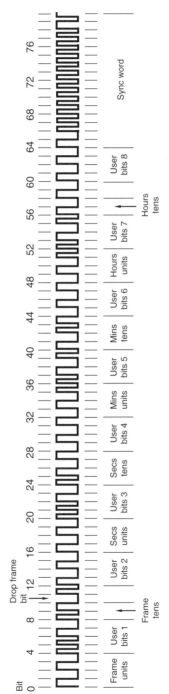

Figure 15.1 The data format of a SMPTE/EBU longitudinal timecode frame

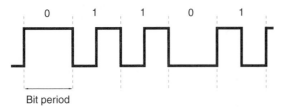

Figure 15.2 Linear timecode data is modulated before recording using a scheme known as 'bi-phase mark' or FM (frequency modulation). A transition from high to low or low to high occurs at every bit-cell boundary, and a binary '1' is represented by an additional transition within a bit cell

it may easily be recorded on an audio machine. The code can be read forwards or backwards, and phase inverted. Readers are available which will read timecode over a very wide range of speeds, from around 0.1 to 200 times play speed. The rise-time of the signal, that is the time it takes to swing between its two extremes, is specified as 25 µs 5 ± µs, and this requires an audio bandwidth of about 10 kHz.

There is another form of timecode known as VITC (Vertical Interval Timecode), used widely in VTRs. VITC is recorded not on an audio track, but in the vertical sync period of a video picture, such that it can always be read when video is capable of being read, such as in slow-motion and pause modes. This code will not be covered further here.

15.2 Recording timecode

Timecode may be recorded or 'striped' on to tape before, during or after the programme material is recorded, depending on the application. In many cases the timecode must be locked to the same speed reference as that used to lock the speed of the tape machine, otherwise a long-term drift can build up between the passage of time on the tape and the measured passage in terms of timecode. Such a reference is usually provided in the form of a video composite sync signal, and video sync inputs are increasingly provided on digital tape recorders for this purpose.

Timecode generators are available in a number of forms, either as stand-alone devices (such as that pictured in Figure 15.3), as part of a synchroniser or editor, or integrally within a tape recorder. In large centres timecode is sometimes centrally distributed and available on a jackfield point. When generated externally, timecode normally appears as an audio signal on an XLR connector or jack, and this should be routed to the track required for timecode on the tape recorder. Most generators allow the user to preset the start time and the frame-rate standard.

Timecode is often recorded on to an outside track of a multitrack tape machine (usually track 24), or a separate timecode or cue track will be provided on digital machines. The signal is recorded at around 10 dB below reference level, and crosstalk between tracks or cables is often a problem due to the very audible mid-frequency nature of timecode. Some quarter-inch analogue machines have a facility for recording timecode in a track which runs down the centre of the guard

Figure 15.3 A stand-alone timecode generator. (Courtesy of Avitel Electronics Ltd)

band in the NAB track format (see section 8.4.1). This is called 'centre-track timecode', and a head arrangement similar to that shown in Figure 15.4 may be used for recording and replay. Normally separate heads are used for recording timecode to those for audio, to avoid crosstalk, although some manufacturers seem to have circumvented this problem and use the same heads. In the former case a delay line is used to synchronise timecode and audio on the tape.

Professional R-DAT machines are often capable of recording timecode, this being converted internally into a DAT running-time code which is recorded in the subcode area of the digital recording. On replay, any frame rate of timecode can be derived, no matter what was used during recording, which is useful in mixed-standard environments.

In mobile film and video work which often employs separate machines for recording sound and picture it is necessary to stripe timecode on both the camera's tape or film and on the audio tape. This can be done by using the same timecode generator to feed both machines, but more usually each machine will carry its own generator and the clocks will be synchronised at the beginning of each day's shooting, both reading absolute time of day. Highly stable crystal control ensures that sync between the clocks will be maintained throughout the day, and it does not then matter whether the two (or more) machines are run at different times or for different lengths of time because each frame has a unique time of day address code which enables successful post-production syncing.

The code should run for around 20 seconds or more before the programme begins in order to give other machines and computers time to lock in. If programme is spread over several reels, the timecode generator should be set and run such that no number repeats itself anywhere throughout the reels, thus avoiding confusion during post-production. Alternatively the reels can be separately numbered.

a)

b)

A = 0.8 mm, B = 2.1 mm, C = 2.0 mm, D = 6.3 mm

Figure 15.4 The centre-track timecode format on quarter-inch tape. (a) Delays are used to record and replay a timecode track in the guard band using separate heads. (Alternatively, specially-engineered combination heads may be used.) (b) Physical dimensions of the centre-track timecode format

15.3 Synchronisers

15.3.1 Overview

A synchroniser is a device which reads timecode from two or more machines and controls the speeds of 'slave' machines so that their timecodes run at the same rate as the 'master' machine. It does so by modifying the capstan speed of the slave machines, using an externally applied speed reference signal, usually in the form of a 19.2 kHz square wave whose frequency is used as a reference in the capstan servo circuit (see Figure 15.5). The synchroniser is microprocessor controlled, and can incorporate offsets between the master and slave machines, programmed by the user. It may also be able to store pre-programmed points for such functions as record drop-in, drop-out, looping and autolocation, for use in post-production.

15.3.2 Chase synchroniser

A simple chase synchroniser could simply be a box with a timecode input for master and slave machines and a remote control interface for each machine (see Figure

Figure 15.5 Capstan speed control is often effected using a servo circuit similar to this one. The frequency of a square wave pulse generated by the capstan tachometer is compared with an externally generated pulse of nominally the same frequency. A signal based on the difference between the two is used to drive the capstan motor faster or slower

15.6). Such a synchroniser is designed to cause the slave to follow the master wherever it goes, like a faithful hound. If the master goes into fast forward so does the slave, the synchroniser keeping the position of the slave as close as possible to the master, and when the master goes back into play the synchroniser parks the slave as close as possible to the master position and then drops it into play, adjusting the capstan speed to lock the two together. A full-featured chase synchroniser is pictured in Figure 15.7.

In fast wind modes, a chase synchroniser will tend not to read timecode, since the tape is not normally in contact with the heads and the timecode reader may not be able to read code at wind speeds, so it reads tachometer pulses from the tape machine's roller guide, transferred over the remote interface. The synchroniser will be programmed so as to count the correct number of tach pulses per second for each machine (they tend to differ considerably) or it may be able to work this out automatically during the first few seconds of operation. When the machine goes back into play it reads timecode again and adjusts its estimation of its position, which should be fairly close to that worked out from the tach pulses. The synchro-

Figure 15.6 A simple chase synchroniser will read timecode, direction and tachometer information from the master, compare it with the slave's position and control the slave accordingly until the two timecodes are identical (plus or minus any entered offset)

Figure 15.7 A modular chase synchroniser with serial bus control facilities. (Courtesy of Audio Kinetics UK Ltd)

niser then uses the difference between the master and slave timecode values, plus or minus any offset, to speed up or slow down the slave in order to lock it closely to the master.

Such a synchroniser could be used to lock two multitrack recorders together, for example in order to increase the number of available tracks, or it could be used to slave a quarter-inch machine to a VTR for laying off or laying back stereo sound tracks in video editing. It should act as an almost invisible link between the machines and should require little attention. The initiation for chasing should not need to come from the user; the slave should start to move as soon as it sees timecode move from the master. Some chase synchronisers of this sort will even work if no remote connection is made to the master, simply chasing the timecode presented to its input (which could have come from anywhere). Systems vary as to what they will do if the master timecode drops out or jumps in time. In the former case most synchronisers wait a couple of seconds or so before stopping the slave, and in the latter case they may try to locate the slave to the new position (this depends on the type of lock employed, as discussed in Fact File 15.2).

Occasionally a machine may be fitted with an internal chase synchroniser which locks to a timecode input on the rear of the machine. It may also have a built-in timecode generator.

15.3.3 Full-featured synchroniser

In post-production operations a controller is often required which offers more facilities than the simple chase synchroniser, such as the example pictured in Figure 15.8. Such a device may allow for multiple machines to be controlled from a single controller, perhaps using a computer network link to communicate commands from the controller to the individual tape machines. In some 'distributed intelligence' systems, each tape machine has a local chase synchroniser which communicates with the controller, the controller not being a synchroniser but a 'command centre' (see Figure 15.9). The ESbus is a remote control bus used increasingly in such applications, designed to act as a remote control bus for audio and video equipment.

The sync controller in such a system will offer facilities for storing full edit decision lists (EDLs) containing the necessary offsets for each slave machine and

FACT FILE 15.2 Types of lock

Frame lock or absolute lock
This term or a similar term is used to describe the mode in which a synchroniser works on the absolute time values of master and slave codes. If the master jumps in time, due to a discontinuous edit for example, then so does the slave, often causing the slave to spool off the end of the reel if it does not have such a value on the tape.

Phase lock or sync lock
These terms are often used to describe a mode in which the synchroniser initially locks to the absolute value of the timecode on master and slaves, switching thereafter to a

mode in which it simply locks to the frame edges of all machines, looking at the sync word in the timecode and ignoring the absolute value. This is useful if discontinuities in the timecode track are known or anticipated, and ensures that a machine will not suddenly drop into a fast spool mode during a programme.

Slow and fast relock
After initial lock is established, a synchroniser may lose lock due to a timecode drop-out or discontinuity in timecode phase. In fast relock mode the synchroniser will attempt to relock the machines as quickly as possible, with no concern for the audible effects of pitch slewing. In slow relock mode, the machines will relock more slowly at a rate intended to be inaudible.

Figure 15.8 A full-featured sync controller: the Audio Kinetics *Eclipse*. (Courtesy of Audio Kinetics UK Ltd)

FACT FILE
15.3

Synchroniser terminology

Pre-roll
The period prior to the required lock point, during which machines play and are synchronised. Typically machines park about 5 seconds before the required lock point and then pre-roll for 5 seconds, after which it is likely that the synchroniser will have done its job. It is rare not to be able to lock machines in 5 seconds, and often it can be faster.

Post-roll
The period after a programmed record drop-out point during which machines continue to play in synchronised fashion.

Loop
A programmed section of tape which is played repeatedly under automatic control, including a pre-roll to lock the machines before each pass over the loop.

Drop-in and drop-out
Points at which the controller or synchroniser executes a pre-programmed record drop-in or drop-out on a selected slave machine. This may be at the start and end of a loop.

Oset
A programmed timecode value which offsets the position of a slave with relation to the master, in order that they lock at an offset. Often each slave may have a separate offset.

Nudge
Occasionally it is possible to nudge a slave's position frame by frame with relation to the master once it has gained lock. This allows for small adjustments to be made in the relative positions of the two machines.

Bit oset
Some synchronisers allow for offsets of less than one frame, with resolution down to one-eightieth of a frame (one timecode bit).

---

the record drop-in and drop-out points for each machine. This can be used for jobs such as automatic dialogue replacement (ADR), in which sections of a programme can be set to loop with a pre-roll (see Fact File 15.3) and drop-in at the point where dialogue on a film or video production is to be replaced. A multitrack recorder may be used as a slave, being dropped in on particular tracks to build up a sound master tape. Music and effects can then be overdubbed.

In locked systems involving video equipment the master machine is normally the video machine, and the slaves are audio machines. This is because it is easier to synchronise audio machines, and because video machines may need to be locked to a separate video reference which dictates their running speed. In cases involving multiple video or digital audio machines, none of the machines is designated the master, and all machines slave to the synchroniser which acts as the master. Its timecode generator is locked to the house video or audio reference, and all machines lock to its timecode generator. This technique is also used in video editing systems.

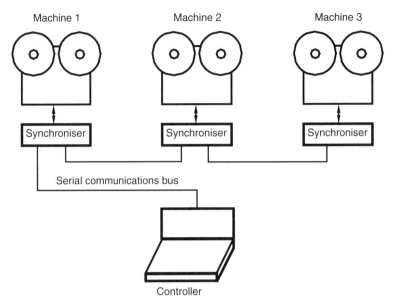

Figure 15.9 In an advanced synchronised system each machine is equipped with its own modular synchroniser, receiving commands and offsets from a serially-linked controller

Recommended further reading

Amyes, T. (1990) *The Technique of Audio Post-Production in Video and Film*. Focal Press
Ratcliff, J. (1995) *Timecode: A User's Guide*. Focal Press

Chapter 16

Two-channel stereo

This chapter covers the principles and practice of two-channel stereophonic recording and reproduction. Two-channel stereophonic reproduction (in international standard terms '2-0 stereo', meaning two front channels and no surround channels) is often called simply 'stereo' as it is the most common way of conveying some spatial content in sound recording and reproduction. In fact 'stereophony' refers to any sound system that conveys three-dimensional sound images, so it is used more generically in this book and includes surround sound. In international standards describing stereo loudspeaker configurations the nomenclature for the configuration is often in the form 'n–m stereo', where n is the number of front channels and m is the number of rear or side channels (the latter only being encountered in surround systems). This distinction can be helpful as it reinforces the slightly different role of the surround channels as explained in the next chapter. (Readers familiar with earlier editions of this book should note that the broadcasting nomenclature of 'A' and 'B', referring to the left and right signals of a stereo pair, has been replaced in this edition by 'L' and 'R'. This is in order to avoid any possible confusion with the American tradition of referring to spaced microphone pairs as 'AB pairs', as contrasted with 'XY' for coincident microphones.)

It might reasonably be supposed that the best stereo sound system would be that which reproduced the sound signal to the ears as faithfully as possible, with all the original spatial cues intact (see Chapter 2). Possibly that should be the aim, and indeed it is the aim of the so-called 'binaural' techniques discussed later in the chapter, but there are many stereo techniques that rely on loudspeakers for reproduction which only manage to provide *some* of the spatial cues to the ears. Such techniques are compromises that have varying degrees of success, and they are necessary for the simple reason that they are reasonably straightforward from a recording point of view and result in subjectively high sound quality. The results can be reproduced in anyone's living room and are demonstrably better than mono (single-channel reproduction). Theoretical correctness is one thing, pragmatism and getting a 'commercial sound' is another. The history of stereo could be characterised as being something of a compromise between the two.

Stereo techniques cannot be considered from a purely theoretical point of view, neither can the theory be ignored, the key being in a proper synthesis of theory and subjective assessment. Some techniques which have been judged subjectively to be good do not always stand up to rigorous theoretical analysis, and those which are held up as theoretically 'correct' are sometimes judged subjectively to be poorer than others. Part of the problem is that the mechanisms of spatial perception are not yet entirely understood. Probably more importantly, most commercial stereo reproduction uses only two loudspeakers so the listening situation already represents a serious departure from natural spatial hearing. (Real sonic experience involves sound arriving from all around the head.) The differences between two-channel stereo reproduction and natural listening may lead listeners to prefer 'distorted' sound fields because of other pleasing artefacts such as 'spaciousness'. Most of the stereo techniques used today combine aspects of imaging accuracy with an attempt to give the impression of spaciousness in the sound field, and to some theorists these two are almost mutually exclusive.

It would be reasonable to surmise that in most practical circumstances, for mainstream consumer applications, one is dealing with the business of creating believable illusions. Sound recording is as much an art as a science. In other words, one needs to create the *impression* of natural spaces, source positions, depth, size and so on, without necessarily being able to replicate the exact sound pressure and velocity vectors that would be needed at each listening position to recreate a sound field accurately. One must remember that listeners rarely sit in the optimum listening position, and often like to move around while listening. While it may be possible to achieve greater spatial accuracy using headphone reproduction, headphones are not always a practical or desirable form of monitoring. Truly accurate soundfield reconstruction covering a wide listening area can only be achieved by using very large numbers of loudspeakers (many thousands) and this is likely to be impractical for most current purposes.

In the following chapters stereo pickup and reproduction is considered from both a theoretical and a practical point of view, recognising that theoretical rules may have to be bent or broken for operational and subjective reasons. Since the subject is far too large even to be summarised in the short space available, a list of recommended further reading is given at the end of the chapter to allow the reader greater scope for personal study.

16.1 Principles of loudspeaker stereo

16.1.1 Historical development

We have become used to stereo sound as a two-channel format, although a review of developments during the last century shows that two channels really only became the norm through economic and domestic necessity, and through the practical considerations of encoding directional sound easily for gramophone records and radio. A two-loudspeaker arrangement is practical in the domestic environment, is reasonably cheap to implement, and provides good phantom images for a central listening position.

Early work on directional reproduction undertaken at Bell Labs in the 1930s involved attempts to recreate the 'sound wavefront' which would result from an infinite number of microphone/loudspeaker channels by using a smaller number of channels, as shown in Figure 16.1(a) and (b). In all cases, spaced pressure response (omnidirectional) microphones were used, each connected via a single amplifier to the appropriate loudspeaker in the listening room. Steinberg and Snow found that when reducing the number of channels from three to two, central sources appeared

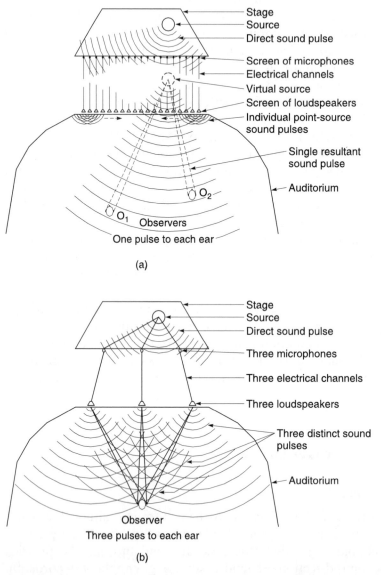

Figure 16.1 Steinberg and Snow's attempt to reduce the number of channels needed to convey a source wavefront to a reproduction environment with appropriate spatial features intact. (a) 'Ideal' arrangement involving a large number of transducers. (b) Compromise arrangement involving only three channels, replying more on the precedence effect

to recede towards the rear of the sound stage and that the width of the reproduced sound stage appeared to be increased. They attempted to make some calculated rather than measured deductions about the way that loudness differences between the channels affected directional perception, apparently choosing to ignore the effects of time or phase difference between channels.

Some twenty years later Snow made comment on those early results, reconsidering the effects of time difference in a system with a small number of channels, since, as he pointed out, there was in fact a marked difference between the multiple-point-source configuration and the small-number-of-channels configuration. It was suggested that in fact the 'ideal' multi-source system re-created the original wavefront very accurately, allowing the ears to use exactly the same binaural perception mechanisms as used in the real-life sound field. The 'wall' of multiple loudspeakers acted as a source of spherical wavelets, re-creating a new plane wave with its virtual source in the same relative place as the original source, thus resulting in a time-of-arrival difference between the listener's ears in the range 0–600 µs, depending on source and listener position. (This is the basis of more recent developments in 'wave field synthesis', developed at the University of Delft in the Netherlands, that also relies on large numbers of closely spaced channels to reconstruct sound fields accurately.)

In the two- or three-channel system, far from this simply being a sparse approximation to the 'wavefront' system, the ears are subjected to two or three discrete arrivals of sound, the delays between which are likely to be in excess of those normally experienced in binaural listening. In this case, the effect of directionality relies much more on the precedence effect and on the relative levels of the channels. Snow therefore begs us to remember the fundamental difference between 'binaural' situations and what he calls 'stereophonic' situations (see Fact File 16.1).

This difference was also recognised by Alan Blumlein, whose now-famous patent specification of 1931 (accepted 1933) allows for the conversion of signals from a binaural format suitable for spaced pressure microphones to a format suitable for reproduction on loudspeakers. His patent also covers other formats of pickup which result in an approximation of the original time and phase differences at the ears when reproduced on loudspeakers. This will be discussed in more detail later on, but it is interesting historically to note how much writing on stereo reproduction even in the early 1950s appears unaware of Blumlein's most valuable work, which appears to have been ignored for some time.

A British paper presented by Clark, Dutton and Vanderlyn (of EMI) in 1957 revives the Blumlein theories, and shows in more rigorous mathematical detail how a two-loudspeaker system may be used to create an accurate relationship between the original location of a sound source and its perceived location on reproduction. This is achieved by controlling only the relative signal amplitudes of the two loudspeakers (derived in this case from a pair of coincident figure-eight microphones). The authors discuss the three-channel system of Bell Labs, and suggest that although it produces convincing results in many listening situations it is uneconomical for domestic use. They also conclude that the two-channel simplification (using microphones spaced about ten feet apart) has a tendency to result in a

| FACT FILE **16.1** | Binaural versus 'stereophonic' localisation | There is a distinct difference between the spatial perception that arises when two ears detect a single wavefront (i.e. from a single source) and that which arises when two |

(a)

(b)

arrivals of a similar sound come from different directions and are detected by both ears. The former, shown at (a), gives rise to spatial perceptions based primarily on what is known as the 'binaural delay' (essentially the time-of-arrival difference that arises between the ears for the particular angle of incidence). The latter, shown at (b) gives rise to spatial perceptions based primarily on various forms of 'precedence effect' (or 'law of the first wavefront'). In terms of sound reproduction, the former may be encountered in the headphone presentation context where sound source positions may be implied by using delays between the ear signals within the interaural delay of about 0.65 ms. Headphones enable the two ears to be stimulated independently of each other.

In loudspeaker listening the precedence effect is more relevant, as a rule. The precedence effect is primarily a feature of transient sounds rather than continuous sounds. In this case there are usually at least

two sound sources in different places, emitting different versions of the same sound, perhaps with a time or amplitude offset to provide directional information. This is what Snow termed the 'stereophonic' situation. Both ears hear both loudspeakers and the brain tends to localise based on the interaural delay arising from the earliest arriving wavefront, the source appearing to come from a direction towards that of the earliest arriving signal. This effect operates over delays between the sources that are somewhat greater than the interaural delay, of the order of a few milliseconds. Similar sounds arriving within up to 50 ms of each other tend to be perceptually fused together, such that one is not perceived as an echo of the other. The time delay over which this fusing effect obtains depends on the source, with clicks tending to separate before complex sounds like music or speech. The timbre and spatial qualities of this 'fused sound', though, may be affected.

'hole-in-the-middle' effect (with which many modern users of spaced microphones may be familiar – sources appearing to bunch towards the left or the right leaving a hole in the centre). They concede that the Blumlein method adapted by them does not take advantage of all the mechanisms of binaural hearing, especially the precedence effect, but that they have endeavoured to take advantage of, and re-create, a few of the directional cues which exist in the real-life situation.

There is therefore a historical basis for both the spaced microphone arrangement which makes use of the time-difference precedence effect (with only moderate level differences between channels), as well as the coincident microphone technique (or any other technique which results in only level differences between channels). There is also some evidence to show that the spaced technique is more effective with three channels than with only two. Later, we shall see that spaced techniques have a fundamental theoretical flaw from a point of view of 'correct' imaging of continuous sounds, which has not always been appreciated, although such techniques may result in subjectively acceptable sounds. Interestingly, three front channels are the norm in cinema sound reproduction, since the central channel has the effect of stabilising the important central image for off-centre listeners, having been used ever since the Disney film *Fantasia* in 1939. (People have often misunderstood the intentions of Bell Labs in the 1930s, since it is not generally realised that they were working on a system suitable for auditorium reproduction with wide-screen pictures, as opposed to a domestic system.)

16.1.2 Creating phantom images

Based on a variety of formal research and practical experience, it has become almost universally accepted that the optimum configuration for two-loudspeaker stereo is an equilateral triangle with the listener located just to the rear of the point of the triangle (the loudspeaker forming the baseline). Wider than this, phantom images (the apparent locations of sound sources in-between the loudspeakers) become less stable, and the system is more susceptible to the effects of head rotation. This configuration gives rise to an angle subtended by the loudspeakers of ±30° at the listening position, as shown in Figure 16.2. In most cases stereo reproduction from two loudspeakers can only hope to achieve a modest illusion of three-dimensional spatiality, since reproduction is from the front quadrant only.

The so-called 'summing localisation' model of stereo reproduction suggests that the best illusion of phantom sources between the loudspeakers will be created when the sound signals present at the two ears are as similar as possible to those perceived in natural listening, or at least that a number of natural localisation cues that are non-contradictory are available. It is possible to create this illusion for sources in the angle between the loudspeakers using only amplitude differences between the

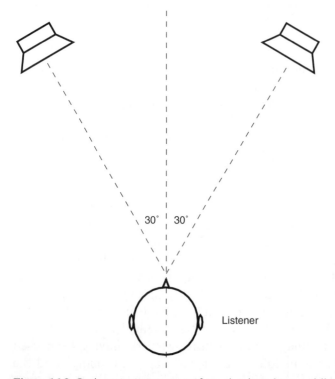

Figure 16.2 Optimum arrangement of two loudspeakers and listener for stereo listening

loudspeakers, where the time difference between the signals is very small («1 ms). To reiterate an earlier point, in loudspeaker reproduction both ears receive the signals from both speakers, whereas in headphone listening each ear only receives one signal channel. The result of this is that the loudspeaker listener seated in a centre seat (see Figure 16.3) receives at his left ear the signal from the left speaker first followed by that from the right speaker, and at his right ear the signal from the right speaker first followed by that from the left speaker. The time ∂t is the time taken for the sound to travel the extra distance from the more distant speaker.

The basis on which 'level-difference' or 'Blumlein' stereo works is to use level differences between two loudspeakers to generate low-frequency phase differences between the ears, based on the summation of the loudspeaker signals at the two ears, as described in Fact File 16.2. Depending on which author one believes, an amplitude difference of between 15 and 18 dB between the channels is needed for a source to be panned either fully left or fully right. A useful summary of experimental data on this issue has been drawn by Hugonnet and Walder and is shown in Figure 16.4. A coincident arrangement of velocity (figure-eight) microphones at ninety degrees to one another produce outputs which differ in amplitude with varying angle over the frontal quadrant by an amount which gives a very close correlation between the true angle of offset of the original source from the centre line and the apparent angle on reproduction, assuming loudspeakers which subtend an angle of 120° to the listening position. This angle of loudspeakers is not found to be very satisfactory for practical purposes for reasons such as the tendency to give rise to a 'hole' in the middle of the image. At smaller loudspeaker angles the change

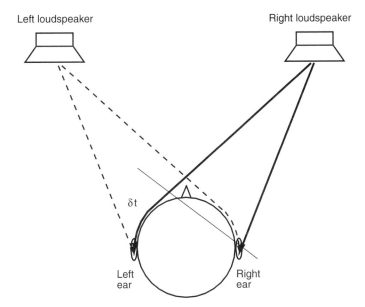

Figure 16.3 An approximation to the situation that arises when listening to sound from two loudspeakers. Both ears hear sound from both loudspeakers, the signal from the right loudspeaker being delyaed by δt at the left ear compared with the time it arrives at the right ear (and reversed for the other ear)

FACT FILE 16.2   Stereo vector summation

If the outputs of the two speakers differ only in amplitude and not in phase (time) then it can be shown (at least for low frequencies up to around 700 Hz) that the vector summation of the signals from the two speakers at each ear results in two signals that, for a given frequency, differ in phase angle proportional to the relative amplitudes of the two signals (the level difference between the ears being negligible at LF). For a given level difference between the speakers, the phase angle changes approximately linearly with frequency, which is the case when listening to a real point source. At higher frequencies the phase

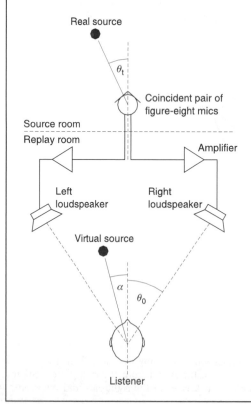

Real source

θ_t

Coincident pair of figure-eight mics

Source room
Replay room

Amplifier

Left loudspeaker

Right loudspeaker

Virtual source

α θ_0

Listener

difference cue becomes largely irrelevant but the shadowing effect of the head results in level differences between the ears. If the amplitudes of the two channels are correctly controlled it is possible to produce resultant phase and amplitude differences for continuous sounds that are very close to those experienced with natural sources, thus giving the impression of virtual or 'phantom' images anywhere between the left and right loudspeakers. This is the basis of Blumlein's (1931) stereophonic system 'invention' although the mathematics is quoted by Clark, Dutton and Vanderyn (1957) and further analysed by others. The result of the mathematical phasor analysis is a simple formula which can be used to determine, for any angle subtended by the loudspeakers at the listener, what the apparent angle of the virtual image will be for a given difference between left and right levels.

Firstly, referring to the diagram, it can be shown that:

$$\sin \alpha = (L - R)/(L + R)) \sin \theta_0$$

where α is the apparent angle of offset from the centre of the virtual image, and θ_0 is the angle subtended by the speaker at the listener. Secondly, it can be shown that:

$$(L - R)/(L + R) = \tan \theta_t$$

where θ_t is the true angle of offset of a real source from the centre-front of a coincident pair of figure-eight velocity microphones. $(L - R)$ and $(L + R)$ are the well-known difference (S) and sum (M) signals of a stereo pair, defined below.

This is a useful result since it shows that it is possible to use positioning techniques such as 'pan-potting' which rely on the splitting of a mono signal source into two components, with adjustment of the relative proportion fed to the left and right channels without affecting their relative timing. It also makes possible the combining of the two channels into mono without cancellations due to phase difference.

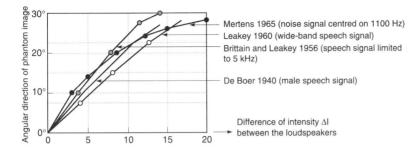

Figure 16.4 A summary of experimental data relating to amplitude differences (here labelled intensity) required between two loundspeaker signals for a particular phantom image location (data compiled by Hugonnet and Walder, 1995). (Courtesy of Christian Hugonnet)

in apparent angle is roughly proportionate as a fraction of total loudspeaker spacing, maintaining a correctly proportioned 'sound stage', so the sound stage with loudspeakers at the more typical 60° angle will tend to be narrower than the original sound stage but still in proportion.

If a time difference also exists between the channels, then transient sounds will be 'pulled' towards the advanced speaker because of the precedence effect, the perceived position depending to some extent on the time delay. If the left speaker is advanced in time relative to the right speaker (or more correctly, the right speaker is delayed!) then the sound appears to come more from the left speaker, although this can be corrected by increasing the level to the right speaker. A delay somewhere between 0.5 and 1.5 ms is needed for a signal to appear fully left or fully right at ±30°, depending on the nature of the signal (see Figure 16.5, after Hugonnet and Walder). With time-difference stereo, continuous sounds may give rise to contradictory phantom image positions when compared with the position implied by transients, owing to the phase differences that are created between the channels. Cancellations may also arise at certain frequencies if the channels are summed to mono.

Figure 16.5 A summary of experimental data relating to time differences required between two loudspeaker signals for a particular phantom image location (Hugonnet and Walder, 1995). (Courtesy of Christian Hugonnet)

Combinations of time and level difference can also be used to create phantom images, as described in Fact File 16.3.

16.2 Principles of binaural or headphone stereo

Binaural recording has fascinated researchers for years but it has received very little commercial attention until recently. Part of the problem has been that it is actually very difficult to get it to work properly for a wide range of listeners over a wide

FACT FILE

16.3

The 'Williams curves'

Stereo microphone technique relies on either interchannel level or time difference or a combination of the two. A trade-off is possible between them, although the exact relationship between time and level differences needed to place a source in a certain position is disputed by different authors and seems to depend to

some extent on the source characteristics. Michael Williams has based an analysis of microphone arrays on some curves of such trade-offs that have generally become known as the 'Williams curves', shown below. These curves represent the time and level difference combinations that may be used between two loudspeakers at ±30° in a typical listening room to obtain certain phantom source positions. The data points marked with circles were determined by a Danish researcher, Simonsen, using speech and maracas for signals.

range of different headphone types, and partly it is related to the limited compatibility between headphone and loudspeaker listening. Conventional loudspeaker stereo is acceptable on headphones to the majority of people, although it creates a strongly 'in-the-head' effect, but binaural recordings do not sound particularly good on loudspeakers unless some signal processing is used, and the stereo image is dubious.

Recent technical developments have made the signal processing needed to synthesise binaural signals and deal with the conversion between headphone and loudspeaker listening more widely available at reasonable cost. It is now possible to create 3D directional sound cues and to synthesise the acoustics of virtual environments quite accurately using digital signal processors (DSP), and it is this area of virtual environment simulation for computer applications that is receiving the most commercial attention for binaural technology today. Flight simulators, computer games, virtual reality applications and architectural auralisation are all areas that are benefiting from these developments.

16.2.1 Basic binaural principles

Binaural approaches to spatial sound representation are based on the premise that the most accurate reproduction of natural spatial listening cues will be achieved if the ears of the listener can be provided with the same signals that they would have experienced in the source environment of during natural listening. In a sense, all stereo reproduction is binaural, but the term is normally taken to mean an approach involving source signals that represent individual ear signals and independent-ear reproduction (such as can be achieved using headphones). Most of the approaches described so far in this chapter have related to loudspeaker reproduction of signals that contain *some* of the necessary information for the brain to localise phantom images and perceive a sense of spaciousness and depth. Much reproduced sound using loudspeakers relies on a combination of accurate spatial cues and believable illusion. In its purest form, binaural reproduction aims to reproduce *all* the cues that are needed for accurate spatial perception, but in practice this is something of a tall order and various problems arise.

An obvious and somewhat crude approach to binaural audio is to place two microphones, one at the position of each ear in the source environment, and to reproduce these signals through headphones to the ears of a listener, as shown in Figure 16.6. For binaural reproduction to work well, the HRTFs of sound sources from the source (or synthesised) environment must be accurately re-created at the listener's ears upon reproduction. This means capturing the time and frequency spectrum differences between the two ears accurately. Since each source position results in a unique HRTF, rather like a fingerprint, one might assume that all that is needed is to ensure the listener hears this correctly on reproduction.

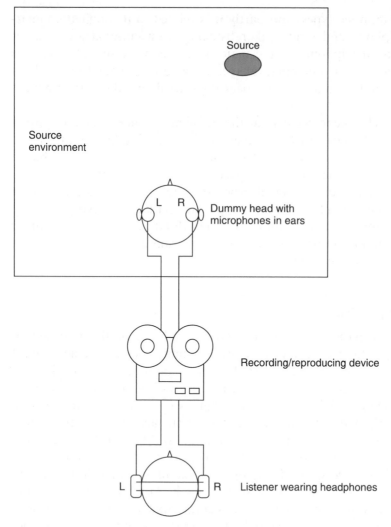

Source

Source
environment

L R   Dummy head with
microphones in ears

Recording/reproducing device

L   R   Listener wearing headphones

Figure 16.6 Basic binaural recording and reproduction

16.2.2 Tackling the problems of binaural systems

The primary problems in achieving an accurate reconstruction of spatial cues can be
summarised as follows:

- People's heads and ears are different (to varying degrees), although there are
  some common features, making it difficult to generalise about the HRTFs
  that should be used for commercial systems that have to serve lots of people
  (see above).
- Head movements that help to resolve directional confusion in natural listening
  are difficult to incorporate in reproduction situations.
- Visual cues are often missing during binaural reproduction and these normally
  have a strong effect on perception.

- Headphones differ in their equalisation and method of mounting, leading to distortions in the perceived HRTFs on reproduction.
- Distortions such as phase and frequency response errors in the signal chain can affect the subtle cues required.

It has been possible to identify the HRTF features that seem to occur in the majority of people and to then create generalised HRTFs that work reasonably well for a wide range of listeners. It has also been found that some people are better at localising sounds than others, and that the HRTFs of so-called 'good localisers' can be used in preference to those of 'poor localisers'. To summarise, it can be said that although a person's own HRTFs provide them with the most stable and reliable directional cues, generalised functions can be used at the expense of absolute accuracy of reproduction for everyone.

The problem of head movements can be addressed in advanced systems by using head tracking to follow the listener's actions and adapt the signals fed to the ears accordingly. This is generally only possible when using synthesised binaural signals that can be modified in real time. The issue of the lack of visual cues commonly encountered during reproduction can only be resolved in full 'virtual reality' systems that incorporate 3D visual information in addition to sound information. In the absence of visual cues, the listener must rely entirely on the sound cues to resolve things like front–back confusions and elevation/ distance estimations.

The issue of headphone equalisation is a thorny one as it depends on the design goal for the headphones. Different equalisation is required depending on the method of recording, unless the equalisation of both ends of the chain is standardised. For a variety of reasons, a diffuse field form of equalisation for headphones, dummy heads and synthesised environments has generally been found preferable to free-field equalisation. This means that the system is equalised to have a flat response to signals arriving from all angles around the head when averaged in a diffuse sound field. Headphones equalised in this way have been found to be quite suitable for both binaural and loudspeaker stereo signals, provided that the binaural signals are equalised in the same way.

Distortions in the signal chain that can affect the timing and spectral information in binaural signals have been markedly reduced since the introduction of digital audio systems. In the days of analogue signal chains and media such as compact cassette and LP records, numerous opportunities existed for interchannel phase and frequency response errors to arise, making it difficult to transfer binaural signals with sufficient integrity for success.

16.3 Loudspeaker stereo over headphones and vice versa

Bauer showed that if stereo signals designed for reproduction on loudspeakers were fed to headphones there would be too great a level difference between the ears compared with the real-life situation, and that the correct interaural delays would not exist. This results in an unnatural stereo image that does not have the expected sense of space and appears to be inside the head. He therefore proposed a network which introduced a measure of delayed crosstalk between the channels to simulate the correct interaural level differences at different frequencies, as well as simulating the interaural time delays which would result from the loudspeaker signals incident at 45° to the listener. He based the characteristics on research done by Weiner which produced graphs for the effects of diffraction around the human head for different angles of incidence. The characteristics of Bauer's circuit are shown in Figure 16.7 (with Weiner's results shown dotted). It may be seen that Bauer chooses to reduce the delay at HF, partially because the circuit design would have been too complicated, and partially because localisation relies more on amplitude difference at HF anyway.

Bauer also suggests the reverse process (turning binaural signals into stereo signals for loudspeakers). He points out that crosstalk must be removed between binaural channels for correct loudspeaker reproduction, since the crossfeed between the channels will otherwise occur twice (once between the pair of binaurally spaced microphones, and again at the ears of the listener), resulting in poor separation and a narrow image. He suggests that this may be achieved using the subtraction of an anti-phase component of each channel from the other channel signal, although he does not discuss how the time difference between the binaural channels may be removed. Such processes are the basis of 'transaural stereo' (see Fact File 16.4).

The idea that unprocessed binaural signals are unsuitable for loudspeaker reproduction has been challenged by Theile. He claims that the brain is capable of associating 'head-related' differences between loudspeakers with appropriate spatial cues for stereo reproduction, provided the timbral quality of head-related signals is equalised for a natural-sounding spectrum (e.g. diffuse field equalisation, as

Figure 16.7 Bauer's filter for processing loudspeaker signals so that they could be reproduced on headphones. The upper graph shows the delay introduced into the crossfeed between channels. The lower graph shows the left and right channel gains needed to imitate the shadowing effect of the head

described above). This theory has led to a variety of companies and recording engineers experimenting with the use of dummy heads such as the Neumann KU100 for generating loudspeaker signals, and created the idea for the Schoeps 'Sphere' microphone described below.

'Spatial equalisation' has been proposed by Griesinger to make binaural recordings more suitable for loudspeaker reproduction. He suggested low-frequency difference channel (L–R) boost of about 15 dB at 40 Hz (to increase the LF width of the reproduction) coupled with overall equalisation for a flat frequency response in the total energy of the recording to preserve timbral quality. This results in reasonably successful stereo reproduction in front of the listener, but the height and front–back cues are not preserved.

16.4 Two-channel signal formats

The two channels of a 'stereo pair' represent the left (L) and the right (R) loudspeaker signals. It is conventional in broadcasting terminology to refer to the left channel of a stereo pair as the 'A' signal and the right channel as the 'B' signal, although this may cause confusion to some who use the term 'AB pair' to refer specifically to a spaced microphone pair. In the case of some stereo microphones or systems the left and right channels are called respectively the 'X' and the 'Y' signals, although some people reserve this convention specifically for coincident microphone pairs. Here we will stick to using L and R for simplicity. In colour coding terms (for meters, cables, etc.), particularly in broadcasting, the L signal is coloured red and the R signal is coloured green. This may be confusing when compared with some domestic hi-fi wiring conventions that use red for the right channel, but it is the same as the convention used for port and starboard on ships. Furthermore there is a German DIN convention which uses yellow for L and red for R.

It is sometimes convenient to work with stereo signals in the so-called 'sum and difference' format, since it allows for the control of image width and ambient signal balance. The sum or main signal is denoted 'M' and is based on the addition of L and R signals. The difference or side signal is denoted 'S' and is based on the subtraction of R from L to obtain a signal which represents the difference between the two channels (see below). The M signal is that which would be heard by someone listening to a stereo programme in mono, and thus it is important in situations where the mono listener must be considered, such as in broadcasting. Colour-coding convention in broadcasting holds that M is coloured white, whilst S is coloured yellow, but it is sometimes difficult to distinguish between these two colours on certain meter types leading to the increasing use of orange for S.

Two-channel stereo signals may be derived by many means. Most simply, they may be derived from a pair of coincident directional microphones orientated at a fixed angle to each other. Alternatively they may be derived from a pair of spaced microphones, either directional or non-directional, with an optional third microphone bridged between the left and right channels. Finally stereo signals may be derived by splitting one or more mono signals into two by means of a 'pan-pot'. A

Transaural stereo

When binaural signals are replayed on loudspeakers there is crosstalk between the signals at the two ears of the listener that does not occur with headphone reproduction. The right ear hears the left channel signal a fraction of a second after it is received by the left ear, with an HRTF corresponding to the location of the left loudspeaker, and vice versa for the other ear. This prevents the correct binaural cues from being established at the listener's

ears and eliminates the possibility for full 3D sound reproduction. Binaural stereo tends to sound excessively narrow at low frequencies when replayed on loudspeakers as there is very little difference between the channels that has any effect at a listener's ears. Furthermore the spectral characteristics of binaural recordings can create timbral inaccuracies when reproduced over loudspeakers unless some form of compromise equalisation is used.

If the full 3D cues of the original binaural recording are to be conveyed over loudspeakers, some additional processing is required. If the left ear is to be presented only with the left channel signal and the right ear with the right channel signal then some means

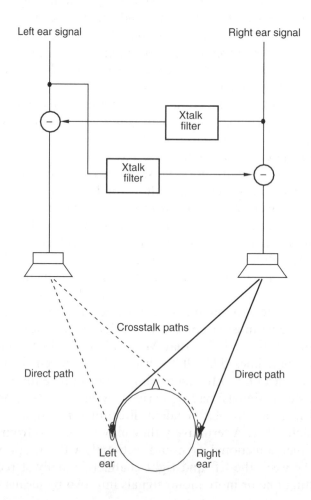

of removing the interaural crosstalk is required. This is often referred to as crosstalk cancelling or 'transaural' processing. Put crudely, transaural crosstalk-cancelling systems perform this task by feeding an anti-phase version of the left channel's signal into the right channel and vice versa, filtered and delayed according to the HRTF characteristic representing the crosstalk path, as shown below.

The effect of this technique can be quite striking, and in the best implementations enables fully three-dimensional virtual sources to be perceived, including behind the listener (from only two loudspeakers located at the front). Crosstalk-cancelling filters are usually only valid for a very narrow range of listening positions. Beyond a few tens of centimetres away from the 'hot spot' the effect often disappears almost completely. The effect is sometimes perceived as unnatural, and some listeners find it fatiguing to listen to for extended periods.

pan-pot is simply a dual-ganged variable resistor that controls the relative proportion of the mono signal being fed to the two legs of the stereo pair, such that as the level to the left side is increased that to the right side is decreased.

MS or 'sum and difference' format signals may be derived by conversion from the LR format using a suitable matrix (see Fact File 4.5 in Chapter 4) or by direct pickup in that format. For every stereo pair of signals it is possible to derive an MS equivalent, since M is the sum of L and R, whilst S is the difference between them. Likewise, signals may be converted from MS to LR formats using the reverse process. Misalignment of signals in either format leads to audible effects as described in Fact File 16.5. In order to convert an LR signal into MS format it is necessary to follow some simple rules. Firstly, the M signal is not usually a simple sum of L and R, as this will result in over-modulation of the M channel in the case where a maximum level signal exists on both L and R (representing a central image). A correction factor is normally applied, ranging between -3 dB and -6 dB (equivalent to a division of the voltage by between $\sqrt{2}$ and 2 respectively):

e.g. $M = (L + R) - 3$ dB   or $(L + R) - 6$ dB

The correction factor will depend on the nature of the two signals to be combined. If identical signals exist on the L and R channels (representing 'double mono' in effect), then the level of the uncorrected sum channel (M) will be two times (6 dB) higher than the levels of either L or R. This requires a correction of -6 dB in the M channel in order for the maximum level of the M signal to be reduced to a satisfactory level. If the L and R signals are non-coherent (random phase relationship), then only a 3 dB rise in the level of M will result when L and R are summed, requiring the -3 dB correction factor to be applied. This is more likely with stereo music signals. As most stereo material has a degree of coherence between the channels, the actual rise in level of M compared with L and R is likely to be somewhat between the two limits for real programme material.

The S signal results from the subtraction of R and L, and is subject to the same correction factor:

FACT FILE

16.5 Stereo misalignment effects

Differences in level, frequency response and phase may arise between signals of a stereo pair, perhaps due to losses in cables, misalignment, and performance limitations of equipment. It is important that these are kept to a minimum for stereo work, as inter-channel anomalies result in various audible side-effects. Differences will also result in poor mono compatibility. These differences and their effects are discussed below.

Frequency response and level
A difference in level or frequency response between L and R channels will result in a stereo image biased towards the channel with the higher overall level or that with the better HF response. Also, an L channel with excessive HF response compared with that of the R channel will result in the apparent movement of sibilant sounds towards the L loudspeaker. Level and response misalignment on MS signals results in increased crosstalk between the equivalent L and R channels, such that if the S level is too low at any

frequency the LR signal will become more monophonic (width narrower), and if it is too high the apparent stereo width will be increased.

Phase
Inter-channel phase anomalies will affect one's perception of the positioning of sound source, and it will also affect mono compatibility. Phase differences between L and R channels will result in 'comb-filtering' effects in the derived M signal due to cancellation and addition of the two signals at certain frequencies where the signals are either out-of- or in-phase.

Crosstalk
It was stated earlier that an inter-channel level difference of only 18 dB was required to give the impression of a signal being either fully left or fully right. Crosstalk between L and R signals is not therefore usually a major problem, since the performance of most audio equipment is far in excess of these requirements. Excessive crosstalk between L and R signals will result in a narrower stereo image, whilst excessive crosstalk between M and S signals will result in a stereo image increasingly biased towards one side.

---

e.g. $S = (L - R) - 3\,dB$ or $(L - R) - 6\,dB$

S can be used to reconstruct L and R when matrixed in the correct way with the M signal (see below), since $(M + S) = 2L$ and $(M - S) = 2R$. It may therefore be appreciated that it is possible at any time to convert a stereo signal from one format to the other and back again.

16.5 Two-channel microphone techniques

This section contains a review of basic two-channel microphone techniques, upon which many spatial recording techniques are based. Panned-spot microphones are often mixed into the basic stereo image created by such techniques.

16.5.1 Coincident-pair principles

The coincident-pair incorporates two directional capsules that may be angled over a range of settings to allow for different configurations and operational requirements. The pair can be operated in either the LR (sometimes known as 'XY') or MS modes (see above), and a matrixing unit is sometimes supplied with microphones which are intended to operate in the MS mode in order to convert the signal to LR format for recording. The directional patterns (polar diagrams) of the two microphones need not necessarily be figure-eight, although if the microphone is used in the MS mode the S capsule must be figure-eight (see below). Directional information is encoded solely in the level differences between the capsule outputs, since the two capsules are mounted physically as close as possible. There are no phase differences between the outputs except at the highest frequencies where inter-capsule spacing may become appreciable in relation to the wavelength of sound.

Coincident pairs are normally mounted vertically in relation to the sound source, so that the two capsules are angled to point symmetrically left and right of the centre of the source stage (see Figure 16.8). The choice of angle depends on the polar response of the capsules used. A coincident pair of figure-eight microphones at 90° provides good correspondence between the actual angle of the source and the apparent position of the virtual image when reproduced on loudspeakers, but there are also operational disadvantages to the figure-eight pattern in some cases, such as the amount of reverberation pickup.

Figure 16.9 shows the polar pattern of a coincident pair using figure-eight mics. Firstly, it may be seen that the fully-left position corresponds to the null point of the right capsule's pickup. This is the point at which there will be maximum level difference between the two capsules. The fully-left position also corresponds to the maximum pickup of the left capsule but it does not always do so in other stereo pairs. As a sound moves across the sound stage from left to right it will result in a gradually decreasing output from the left mic, and an increasing output from the

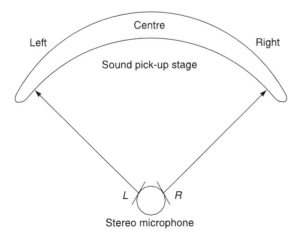

Figure 16.8 A coincident pair's capsules are oriented so as to point left and right of the centre of the sound stage

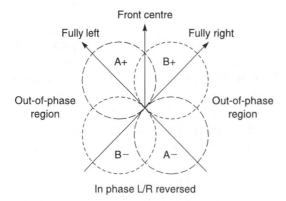

Figure 16.9 Polar pattern of a coincident pair using figure-eight microphones

right mic. Since the microphones have cosine responses, the output at 45° off axis is $\sqrt{2}$ times the maximum output, or 3 dB down in level, thus the takeover between left and right microphones is smooth for music signals. Fact File 16.6 goes into greater detail concerning the relationship between capsule angle and stereo width.

The second point to consider with this pair is that the rear quadrant of pickup suffers a left–right reversal, since the rear lobes of each capsule point in the opposite direction to the front. This is important when considering the use of such a microphone in situations where confusion may arise between sounds picked up on the rear and in front of the mic, such as in television sound where the viewer can also see the positions of sources. The third point is that pickup in both side quadrants results in out-of-phase signals between the channels, since a source further round than 'fully left' results in pickup by both the negative lobe of the right capsule and the positive lobe of the left capsule. There is thus a large region around a crossed pair of figure-eights that results in out-of-phase information, this information often being reflected or reverberant sound. Any sound picked up in this region will suffer cancellation if the channels are summed to mono, with maximum cancellation occurring at 90° and 270°, assuming 0° as the centre-front.

The operational advantages of the figure-eight pair are the crisp and accurate phantom imaging of sources, together with a natural blend of ambient sound from the rear. Some cancellation of ambience may occur, especially in mono, if there is a lot of reverberant sound picked up by the side quadrants. Disadvantages lie in the large out-of-phase region, and in the size of the rear pickup which is not desirable in all cases and is left–right reversed. Stereo pairs made up of capsules having less rear pickup may be preferred in cases where a 'drier' or less reverberant balance is required, and where frontal sources are to be favoured over rear sources. In such cases the capsule responses may be changed to be nearer the cardioid pattern, and this requires an increased angle between the capsules to maintain good correlation between actual and perceived angle of sources.

The cardioid crossed pair shown in Figure 16.10 is angled at approximately 131°, although angles of between 90° and 180° may be used to good effect depending on the width of the sound stage to be covered. At an angle of 131° a centre source is

65.5° off-axis from each capsule, resulting in a 3 dB drop in level compared with the maximum on-axis output (the cardioid mic response is equivalent to $0.5(1 + \cos \vartheta)$, where ϑ is the angle off-axis of the source, and thus the output at 65.5° is $\sqrt{2}$ times that at 0°). A departure from the theoretically correct angle is often necessary in practical situations, and it must be remembered that the listener will not necessarily be aware of the 'correct' location of each source, neither may it matter that the true and perceived positions are different. A pair of 'back-to-back' cardioids has often been used to good effect (see Figure 16.11), since it has a simple MS equivalent of an omni and a figure-eight, and has no out-of-phase region. Although the maximum level difference between the channels is at 90° off-centre there will in fact be a satisfactory level difference for a phantom image to appear fully left or right at a substantially smaller angle than this.

XY or LR coincident pairs in general have the possible disadvantage that central sounds are off-axis to both mics, perhaps considerably so in the case of crossed cardioids. This may result in a central signal with a poor frequency response and possibly an unstable image if the polar response is erratic. Whether or not this is important depends on the importance of the central image in relation to that of offset images, and will be most important in cases where the main source is central (such as in television, with dialogue). In such cases the MS technique described in the next section is likely to be more appropriate, since central sources will be on-axis to the M microphone. For music recording it would be hard to say whether central sounds are any more important than offset sources, so either technique may be acceptable.

16.5.2 Using MS processing on coincident pairs

Although some stereo microphones are built specifically to operate in the MS mode, it is possible to take any coincident pair capable of at least one capsule being switched to figure-eight, and orientate it so that it will produce suitable signals. The S component (being the difference between left and right signals) is always a sideways-facing figure-eight with its positive lobe facing left. The M (middle) component may be any polar pattern facing to the centre-front, although the choice of M pattern depends on the desired equivalent pair, and will be the signal that a mono listener would hear. True MS mics usually come equipped with a control box that matrixes the MS signals to LR format if required. A control for varying S gain is often provided as a means of varying the effective acceptance angle between the equivalent LR pair.

MS signals are not suitable for direct stereo monitoring, they are sum and difference components and must be converted to a conventional loudspeaker format at a convenient point in the production chain. The advantages of keeping a signal in the MS format until it needs to be converted will be discussed below, but the major advantage of pickup in the MS format is that central signals will be on-axis to the M capsule, resulting in the best frequency response. Furthermore, it is possible to operate an MS mic in a similar way to a mono mic which may be useful in television operations where the MS mic is replacing a mono mic on a pole or in a boom.

FACT FILE
16.6

Stereo width issues

With any coincident pair, fully left or fully right corresponds to the null point of pickup of the opposite channel's microphone, although psychoacoustically this point may be reached before the maximum level difference is arrived at. This also corresponds to the point where the M signal equals the S signal (where the sum of the channels is the same as the difference between them). As the angle between the capsules is made larger, the angle between the null points will become smaller, as shown below. Operationally, if one wishes to widen the reproduced sound stage one will widen the angle between the microphones which is intuitively the right thing to do. This results in a narrowing of the angle between fully left and fully right, so sources

which had been, say, half left in the original image will now be further towards the left. A narrow angle between fully left and fully right results in a very wide sound stage, since sources have only to move a small distance to result in large changes in reproduced position. This corresponds to a wide angle between the capsules.

Further coincident pairs are possible using any polar pattern between figure-eight and omni, although the closer that one gets to omni, the greater the required angle to achieve adequate separation between the channels. The hypercardioid pattern is often chosen for its smaller rear lobes than the figure-eight, allowing a more distant placement from the source for a given direct-to-reverberant ratio (although in practice hypercardioid pairs tend to be used closer to make the image width similar to that of a figure-eight pair). Since the hypercardioid pattern lies between figure-eight and cardioid, the angle required between the capsules is correctly around 110°.

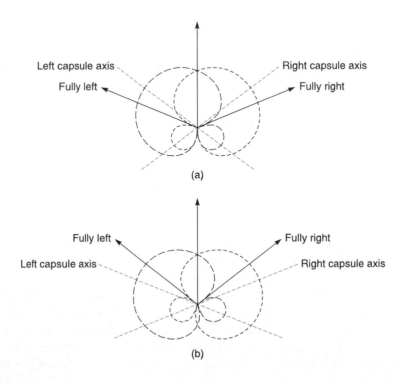

(a)

(b)

Psychoacoustic requirements suggest the need for an electrical narrowing of the image at high frequencies in order to preserve the correct angular relationships between low- and high-frequency signals, although this is rarely implemented in practice with coincident pair recording. A further consideration to do with the theoretical versus the practical is that although microphones tend to be referred to as having a particular polar pattern, this pattern is unlikely to be consistent across the frequency range and this will have an effect on the stereo image. Cardioid crossed pairs should theoretically exhibit no out-of-phase region (there should be no negative rear lobes), but in practice most cardioid capsules become more omni at LF and narrower at HF. As a result some out-of-phase components may be noticed in the HF range while the width may appear too narrow at LF. Attempts have been made to compensate for this in some stereo microphone designs.

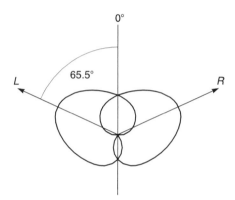

Figure 16.10 A coincident pair of cardioid microphones should theoretically be angled at 131°, but deviations either side of this may be acceptable in practice

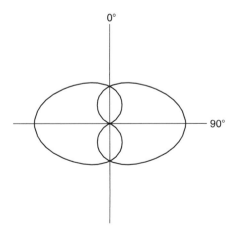

Figure 16.11 Back-to-back cardioids have been found to work well in practice and should have no out-of-phase region

To see how MS and LR pairs relate to each other, and to draw some useful conclusions about stereo width control, it is informative to consider a coincident pair of figure-eight mics again. For each MS pair there is an LR equivalent. The polar pattern of the LR equivalent to any MS pair may be derived by plotting the level of $(M + S)/2$ and $(M - S)/2$ for every angle around the pair. Taking the MS pair of figure-eight mics shown in Figure 16.12, it may be seen that the LR equivalent is simply another pair of figure-eights, but rotated through 45°. Thus the correct MS arrangement to give an equivalent LR signal where both 'capsules' are oriented at 45° to the centre-front (the normal arrangement) is for the M capsule to face forwards and the S capsule to face sideways.

A number of interesting points arise from a study of the LR/MS equivalence of these two pairs, and these points apply to all equivalent pairs. Firstly, fully left or right in the resulting stereo image occurs at the point where $S = M$ (in this case at 45° off-centre). This is easy to explain, since the fully left point is the point at which the output from the right capsule is zero. Therefore $M = L + 0$, and $S = L - 0$, both of which equal L. Secondly, at angles of incidence greater than 45° off-centre in either

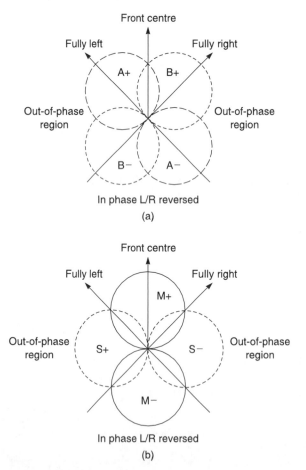

Figure 16.12 Every coincident pair has an MS equivalent. The conventional left–right arrangement is shown in (a), and the MS equivalent in (b)

direction the two channels become out-of-phase, as was seen above, and this corresponds to the region in which S is greater than M. Thirdly, in the rear quadrant where the signals are in phase again, but left–right reversed, the M signal is greater than S again. The relationship between S and M levels, therefore, is an excellent guide to the phase relationship between the equivalent LR signals. If S is lower than M, then the LR signals will be in phase. If S = M, then the source is either fully left or right, and if S is greater than M, then the LR signals will be out-of-phase.

To show that this applies in all cases, and not just that of the figure-eight pair, look at the MS pair in Figure 16.13 together with its LR equivalent. This MS pair is made up of a forward-facing cardioid and a sideways-facing figure-eight (a popular arrangement). Its equivalent is a crossed pair of hypercardioids, and again the extremes of the image (corresponding to the null points of the LR hypercardioids) are the points at which S equals M. Similarly, the signals go out-of-phase in the region where S is greater than M, and come back in phase again for a tiny angle round the back, due to the rear lobes of the resulting hypercardioids. Thus the angle of acceptance (between fully left and fully right) is really the frontal angle between the two points on the MS diagram where M equals S.

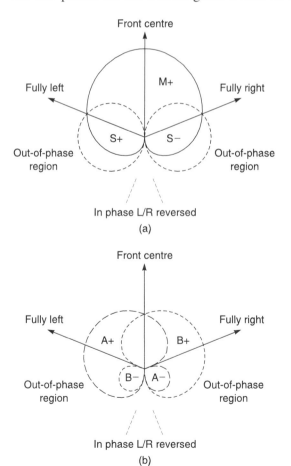

Figure 16.13 The MS equivalent of a forward facing cardioid and sidways figure-eight, as shown in (a), is a pair of hypercardioids whose effective angle depends on S gain, as shown in (b)

Now, consider what would happen if the gain of the S signal was raised (imagine expanding the lobes of the S figure-eight). The result of this would be that the points where S equalled M would move inwards, making the acceptance angle smaller. As explained earlier, this results in a wider stereo image, since off-centre sounds will become closer to the extremes of the image, and is equivalent to increasing the angle between the equivalent LR capsules. Conversely, if the S gain is reduced, the points at which S equals M will move further out from the centre, resulting in a narrower stereo image, equivalent to decreasing the angle between the equivalent LR capsules. This helps to explain why Blumlein-style shufflers work by processing the MS equivalent signals of stereo pairs, as one can change the effective stereo width of pairs of signals, and this can be made frequency dependent if required.

This is neatly exemplified in a commercial example, the Neumann RSM 191i, which is an MS mic in which the M capsule is a forward-facing short shotgun mic with a polar pattern rather like a hypercardioid. The polar pattern of the M and S capsules and the equivalent LR pair is shown in Figure 16.14 for three possible

Figure 16.14 Polar patterns on the Neumann RSM191i microphone. (a) M capsule, (b) S capsule, (c) LR equivalent with −6 dB S gain, (d) 0 dB S gain, (e) +6 dB S gain

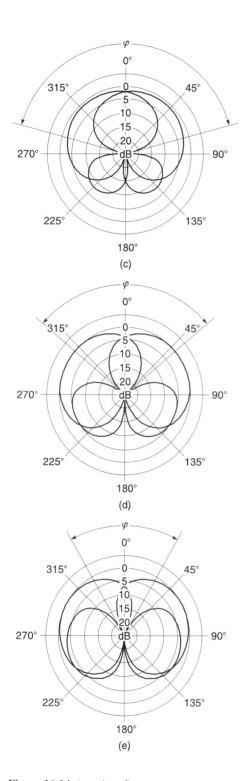

Figure 16.14 (*continued*)

gains of the S signal with relation to M (−6 dB, 0 dB and +6 dB). It will be seen that the acceptance angle (M) changes from being large (narrow image) at −6 dB, to small (wide image) at +6 dB. Changing the S gain also affects the size of the ear lobes of the LR equivalent. The higher the S gain, the larger the rear lobes. Not only does S gain change stereo width, it also affects rear pickup, and thus the ratio of direct to reverberant sound.

Any stereo pair may be operated in the MS configuration, simply by orientating the capsules in the appropriate directions and switching them to an appropriate polar pattern, but certain microphones are dedicated to MS operation simply by the physical layout of the capsules (see Fact File 16.7).

FACT FILE 16.7 — End-fire and side-fire configurations

There are two principal ways of mounting the capsules in a coincident stereo microphone, be it MS or LR format: either in the 'end-fire' configuration where the capsules 'look out' of the end of the microphone, such that the microphone may be pointed at the source (see the diagram), or in the 'side-fire' configuration where the capsules look out of the sides of the microphone housing. It is less easy to see the direction in which the capsules are pointing in a side-fire microphone, but such a microphone makes it possible to align the capsules vertically above each other so as to be time-coincident in the horizontal plane, as well as allowing for the rotation of one capsule with relation to the other. An end-fire configuration is more suitable for the MS capsule arrangement (see diagram below), since the S capsule may be mounted sideways behind the M capsule, and no rotation of the capsules is required. There is a commercial example of an LR end-fire microphone for television ENG (electronic news gathering) use which houses two fixed cardioids side-by-side in an enlarged head.

16.5.3 Operational considerations with coincident pairs

The control of S gain is an important tool in determining the degree of width of a stereo sound stage, and for this reason the MS output from a microphone might be brought (unmatrixed) into a mixing console, so that the engineer has control over the width. This in itself can be a good reason for keeping a signal in MS form during the recording process, although M and S can easily be derived at any stage using a conversion matrix.

Although some mixers have MS matrixing facilities on board, the diagram in Figure 16.15 shows how it is possible to derive an LR mix with variable width from an MS microphone using three channels on a mixer without using an external MS matrix. M and S outputs from the microphone are fed in phase through two mixer channels and faders, and a post-fader feed of S is taken to a third channel line input, being phase-reversed on this channel. The M signal is routed to both left and right mix buses (panned centrally), whilst the S signal is routed to the left mix box (M + S = 2L) and the −S signal (the phase-reversed version) is routed to the right mix bus (M − S = 2R). It is important that the gain of the −S channel is matched very closely with that of the S channel. (A means of deriving M and S from an LR format input is to mix L and phase-reversed R together to get S, and without the phase reverse to get M.)

Outdoors, coincident pairs will be susceptible to wind noise and rumble, as they incorporate velocity-sensitive capsules which always give more problems in this respect than omnis. Most of the interference will reside in the S channel, since this has always a figure-eight pattern, and thus would not be a problem to the mono listener. Similarly, physical handling of the stereo microphone, or vibration picked up through a stand, will be much more noticeable than with pressure microphones. Coincident pairs should not generally be used close to people speaking, as

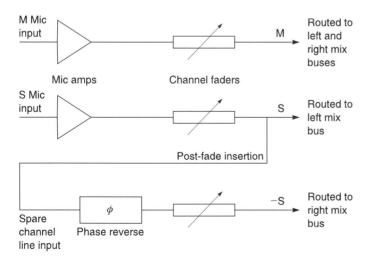

Figure 16.15 An LR mix with variable width can be derived from an MS microphone connected to three channels of a mixer as shown. The S faders should be ganged together and used as a width control

small movements of their heads can cause large changes in the angle of incidence, leading to considerable movement in their apparent position in the sound stage.

16.5.4 Near-coincident microphone configurations

'Near-coincident' pairs of directional microphones introduce small additional timing differences between the channels which may help in the localisation of transient sounds and increase the spaciousness of a recording, which at the same time remaining nominally coincident at low frequencies and giving rise to suitable amplitude differences between the channels. Headphone compatibility is also quite good owing to the microphone spacing being similar to ear spacing. The family of near-coincident (or closely spaced) techniques relies on a combination of time and level differences between the channels that can be traded off for certain widths of sound stage and microphone pattern.

Subjective evaluations often seem to show good results for such techniques. One comprehensive subjective assessment of stereo microphone arrangements, performed at the University of Iowa, consistently resulted of the near-coincident pairs scoring among the two few performers for their sense of 'space' and realism. Critics have attributed these effects to 'phasiness' at high frequencies (which some people may like, nonetheless), and argued that truly coincident pairs were preferable.

A number of examples of near-coincident pairs exist as 'named' arrangements, although there is a whole family of possible near-coincident arrangements using combinations of spacing and angle. Some near-coincident pairs of different types, based on the 'Williams curves' (see Fact File 16.3) are given in Table 16.1. The so-called 'ORTF pair' is an arrangement of two cardioid mics, deriving its name from the organisation which first adopted it (the Office de Radiodiffusion-Television Française). The two mics are spaced apart by 170 mm, and angled at 110°. The 'NOS' pair (Nederlande Omroep Stichting, the Dutch Broadcasting Company), uses a cardioid mics spaced apart by 300 mm and angled at 90°. Figure 16.16 illustrates these two pairs, along with a third pair of figure-eight microphones spaced apart by 200 mm, which has been called a 'Faulkner' pair, after the British recording engineer who first adopted it (this is not strictly based on the Williams curves). This latter pair has been found to offer good image focus on a

Table 16.1 Some near-coincident pairs based on the 'Williams curves'

Designation	Polar pattern	Mic angle	Spacing	Recording angle
NOS	Cardioid	±45°	30 cm	80°
RAI	Cardioid	±50°	21 cm	90°
ORTF	Cardioid	±55°	17 cm	95°
DIN	Cardioid	±45°	20 cm	100°
–	Omni	0°	50 cm	130°
–	Omni	0°	35 cm	160°

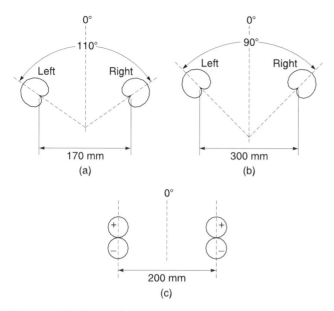

Figure 16.16 Near-coincident pairs (a) ORTF, (b) NOS, (c) Faulkner

small-to-moderate-sized central ensemble with the mics placed further back than would normally be expected.

16.5.5 Pseudo-binaural techniques

Binaural techniques could be classed as another form of near-coincident technique. The spacing between the omni microphones in a dummy head is not great enough to fit any of the Williams models described above for near-coincident pairs, but the shadowing effect of the head makes the arrangement more directional at high frequencies. Low-frequency width is likely to need increasing to make the approach more loudspeaker-compatible, as described earlier, unless one adheres to Theile's association theory of stereo in which case little further processing is required except for equalisation.

The Schoeps KFM6U microphone, pictured in Figure 16.17, was designed as a head-sized sphere with pressure microphones mounted on the surface of the sphere, equalised for a flat response to frontal incidence sound and suitable for generating signals that could be reproduced on loudspeakers. This is in effect a sort of dummy head without ears. Dummy heads also exist that have been equalised for a reasonably natural timbral quality on loudspeakers, such as the Neumann KU100. The use of unprocessed dummy head techniques for stereo recording intended for loudspeakers has found favour with some recording engineers because they claim to like the spatial impression created, although others find the stereo image somewhat unfocused or vague.

Figure 16.17 The Schoeps KFM6U microphone consists of two presssure microphones mounted on the surface of a sphere. (Courtesy of Schalltechnik Dr. -Ing. Schoeps GmbH)

16.5.6 Spaced microphone configurations

Spaced arrays have a historical precedent for their usage, since they were the first to be documented (in the work of Clement Ader at the Paris Exhibition in 1881), were the basis of the Bell Labs stereo systems in the 1930s, and have been widely used since then. They are possibly less 'correct' theoretically, from a standpoint of sound-field representation, but they can provide a number of useful spatial cues that give rise to believable illusions of natural spaces. Many recording engineers prefer spaced arrays because the omni microphones often used in such arrays tend to have a flatter and more extended frequency response than their directional counter-parts, although it should be noted that spaced arrays do not have to be made up of omni mics (see below).

Spaced arrays rely principally on the precedence effect. The delays that result between the channels tend to be of the order of a number of milliseconds. With spaced arrays the level and time difference resulting from a source at a particular left–right position on the sound stage will depend on how far the source is from the microphones (see Figure 16.18), with a more distant source resulting in a much smaller delay and level difference. In order to calculate the time and level differences that will result from a particular spacing it is possible to use the following two formulae:

$$\Delta t = (d_1 - d_2)/c \qquad \Delta L = 20 \log_{10}(d_1/d_2)$$

where Δt is the time difference and ΔL the pressure level difference which results from a source whose distance is d_1 and d_2 respectively from the two microphones, and c is the speed of sound (340 m/s).

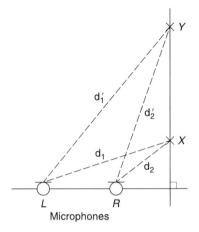

Figure 16.18 With spaced omnis a source at position X results in path lengths d_1 and d_2 to each microphone respectively, whilst for a source in the same LR position but at a greater distance (source Y) the path length difference is smaller, resulting in smaller time difference than for X

When a source is very close to a spaced pair there may be a considerable level difference between the microphones, but this will become small once the source is more than a few metres distant. The positioning of spaced microphones in relation to a source is thus a matter of achieving a compromise between closeness (to achieve satisfactory level and time differences between channels), and distance (to achieve adequate reverberant information relative to direct sound). When the source is large and deep, such as a large orchestra, it will be difficult to place the microphones so as to suit all sources. It may therefore be found necessary to raise the microphones somewhat so as to reduce the differences in path length between sources at the front and rear of the orchestra.

Spaced microphone arrays do not stand up well to theoretical analysis when considering the imaging of continuous sounds, the precedence effect being related principally to impulsive or transient sounds. Because of the phase differences between signals at the two loudspeakers created by the microphone spacing, interference effects at the ears at low frequencies may in fact result in a contradiction between level and time cues at the ears. It is possible in fact that the ear on the side of the earlier signal may not experience the higher level, thus producing a confusing difference between the cues provided by impulsive sounds and those provided by continuous sounds. The lack of phase coherence in spaced-array stereo is further exemplified by phase inverting one of the channels on reproduction, an action which does not always appear to affect the image particularly, as it would with coincident stereo, showing just how uncorrelated the signals are. (This is most noticeable with widely spaced microphones.)

Accuracy of phantom image positioning is therefore lower with spaced arrays, although many convincing recordings have resulted from their use. It has been suggested that the impression of spaciousness that results from the use of spaced arrays is in fact simply the result of phasiness and comb-filtering effects. Others suggest that there is a place for the spaciousness that results from spaced techniques, since the highly decorrelated signals which result from spaced techniques are also a feature of concert hall acoustics.

Griesinger has often claimed informally that spacing the mics apart by at least the reverberation radius (critical distance) of a recording space gives rise to adequate decorrelation between the microphones to obtain good spaciousness, and that this might be a suitable technique for ambient sound in surround recording. Mono compatibility of spaced pairs is variable, although not always as poor in practice as might be expected.

The so-called 'Decca Tree' is a popular arrangement of three spaced omnidirectional mics. The name derives from the traditional usage of this technique by the Decca Record Company, although even that company did not adhere rigidly to this arrangement. A similar arrangement is described by Grignon (1949). Three omnis are configured according to the diagram in Figure 16.19, with the centre microphone spaced so as to be slightly forward of the two outer mics, although it is possible to vary the spacing to some extent depending on the size of the source stage to be covered. The reason for the centre microphone and its spacing is to stabilise the central image which tends otherwise to be rather imprecise, although the existence of the centre mic will also complicate the phase relationships between the channels, thus exacerbating the comb-filtering effects that may arise with spaced pairs. The advance in time experienced by the forward mic will tend to solidify the central image, due to the precedence effect, avoiding the hole-in-the-middle often resulting from spaced pairs. The outer mics are angled outwards slightly, so that the axes of best HF response favour sources towards the edges of the stage whilst central sounds are on-axis to the central mic.

A pair of omni outriggers are often used in addition to the tree, towards the edges of wide sources such as orchestras and choirs, in order to support the extremes of the sound stage that are some distance from the tree or main pair (see Figure 16.20). This is hard to justify on the basis of any conventional imaging theory, and is beginning to move toward the realms of multi-microphone pickup, but can be used to produce a commercially acceptable sound. Once more than around three microphones are used to cover a sound stage one has to consider a combination of theories, possibly suggesting conflicting information between the outputs of the different microphones. In such cases the sound balance will be

Figure 16.19 The classic 'Decca Tree' involved three omnis, with the centre microphone spaced slighly forward of the outer mics

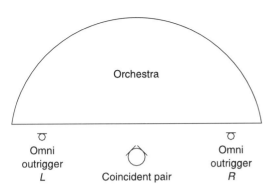

Figure 16.20 Omni outriggers may be used in addition to a coincident pair or Decca Tree, for wide sources

optimised on a mixing console, subject to the creative control of the recording engineer.

Spaced microphones with either omnidirectional or cardioid patterns may be used in configurations other than Decca Tree described above, although the 'tree' has certainly proved to be the more successful arrangement in practice. The precedence effect begins to break down for delays greater than around 40 ms, because the brain begins to perceive the two arrivals of sound as being discrete rather than integrated. It is therefore reasonable to assume that spacings between microphones which give rise to greater delays than this between channels should be avoided. This maximum delay, though, corresponds to a mic spacing of well over ten metres. Such extremes have not proved to work well in practice due to the great distance of central sources from either microphone compared with the closeness of sources at the extremes, resulting in a considerable level drop for central sounds and thus a hole in the middle.

Dooley and Streicher have shown that good results may be achieved using spacings of between one-third and one-half of the width of the total sound stage to be covered (see Figure 16.21), although closer spacings have also been used to good

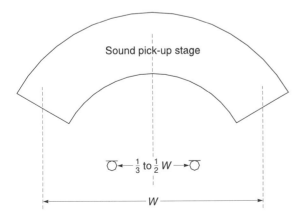

Figure 16.21 Dooley and Streicher's proposal for omni spacing

effect. Bruel and Kjaer manufacture matched stereo pairs of omni microphones together with a bar which allows variable spacing, as shown in Figure 16.22, and suggest that the spacing used is smaller than one-third of the stage width (they suggest between 5 cm and 60 cm, depending on stage width). Their principal rule is that the distance between the microphones should be small compared with the distance from microphones to source.

16.6 Binaural recording and 'dummy head' techniques

While it is possible to use a real human head for binaural recording (generally attached to a live person), it can be difficult to mount high-quality microphones in the ears and the head movements and noises of the owner can be obtrusive. Sometimes heads are approximated by the use of a sphere or a disk separating a pair of microphones, and this simulates the shadowing effect of the head but it does not give rise to the other spectral filtering effects of the outer ear. Recordings made using such approaches have been found to have reasonable loudspeaker compatibility as they do not have the unusual equalisation that results from pinna filtering. (Unequalised true binaural recordings replayed on loudspeakers will typically suffer two stages of pinna filtering – once on recording and then again on reproduction – giving rise to distorted timbral characteristics.)

Dummy heads are models of human heads with pressure microphones in the ears that can be used for originating binaural signals suitable for measurement or reproduction. A number of commercial products exist, some of which also include either shoulders or a complete torso. A complete head-and-torso simulator is often referred to as a 'HATS', and an example is shown in Figure 16.23. The shoulders

Figure 16.22 B&K omni microphones mounted on a stereo bar that allows variable spacing

Figure 16.23 Head and torso simulator (HATS) from B&K

and torso are considered by some to be important owing to the reflections that result from them in natural listening, which can contribute to the HRTF. This has been found to be a factor that differs quite considerably between individuals and can therefore be a confusing cue if not well matched to the listener's own torso reflections.

Some dummy heads or ear inserts are designed specifically for recording purposes whereas others are designed for measurement. As a rule, those designed for recording tend to have microphones at the entrances of the ear canals, whereas those designed for measurement have the mics at the ends of the ear canals, where the ear drum should be. (Some measurement systems also include simulators for the transmission characteristics of the inner parts of the ear.) The latter types will therefore include the ear canal resonance in the HRTF, which would have to be equalised out for recording/reproduction purposes in which headphones were located outside the ear canal. The ears of dummy heads are often interchangeable in order to vary the type of ear to be simulated, and these ears are modelled on 'average' or 'typical' physical properties of human ears, giving rise to the same problems of HRTF standardisation as mentioned above.

The equalisation of dummy heads for recording has received much attention over the years, mainly to attempt better headphone/loudspeaker compatibility. Equalisation can be used to modify the absolute HRTFs of the dummy head in such a way that the overall spatial effect is not lost, partly because the differences between the ears are maintained. Just as Theile has suggested using diffuse field equalisation for headphones as a good means of standardising their response, he and others have also suggested diffuse field equalisation of dummy heads so that recordings made on such heads replay convincingly on such headphones and sound reasonably natural on loudspeakers. This essentially means equalising the dummy head microphone so that it has a near-flat response when measured in one-third octave bands in a diffuse sound field. The Neumann KU100, pictured in Figure 16.24, is a dummy head that is designed to have good compatibility between loudspeaker and headphone reproduction, and uses equalisation that is close to Theile's proposed diffuse field response.

Figure 16.24 Neumann KU100 dummy head. (Courtesy of Georg Neumann GmbH, Berlin)

Binaural cues do not have to be derived from dummy heads. Provided the HRTFs are known, or can be approximated for the required angle of sound incidence, signals can be synthesised with the appropriate time delays and spectral characteristics. Such techniques are increasingly used in digital signal processing applications that aim to simulate natural spatial cues, such as flight simulators and virtual reality. Accurate sets of HRTF data for all angles of incidence and elevation have been hard to come by until recently, and they are often quite closely guarded intellectual property as they can take a long time and a lot of trouble to measure. The question also arises as to how fine an angular resolution is required in the data set. For this reason a number of systems base their HRTF implementation on relative coarse resolution data and interpolate the points in between.

16.7 Spot microphones and two-channel panning laws

We have so far considered the use of a small number of microphones to cover the complete sound stage. It is also possible to make use of a large number of mono microphones or other mono sources, each covering a small area of the sound stage and intended to be as independent of the others as possible. This is the normal basis of most studio pop music recording, with the sources often being recorded at separate times using overdubbing techniques. In the ideal world, each mic in such an arrangement would pick up sound only from the desired sources, but in reality there is usually considerable spill from one to another. It is not the intention in this chapter to provide a full resumé of studio microphone technique, and thus discussion will be limited to an overview of the principles of multi-mic pickup as distinct from the more simple techniques described above.

In multi-mic recording each source feeds a separate channel of a mixing console, where levels are individually controlled and the mic signal is 'panned' to a virtual position somewhere between left and right in the sound stage. The pan control takes the monophonic signal and splits it two ways, controlling the proportion of the signal fed to each of the left and right mix buses. Typical pan control laws follow a curve which gives rise to a 3 dB drop in the level sent to each channel at the centre, resulting in no perceived change in level as a source is moved from left to right (see Fact File 6.2). This has often been claimed to be due to the way signals from left and right loudspeakers sum acoustically at the listening position, which includes a diffuse field component of the room. The -3 dB pan-pot law is not correct if the stereo signal is combined electrically to mono, since the summation of two equal signal voltages would result in a 6 dB rise in level for signals panned centrally. A -6 dB law is more appropriate for mixers whose outputs will be summed to mono (eg: radio and TV operations) as well as stereo, although this will then result in a drop in level in the centre for stereo signals. A compromise law of -4.5 dB is sometimes adopted by manufacturers for this reason.

Panned mono balances rely on channel level differences, separately controlled for each source, to create phantom images on a synthesised sound stage, with relative level between sources used to adjust the prominence of a source in a mix. Time delay is hardly ever used as a panning technique, for reasons of poor mono compatibility and technical complexity. Artificial reverberation may be added to restore a sense of space to a multi-mic balance. Source distance can be simulated by the addition of reflections and reverberation, as well as by changes in source spectrum and overall level (e.g. HF roll-off can simulate greater distance).

It is common in classical music recording to use close mics in addition to a coincident pair or spaced pair in order to reinforce sources that appear to be weak in the main pickup. These close mics are panned to match the true position of the source. The results of this are variable and can have the effect of flattening the perspective, removing any depth which the image might have had, and thus the use of close mics must be handled with subtlety. David Griesinger has suggested that the use of stereo pairs of mics as spots can help enormously in removing this flattening effect, because the spill that results between spots is now in stereo rather than in mono and is perceived as reflections separated spatially from the main signal.

The recent development of cheaper digital signal processing (DSP) has made possible the use of delay lines, sometimes as an integral feature of digital mixer channels, to adjust the relative timing of spot mics in relation to the main pair. This can help to prevent the distortion of distance, and to equalise the arrival times of distant mics so that they do not exert a precedence 'pull' over the output of the main pair. It is also possible to process the outputs of multiple mono sources to simulate binaural delays and head-related effects in order to create the effect of sounds at any position around the head when the result is monitored on headphones or on loudspeaker using crosstalk cancelling, as described earlier.

Recommended further reading

Bartlett, B. (1991) *Stereo Microphone Techniques*. Focal Press

Eargle, J. (ed.) (1986) *Stereophonic Techniques – An Anthology*. Audio Engineering Society

Hugonnet, C. and Walder, P. (1998) *Stereophonic Sound Recording – Theory and Practice*. John Wiley

Rumsey, F. (2001) *Spatial Audio*. Focal Press

Chapter 17

Surround sound

This chapter is concerned with the most commonly encountered multichannel (i.e. more than two channels) stereo reproduction configurations, most of which are often referred to as surround sound. Standards or conventions that specify basic channel or loudspeaker configurations are distinguished from proprietary systems such as Dolby Digital and DTS whose primary function is the coding and delivery of multi-channel audio signals. The latter are discussed in the second part of the chapter, in which is also contained an explanation of the Ambisonic system for stereo signal representation. Surround sound standards often specify little more than the channel configuration and the way the loudspeakers should be arranged. This leaves the business of how to create or represent a spatial sound field entirely up to the user.

17.1 Three-channel (3-0) stereo

It is not proposed to say a great deal about the subject of three-channel stereo here, as it is rarely used on its own. Nonetheless it does form the basis of a lot of surround sound systems. It requires the use of a left (L), centre (C) and right (R) channel, the loudspeakers arranged equidistantly across the front sound stage, as shown in Figure 17.1. It has some precedents in historical development, in that the stereo-phonic system developed by Steinberg and Snow in the 1930s used three channels (see Chapter 16). Three front channels have also been commonplace in cinema stereo systems, mainly because of the need to cover a wide listening area and because wide screens tend to result in a large distance between left and right loud-speakers. Two channels only became the norm in consumer systems for reasons of economy and convenience, and particularly because it was much more straight-forward to cut two channels onto an analogue disk than three.

There are various advantages of three-channel stereo. Firstly, it allows for a some-what wider front sound stage than two-channel stereo, if desired, because the centre channel acts to 'anchor' the central image and the left and right loudspeakers can be placed further out to the sides (say $\pm 45°$). (Note, though, that in the current five-channel surround sound standard the L and R loudspeakers are in fact placed at $\pm 30°$, for compatibility with two-channel stereo material.) Secondly, the centre

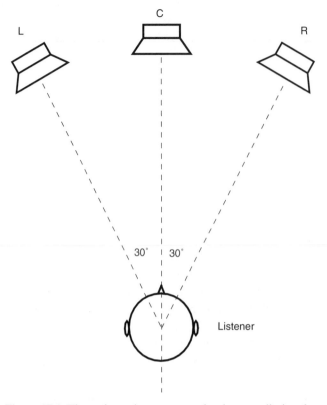

Figure 17.1 Three-channel stereo reproduction usually involves three equally spaced loudspeakers in front of the listener. The angle between the outer loudspeakers is 60° in the ITU standard configuration, for compatibility with two-channel reproduction, but the existence of a centre loudspeaker makes wider spacings feasible if compatibility is sacrificed

loudspeaker enables a wider range of listening positions in many cases, as the image does not collapse quite as readily into the nearest loudspeaker. It also anchors dialogue more clearly in the middle of the screen in sound-for-picture applications. Thirdly, the centre image does not suffer the same timbral modification as the centre image in two-channel stereo, because it emanates from a real source.

A practical problem with three-channel stereo is that the centre loudspeaker position is often very inconvenient. Although in cinema reproduction it can be behind an acoustically transparent screen, in consumer environments, studios and television environments it is almost always just where one wants a television monitor or a window. Consequently the centre channel has to be mounted above or below the object in question, and possibly made smaller than the other loudspeakers.

17.2 Four-channel surround (3-1 stereo)

In this section the form of stereo called '3-1 stereo' in some international standards, or 'LCRS surround' in some other circles, is briefly described. Proprietary encoding and decoding technology from Dolby relating to this format is described later.

'Quadraphonic' reproduction using four loudspeakers in a square arrangement is not covered further here (it was mentioned in the Introduction), as it has little relevance to current practice.

17.2.1 Purpose of four-channel systems

The merits of three front channels have already been introduced in the previous section. In the 3-1 approach, an additional 'effects' channel or 'surround' channel is added to the three front channels, routed to a loudspeaker or loudspeakers located behind (and possibly to the sides) of listeners. It was developed first for cinema applications, enabling a greater degree of audience involvement in the viewing/listening experience by providing a channel for 'wrap-around' effects. This development is attributed to 20th Century Fox in the 1950s, along with wide-screen Cinemascope viewing, being intended to offer effective competition to the new television entertainment.

There is no specific intention in 3-1 stereo to use the effects channel as a means of enabling 360° image localisation. In any case, this would be virtually impossible with most configurations as there is only a single audio channel feeding a larger number of surround loudspeakers, effectively in mono.

17.2.2 Loudspeaker configuration

Figure 17.2 shows the typical loudspeaker configuration for this format. In the cinema there are usually a large number of surround loudspeakers fed from the single S channel ('surround channel', not to be confused with the 'S' channel in sum-and-difference stereo), in order to cover a wide audience area. This has the tendency to create a relatively diffuse or distributed reproduction of the effects signal. The surround speakers are sometimes electronically decorrelated to increase the degree of spaciousness or diffuseness of surround effects, in order that they are not specifically localised to the nearest loudspeaker or perceived inside the head.

In consumer systems reproducing 3-1 stereo, the mono surround channel is normally fed to two surround loudspeakers located in similar positions to the 3-2 format described below. The gain of the channel is usually reduced by 3 dB so that the summation of signals from the two speakers does not lead to a level mismatch between front and rear.

17.2.3 Limitations of four-channel reproduction

The mono surround channel is the main limitation in this format. Despite the use of multiple loudspeakers to reproduce the surround channel, it is still not possible to create a good sense of envelopment of spaciousness without using surround signals that are different on both sides of the listener. Most of the psychoacoustic research suggests that the ears need to be provided with decorrelated signals to create the best sense of envelopment and effects can be better spatialised using stereo surround channels.

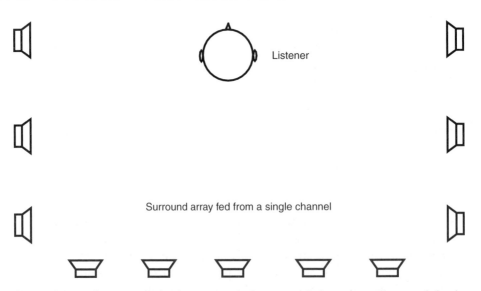

Figure 17.2 3-1 format reproduction uses a single surround channel usually routed (in cinema environments) to an array of loudspeakers to the sides and rear of the listening area. In consumer reproduction the mono surround channel may be reproduced through only two surround loudspeakers, possibly using artificial decorrelation and/or dipole loudspeakers to emulate the more diffused cinema experience

17.3 5.1 channel surround (3-2 stereo)

This section deals with the 3-2 configuration that has been standardised for numerous surround sound applications, including cinema, television and consumer applications. Because of its wide use in general parlance, the term '5.1 surround' will be used below. While without doubt a compromise, it has become widely adopted in professional and consumer circles and is likely to form the basis for consumer surround sound for the foreseeable future.

Various international groups have worked on developing recommendations for common practice and standards in this area, and some of the information below is based on the effort of the AES Technical Committee on Multichannel and Binaural Audio Technology to bring together a number of proposals.

17.3.1 Purpose of 5.1-channel systems

Four-channel systems have the disadvantage of a mono surround channel, and this limitation is removed in the 5.1-channel system, enabling the provision of stereo effects or room ambience to accompany a primarily front-orientated sound stage. This front-oriented paradigm is a most important one as it emphasises the intentions of those that finalised this configuration, and explains the insistence in some standards on the use of the term '3-2 stereo' rather than 'five-channel surround'. Essentially the front three channels are intended to be used for a conventional three-channel stereo sound image, while the rear/side channels are only intended for generating supporting ambience, effects or 'room impression'. In this sense, the standard does not directly support the concept of 360° image localisation, although it may be possible to arrive at recording techniques or signal processing methods that achieve this to a degree.

The front–rear distinction is a conceptual point often not appreciated by those that use the format. Two-channel stereo can be relatively easily modelled and theoretically approached in terms of localisation vectors etc. for sounds at any angle between the loudspeakers. It is more difficult, though, to come up with such a model for the five-channel layout described below, as it has unequal angles between the loudspeakers and a particularly large angle between the two rear loudspeakers. It is possible to arrive at gain and phase relationships between these five loudspeakers that are similar to those used in Ambisonics for representing different source angles, but the varied loudspeaker angles make the imaging stability less reliable in some sectors than others. For those who do not have access to the sophisticated panning laws or psychoacoustic matrices required to feed five channels accurately for all-round localisation it may be better to treat the format in 'cinema style' – in other words with a three-channel front image and two surround effect channels. With such an approach it is still possible to create very convincing spatial illusions, with good envelopment and localisation qualities.

One cannot introduce the 5.1 surround system without explaining the meaning of the '.1' component. This is a dedicated low-frequency effects (LFE) channel or sub-bass channel. It is called '.1' because of its limited bandwidth. Strictly, the international standard nomenclature for 5.1 surround should be '3-2-1', the last digit indicating the number of LFE channels.

17.3.2 International standards and configurations

The loudspeaker layout and channel configuration is specified in the ITU-R BS.775 standard. This is shown in Figure 17.3 and Fact File 17.1 A display screen is also shown in the figure for sound with picture applications, and there are recommenda-

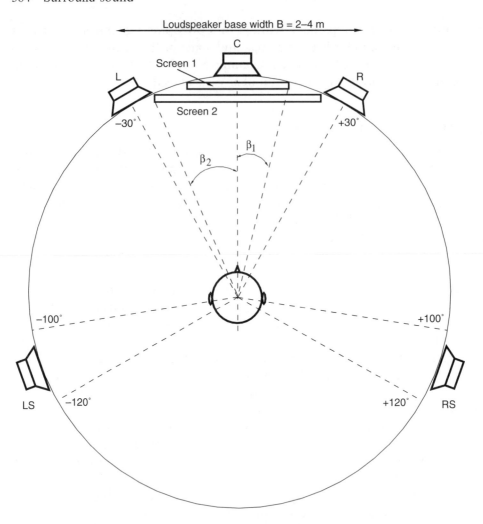

Screen 1: Listening distance = 3H (2 β_1 = 33°) (possibly more suitable for TV screen)
Screen 2: Listening distance = 2H (2 β_2 = 48°) (more suitable for projection screen)
H: Screen height

Figure 17.3 3-2 format reproduction according to the ITU-R BS.775 standard uses two independent surround channels routed to one or more loudspeakers per channel

tions concerning the relative size of the screen and the loudspeaker base width shown in the accompanying table. The left and right loudspeakers are located at ±30° for compatibility with two-channel stereo reproduction. In many ways this need for compatibility with 2/0 is a pity, because the centre channel unavoidably narrows the front sound stage in many applications, and the front stage could otherwise take advantage of the wider spacing facilitated by three-channel reproduction. It was none the less considered crucial for the same loudspeaker configuration to be usable for all standard forms of stereo reproduction, for reasons most people will appreciate.

FACT FILE
17.1

Track allocations in 5.1

Standards recommend the track allocations to be used for 5.1 surround on eight-track recording formats, as shown in the table below.

Although other configurations are known to exist there is a strong move to standardise on this arrangement (see also the notes below the table).

Track[1]	Signal		Comments	Colour[2]
1	L	Left		Yellow
2	R	Right		Red
3	C	Centre		Orange
4	LFE	Low frequency enhancement	Additional sub-bass and effects signal for subwoofer, optional[3]	Grey
5	LS	Left surround	−3 dB in the case of mono surround	Blue
6	RS	Right surround	−3 dB in the case of mono surround	Green
7	Free use in programme exchange[4]		Preferably left signal of a 2/0 stereo mix	Violet
8	Free use in programme exchange[4]		Preferably right signal of a 2/0 stereo mix	Brown

1 The term 'track' is used to mean either tracks on magnetic tape or virtual tracks on other storage media where no real tracks exist.
2 This colour coding is only a proposal of the German Surround Sound Forum at present, and not internationally standardised.
3 Preferably used in film sound, but is optional for home reproduction. If no LFE signal is being used, track 4 can be used freely, e.g. for commentary. In some regions a mono surround signal $MS = LS + RS$ is applied, where the levels of LS and RS are decreased by 3 dB before summing.
4 Tracks 7 and 8 can be used alternatively, for example for commentary, for additional surround-signals, or for half-left/half-right front signal (e.g. for special film formats), or rather for the matrix format sum signal L_t/R_t.

The surround loudspeaker locations, at approximately ±110°, are placed so as to provide a compromise between the need for effects panning behind the listener and the lateral energy important for good envelopment. In this respect they are more like 'side' loudspeakers than rear loudspeakers, and in many installations this is an inconvenient location causing people to mount them nearer the rear than the standard suggests. (Some have said that a 150° angle for the rear loudspeakers provides a more exciting surround effect.) In the 5.1 standard there are normally no loudspeakers directly behind the listener, which can make for creative difficulties. This has led to a Dolby proposal called EX (described below) that places an additional speaker at the centre-rear location. (This is not part of the current standard, though.) The ITU standard allows for additional surround loudspeakers to cover the region around listeners, similar to the 3-1 arrangement described earlier. If these are used then they are expected to be distributed evenly in the angle between ±60° and ±150°.

Surround loudspeakers should be the same as front loudspeakers where possible, in order that uniform sound quality can be obtained all around. That said, there are arguments for use of dipole loudspeakers in these positions. Dipoles radiate sound in more of a figure-eight pattern and one way of obtaining a diffuse surround impression is to orient these with the nulls of the figure-eight towards the listening position. In this way the listener experiences more reflected than direct sound and this can give the impression of a more spacious ambient soundfield that may better emulate the cinema listening experience in small rooms. Dipoles make it correspondingly more difficult to create defined sound images in rear and side positions, though.

17.3.3 The LFE channel and use of subwoofers

The low-frequency effects channel is a separate sub-bass channel with an upper limit extending to a maximum of 120 Hz (see Fact File 17.2). It is intended for conveying special low-frequency content that requires greater sound pressure levels and headroom than can be handled by the main channels. It is not intended for conveying the low-frequency component of the main channel signals, and its application is likely to be primarily in sound-for-picture applications where explosions and other high-level rumbling noises are commonplace, although it may be used in other circumstances.

In consumer audio systems, reproduction of the LFE channel is considered optional. Because of this, recordings should normally be made so that they sound satisfactory even if the LFE channel is not reproduced. The EBU (European Broadcasting Union) comments on the use of the LFE channel as follows:

When an audio programme originally produced as a feature film for theatrical release is transferred to consumer media, the LFE channel is often derived from the dedicated theatrical subwoofer channel. In the cinema, the dedicated subwoofer channel is always reproduced, and thus film mixes may use the subwoofer channel to convey important low frequency programme content. When transferring programmes originally produced for the cinema over television media (e.g. DVD), it may be necessary to re-mix some of the content of the subwoofer channel into the main full bandwidth channels. It is important that any low frequency audio which is very significant to the integrity of the programme content is not placed into the LFE channel. The LFE channel should be reserved for extreme low frequency, and for very high level < 120 Hz programme content which, if not reproduced, will not compromise the artistic integrity of the programme.

With cinema reproduction the in-band gain of this channel is usually 10 dB higher than that of the other individual channels. This is achieved by a level increase of the reproduction channel, not by an increased recording level. (This does not mean that

FACT FILE
17.2

Bass
management
in 5.1

It is a common misconception that any sub-bass or subwoofer loudspeaker(s) that may be used on reproduction must be fed directly from the LFE channel in all circumstances. While this may be the case in the cinema, bass management in the consumer reproducing system is not specified in the standard and is entirely system dependent. It is not mandatory to feed low-frequency information to the LFE channel during the recording process, neither is it mandatory to use a subwoofer, indeed it has been suggested that restricting extreme low-frequency information to a monophonic channel may limit the potential for low-frequency spaciousness in balances. In music

mixing it is likely to be common to send the majority of full-range LF information to the main channels, in order to retain the stereo separation between them.

In practical systems it may be desirable to use one or more subwoofers to handle the low-frequency content of a mix on reproduction. The benefit of this is that it enables the size of the main loudspeakers to be correspondingly reduced, which may be useful practically when it comes to finding places to put them in living rooms or sound control rooms. In such cases crossover systems split the signals between main loudspeakers and subwoofer(s) somewhere between 80 Hz and 160 Hz. In order to allow for reproduction of the LFE channel and/or the low-frequency content from the main channels through subwoofer loudspeakers, a form of bass management akin to that shown below is typically employed.

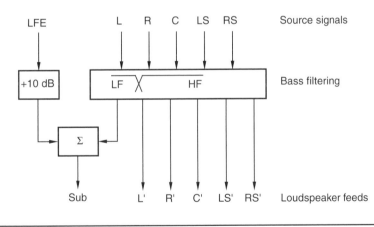

the broadband or weighted SPL of the LFE loudspeaker should measure 10 dB higher than any of the other channels – in fact it will be considerably less than this as its bandwidth is narrower.)

17.3.4 Limitations of 5.1-channel reproduction

The main limitations of the 5.1 surround format are firstly, that it was not intended for accurate 360° phantom imaging capability, as explained above. While it may be possible to achieve a degree of success in this respect, the loudspeaker layout is not ideally suited to it. Secondly, the front sound stage is narrower than it could be if compatibility with 2/0 reproduction was not a requirement. Thirdly, the centre channel can prove problematic for music balancing, as conventional panning laws and coincident microphone techniques are not currently optimised for three loudspeakers, having been designed for two-speaker stereo. Simple bridging of the centre loudspeaker between left and right signals has the effect of narrowing the front image compared with a two-channel stereo reproduction of the same material. This may be resolved over time as techniques suited better to three-channel stereo are resurrected or developed. Fourthly, the LS and RS loudspeakers are located in a compromise position, leading to a large hole in the potential image behind the listener and making it difficult to find physical locations for the loudspeakers in practical rooms.

These various limitations of the format, particularly in some peoples' view for music purposes, have led to various non-standard uses of the five or six channels available on new consumer disk formats such as DVD-A (Digital Versatile Disk – Audio) and SACD (Super Audio Compact Disc). For example, some are using the sixth channel (that would otherwise be LFE) in its full bandwidth form on these media to create a height channel. Others are making a pair out of the 'LFE' channel and the centre channel so as to feed a pair of front-side loudspeakers, enabling the rear loudspeakers to be further back. These are non-standard uses and should be clearly indicated on any recordings.

17.3.5 Signal levels in 5.1 surround

In film sound environments it is the norm to increase the relative recording level of the surround channels by 3 dB compared with that of the front channels. This is in order to compensate for the −3 dB acoustic alignment of each surround channel's SPL with respect to the front that takes place in dubbing stages and movie theatres. It is important to be aware of this discrepancy between practices, as it is the norm in music mixing and broadcasting to align all channels for equal level both on recording media and for acoustical monitoring. Transfers from film masters to consumer or broadcast media may require 3 dB alteration in the gain of the surround channels.

17.4 Other multichannel configurations

Although the 5.1 surround standard is becoming widely adopted as the norm for the majority of installations, other proposals and systems exist, typically involving more channels to cover a large listening area more accurately. It is reasonable to assume that the more real loudspeakers exist in different locations around the listener, the less one has to rely on the formation of phantom images to position sources accurately, and the more freedom one has in listener position. The added complication of mixing for such larger numbers of channels must be considered as a balancing factor.

The reader is also referred to the discussion of Ambisonics, as this system can be used with a wide range of different loudspeaker configurations depending on the decoding arrangements used.

17.4.1 7.1 channel surround

Deriving from widescreen cinema formats, the 7.1 channel configuration normally adds two further loudspeakers to the 5.1 channel configuration, located at centre-left (CL) and centre-right (CR), as shown in Figure 17.4. This is not a format primarily intended for consumer applications, but for large cinema auditoria where the screen width is such that the additional channels are needed to cover

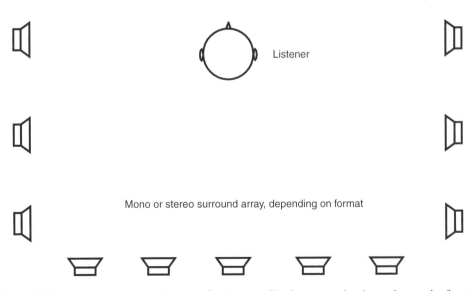

Figure 17.4 Some cinema sound formats for large auditorium reproduction enhance the front imaging accuracy by the addition of two further loudspeakers, centre-left and centre-right

the angles between the loudspeakers satisfactorily for all the seats in the auditorium. Sony's SDDS cinema system is a common proprietary implementation of this format, as is the original 70 mm Dolby Stereo format (see below), although the original 70 mm analogue format only used one surround channel.

Lexicon and Meridian have also implemented a seven-channel mode in their consumer surround decoders, but the recommended locations for the loudspeakers are not quite the same as in the cinema application. The additional channels are used to provide a wider side-front component and allow the rear speakers to be moved round more to the rear than in the 5.1 arrangement.

17.4.2 10.2 channel surround

Tomlinson Holman has spent considerable effort promoting a 10.2-channel surround sound system as 'the next step' in spatial reproduction, but this has not yet been adopted as standard. To the basic five-channel array he adds wider side-front loudspeakers and a centre-rear channel to 'fill in the holes' in the standard layout. He also adds two height channels and a second LFE channel. The second LFE channel is intended to provide lateral separation of decorrelated low bass content to either side of the listening area, as suggested by Griesinger to enhance low-frequency spaciousness.

17.5 Surround sound systems

This part of the chapter concerns what will be called surround sound 'systems', which includes proprietary formats for the coding and transfer of surround sound. These are distinguished from the generic configurations and international standards discussed already. Most of the systems covered here are the subject of patents and intellectual property rights. In some proprietary systems the methods of signal coding or matrixing for storage and delivery are defined (e.g. Dolby Stereo), whilst others define a full source-receiver signal representation system (e.g. Ambisonics).

17.6 Matrixed surround sound systems

While ideally one would like to be able to transfer or store all the channels of a surround sound mix independently and discretely, it may be necessary to make use of existing two-channel media for compatibility with other systems. The systems described in the following sections all deal with multichannel surround sound in a matrixed form (in other words, using an algorithm that combines the channels in such a way that they can be subsequently extracted using a suitable decoder). By matrixing the signals they can be represented using fewer channels than the source material contains. This gives rise to some side-effects and the signals require careful dematrixing, but the approach has been used widely for many years, mainly because of the unavailability of multichannel delivery media in many environments.

17.6.1 Dolby Stereo, Surround and Prologic

Dolby Labs was closely involved with the development of cinema surround sound systems, and gradually moved into the area of surround sound for consumer applications.

The original Dolby Stereo system involved a number of different formats for film sound with three to six channels, particularly a 70 mm film format with six discrete tracks of magnetically recorded audio, and a 35 mm format with two optically recorded audio tracks onto which were matrixed four audio channels in the 3-1 configuration (described above). The 70 mm format involved L, LC, C, RC, R and S channels, whereas the 35 mm format involved only L, C, R and S. Both clearly only involved mono surround information. The four-channel system is the one most commonly known today as Dolby Stereo, having found widespread acceptance in the cinema world and used on numerous movies. Dolby Surround was introduced in 1982 as a means of emulating the effects of Dolby Stereo in a consumer environment. Essentially the same method of matrix decoding was used, so movies transferred to television formats could be decoded in the home in a similar way to the cinema. Dolby Stereo optical sound tracks for the cinema were Dolby A noise-reduction encoded and decoded, in order to improve the signal-to-noise ratio, but this is not a feature of consumer Dolby Surround (more recent cinema formats have used Dolby SR-type noise reduction, alongside a digital soundtrack).

The Dolby Stereo matrix (see Figure 17.5) is a form of '4-2-4' matrix that encodes the mono surround channel so that it is added out of phase into the left and right channels ($+90°$ in one channel and $-90°$ in the other). The centre channel signal is added to left and right in phase. The resulting sum is called L_t/R_t (left total and right total). By doing this the surround signal can be separated from the front signals upon decoding by summing the L_t/R_t signals out of phase (extracting the stereo difference signal), and the centre channel can be extracted by summing L_t/R_t in phase. In consumer systems using passive decoding the centre channel is not always fed to a separate loudspeaker but can be heard as a phantom image between left and right. A decoder block diagram for the consumer version (Dolby Surround) is shown in Figure 17.6. Here it can be seen that in addition to the sum-and-difference-style decoding, the surround channel is subject to an additional delay, band-limiting between 100 Hz and 7 kHz and a modified form of Dolby B noise reduction. The low-pass filtering and the delay are both designed to reduce matrix

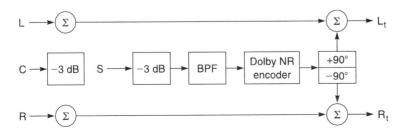

Figure 17.5 Basic components of the Dolby Stereo matrix encoding process

Figure 17.6 Basic components of the passive Dolby surround decoder

side-effects that could otherwise result in front signals appearing to come from behind. Crosstalk between channels and effects of any misalignment in the system can cause front signals to 'bleed' into the surround channel, and this can be worse at high frequencies than low. The delay (of the order of 20–30 ms in consumer systems, depending on the distance of the rear speakers) relies on the precedence effect (see Chapter 2) to cause the listener to localise signals according to the first arriving wavefront which will now be from the front rather than the rear of the sound stage. The rear signal then becomes psychoacoustically better separated from the front and localisation of primary signals is biased more towards the front. The modified B-type NR reduces surround channel noise and also helps to reduce the effects of decoding errors and interchannel crosstalk, as some distortions introduced between encoding and decoding will be reduced by B-type decoding.

A problem with passive Dolby Surround decoding is that the separation between adjacent channels is relatively modest, although the separation of left/right and centre/surround remains high. When a signal is panned fully left it will tend to appear only 3 dB down in the centre, and also in the surround, for example. The effects of this can be ameliorated in passive consumer systems by the techniques described above (phantom centre and surround delay/filtering). Dolby's ProLogic system, based on principles employed in the professional decoder, attempts to resolve this problem by including sophisticated 'steering' mechanisms into the decoder circuit to improve the perceived separation between the channels. A basic block diagram is shown in Figure 17.7. This enables a real centre loudspeaker to be employed. Put crudely, ProLogic works by sensing the location of 'dominant' signal components and selectively attenuating channels away from the dominant component. (A variety of other processes are involved as well as this.) So, for example, if a dialogue signal is predominantly located in the centre, the control circuit will reduced the output of the other channels (L, R, S) in order that the signal comes mainly from the centre loudspeaker. (Without this it would also have appeared at quite high level in left and right as well.) A variety of algorithms are used to determine how quickly the system should react to changes in dominant signal position, and what to do when no signal appears dominant.

Figure 17.7 Basic components of the active Dolby ProLogic decoder

Dolby has recently introduced an enhancement to ProLogic, entitled ProLogic 2, that adds support for full-bandwidth stereo rear channels, with various options that make it more suitable for music programmes. It is also claimed to be effective in the up-conversion of unencoded two-channel material to five-channel surround.

Mixes that are to be matrix encoded using the Dolby system should be monitored via the encode–decode chain in order that the side-effects of the process can be taken into account by the balance engineer. Dolby normally licenses the system for use on a project, and will assist in the configuration and alignment of their equipment during the project.

Dolby Stereo/Surround can be complemented by the THX system, as described in Fact File 17.3.

17.6.2 Circle Surround

Circle Surround was developed by the Rocktron Corporation (RSP Technologies) as a matrix surround system capable of encoding stereo surround channels in addition to the conventional front channels. They proposed the system as more appropriate than Dolby Surround for music applications, and claimed that it should be suitable for use on material that had not been encoded as well as that which had.

The Circle Surround encoder is essentially a sum and difference L_t/R_t process (similar to Dolby but without the band limiting and NR encoding of the surround channel). One incarnation of this involves 5-2 encoding, intended for decoding back to five channels (the original white paper on the system described a 4-2 encoder). Among other methods, the Circle decoder steers the rear channels separately according to a split-band technique that steers low- and high-frequency components independently from each other. In this way they claim to avoid the broad-band 'pumping' effects associated with some other systems. They also decode the rear channels slightly differently, using L–R for the left rear channel and R–L for the right rear channel, which it is claimed allows side-images to be created on either

---

FACT FILE

17.3

What is THX?

The THX system was developed by Tomlinson Holman at Lucasfilm (THX is derived from 'Tomlinson Holman Experiment'). The primary aim of the system was to improve the sound quality in movie theatres and make it closer to the sound experienced by sound mixers during post-production. It was designed to complement the Dolby Stereo system, and does not itself deal with the encoding or representation of surround sound. In fact THX is more concerned with the acoustics of cinemas and the design of loudspeaker systems, optimising the acoustic characteristics and noise levels of the theatre, as well as licensing a particular form of loudspeaker system and crossover network. THX licenses the system to theatres and requires that the installation is periodically tested to ensure that it continues to meet the specification.

Home THX was developed, rather like Dolby Surround, in an attempt to convey the cinema experience to the home. Through the use of a specific controller, amplifiers and speakers, the THX system enhances the decoding of Dolby Surround and can also be used with digital surround sound signals. The mono surround signal of Dolby Surround is subject to decorrelation of the signals sent to the two surround loudspeakers in order that the surround signal is made more diffuse and less 'mono'. It is claimed that this has the effect of preventing surround signals from collapsing into the nearest loudspeaker. Signals are re-equalised to compensate for the excessive high-frequency content that can arise when cinema balances are replayed in small rooms, and the channels are 'timbre matched' to compensate for the spectral changes that arise when sounds are panned to different positions around the head.

In terms of hardware requirements, the Home THX system also specifies certain aspects of amplifier performance, as well as controlling the vertical and horizontal directivity of the front loudspeakers. Vertical directivity is tightly controlled to increase the direct sound component arriving at listeners, while horizontal directivity is designed to cover a reasonably wide listening area. Front speakers should have a frequency response from 80 Hz to 20 kHz and all speakers must be capable of radiating an SPL of 105 dB without deterioration in their response or physical characteristics. The surround speakers are unusual in having a bipolar radiation pattern, arranged so that the listener hears reflected sound rather than direct sound from these units. These have a more relaxed frequency response requirement of 125 Hz to 8 kHz. A subwoofer feed is usually also provided.

---

side. They avoid the use of a delay in the rear channels for the 'Music' mode of the system and do not band-limit the rear channels as Dolby Surround does.

17.6.3 Lexicon Logic 7

Logic 7 is another surround matrix decoding process that can be used as an alternative for Dolby Surround decoding. Variants on this algorithm (such as the so-called Music Logic and Music Surround modes) can also be used for generating a good surround effect from ordinary two-channel material. Lexicon developed the algorithm for its high-end consumer equipment, and it is one of a family of steered decoding processes that distributes sound energy appropriately between a number of loudspeakers depending on the gain and phase relationships in the source material. In this case seven loudspeaker feeds are provided rather than five, adding two 'side' loudspeakers to the array, as shown in Figure 17.8. The rear speakers can then be further to the rear than would otherwise be desirable. The side loudspeakers can be used for creating an enhanced envelopment effect in music modes and more accurate side panning of effects in movie sound decoding.

In Logic 7 decoding of Dolby matrix material the front channel decoding is almost identical to Dolby ProLogic, with the addition of a variable centre channel delay to compensate for non-ideal locations of the centre speaker. The rear channels operate differently depending on whether the front channel content is primarily steered dialogue/effects or music/ambience. In the former case the front signals are cancelled from the rear channels and panned effects behave as they would with ProLogic, with surround effects panned 'full rear' appearing in mono on both rear channels. In the latter case the rear channels work in stereo, but reproducing the front left and right channels with special equalisation and delay to create an

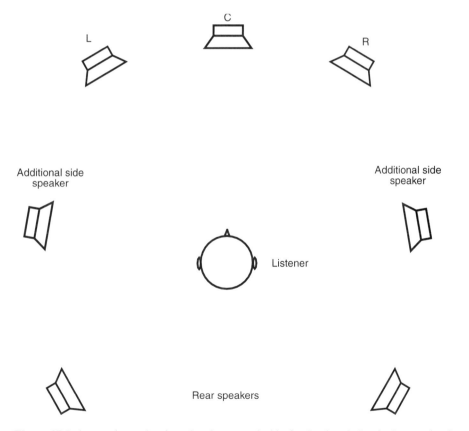

Figure 17.8 Approximate loudspeaker layout suitable for Lexicon's Logic 7 reproduction. Notice the additional side loudspeakers that enable a more enveloping image and may enable rear loudspeakers to be placed further to the rear

enveloping spatial effect. The side channels carry steered information that attempts to ensure that effects which pan from left to rear pass through the left-side on the way, and similarly for the right side with right-to-rear pans.

It is claimed that by using these techniques the effect of decoding a 3-1 matrix surround version of a 3-2 format movie can be brought close to that of the original 3-2 version. Matrix encoding of five channels to L_t/R_t is also possible with a separate algorithm, suitable for decoding to five or more loudspeakers using Logic 7.

17.6.4 Dolby EX

In 1998 Dolby and Lucasfilm THX joined forces to promote an enhanced surround system that added a centre rear channel to the standard 5.1-channel setup. They introduced it, apparently, because of frustrations felt by sound designers for movies in not being able to pan sounds properly to the rear of the listener – the surround effect typically being rather diffuse. This system was christened 'Dolby Digital – Surround EX', and apparently uses matrix-style centre channel encoding and decoding between the left and right surround channels of a 5.1-channel mix. The loudspeakers at the rear of the auditorium are then driven separately from those on the left and right sides, using the feed from this near 'rear centre' channel, as shown in Figure 17.9.

17.7 Digital surround sound formats

Matrix surround processes are gradually giving way to digital formats that enable multiple channels to be delivered discretely, bypassing the two-channel restriction of most previous delivery formats. While it is desirable to be able to store or transfer multichannel surround sound signals in a discrete, full-resolution digital PCM format, this can occupy considerable amounts of storage space or transmission bandwidth (somewhere between about 0.75 and 2 Mbit/s per channel, depending on the resolution). This is too high for practical purposes in broadcasting, film sound and consumer systems, using current technology. Consequently a number of approaches have been developed whereby the information can be digitally encoded at a much lower bit rate than the source material, with minimal loss of sound quality (see Chapter 10). The sections below briefly describe some of these systems, used in cinema sound, digital consumer formats and broadcasting systems.

17.7.1 Dolby Digital

Dolby Digital or AC-3 encoding was developed as a means of delivery 5.1-channel surround to cinemas or the home without the need for analogue matrix encoding. It is likely to replace Dolby Stereo/Surround gradually as digital systems replace analogue ones. It relies on a digital low-bit-rate encoding and decoding process that enables the multiple channels of the surround mix to be conveyed without the separation and steering problems inherent in matrixed surround. Dolby Digital

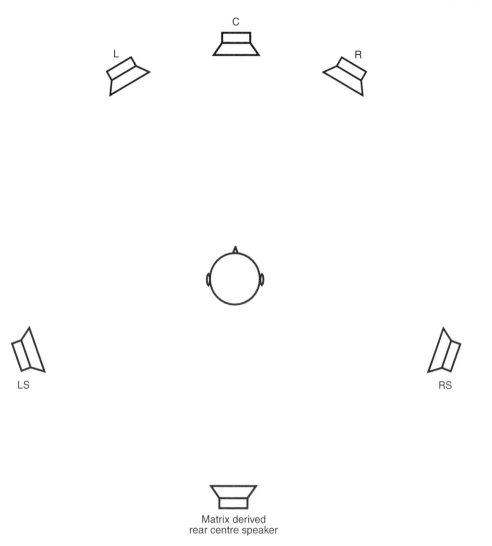

Figure 17.9 Dolby EX adds a centre-rear channel fed from a matrix-decoded signal that was originally encoded between left and right surround channels in a manner similar to the conventional Dolby Stereo matrix process

can code signals based on the ITU-standard 3-2-1 surround format of loudspeakers and it should be distinguished from such international standards since it is primarily a signal coding and representation method. In fact, the AC-3 coding algorithm can be used for a wide range of different audio signal configurations and bit rates from 32 kbit/s for a single mono channel up to 640 kbit/s for surround signals. It is used widely for the distribution of digital sound tracks on 35 mm movie films, the data being stored optically in the space between the sprocket holes on the film, as shown in Figure 17.10. In this way, the analogue optical soundtracks can be retained in their normal place alongside the picture for compatibility purposes. In this format it is combined with a Dolby-SR encoded analogue Dolby Stereo mix, and the

Figure 17.10 Dolby Digital data is stored optically in an area between the sprocket holes of 35 mm film. (Courtesy of Dolby Laboratories, Inc.)

combined format is called Dolby SR-D. Dolby Digital is also used for surround sound on DVD video releases, and for certain digital broadcasting applications.

The Dolby Digital encoding process can be controlled by a software application that enables various parameters of the encoding process to be varied, as shown in Figure 17.11. Dolby Digital can operate at sampling rates of 32, 44.1 or 48 kHz, and the LFE channel is sampled at 240 Hz (because of its limited bandwidth). A 90° phase shift is normally introduced into each of the surround channels during encoding, which apparently improves the smoothness of front–back panning and reduces crosstalk between centre and surround channels when decoded to Dolby Surround. For this reason it is important to monitor recordings via the encode–decode process to ensure that this phase shift does not affect the spatial intention of the producer.

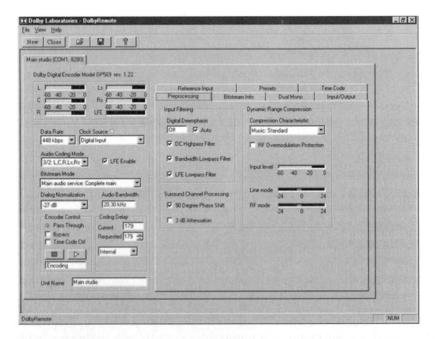

Figure 17.11 Screen display of Dolby Digital encoding software options. (Courtesy of Dolby Laboratories, Inc.)

Aside from the representation of surround sound in a compact digital form, Dolby Digital includes a variety of operational features that enhance system flexibility and help adapt replay to a variety of consumer situations. These include dialogue normalisation ('dialnorm') and the option to include dynamic range control information alongside the audio data for use in environments where background noise prevents the full dynamic range of the source material to be heard. Downmix control information can also be carried alongside the audio data in order that a two-channel version of the surround sound material can be reconstructed in the decoder. As a rule, Dolby Digital data is stored or transmitted with the highest number of channels needed for the end product to be represented, and any compatible downmixes are created in the decoder. This differs from some other systems where a two-channel downmix is carried alongside the surround information.

Dialnorm indication can be used on broadcast and other material to ensure that the dialogue level remains roughly constant from programme to programme. It is assumed that dialogue level is the main factor governing the listening level used in peoples' homes, and that they do not want to keep changing this as different programmes come on the air (e.g. from advertising to news programmes). The dialnorm level is the average dialogue level over the duration of the programme compared to the maximum level that would be possible, measured using an A-weighted L_{EQ} reading (this averages the level linearly over time). So, for example, if the dialogue level averaged at 70 dBA over the programme, and the SPL corresponding to peak recording level was 100 dBA, the dialnorm setting would be -30 dB.

17.7.2 DTS

The DTS (Digital Theater Systems) 'Coherent Acoustics' system is another digital signal coding format that can be used to deliver surround sound in consumer or professional applications, using low bit rate coding techniques to reduce the data rate of the audio information. The DTS system can accommodate a wide range of bit rates from 32 kbits/up to 4.096 Mbit/s (somewhat higher than Dolby Digital), with up to eight source channels and with sampling rates up to 192 kHz. Variable bit rate and lossless coding are also optional. Downmixing and dynamic range control options are provided in the system.

DTS data is found on some film releases and occupies a different area of the film from Dolby Digital and SDDS data (see below). In fact it is possible to have film release prints in a multi-format version with all three digital sound formats plus the analogue Dolby Stereo tracks on one piece of film, making it almost universally compatible. DTS is also used on a number of surround CD releases and is optional on DVD, requiring a special decoder to replay the data signal from the disk. Because the maximum data rate is typically somewhat higher than that of Dolby Digital or MPEG, a greater margin can be engineered between the signal and any artefacts of low bit rate coding, leading to potentially higher sound quality. Such judgements, though, are obviously up to the individual and it is impossible to make blanket statements about comparative sound quality between systems.

17.7.3 SDDS

SDDS stands for Sony Dynamic Digital Sound, and is the third of the main competing formats for digital film sound. Using Sony's ATRAC data reduction system, it too encodes audio data with a substantial saving in bit rate compared with the original PCM (about 5:1 compression). The SDDS system employs 7.1 channels rather than 5.1, as described earlier in this chapter, providing detailed positional coverage of the front sound stage. It is not common to find SDDS data on anything but film release prints, although it could be included on DVD as a proprietary format if required. Consumer decoders are not currently available, to the author's knowledge.

17.7.4 MPEG surround modes

The MPEG (Moving Pictures Expert Group) standards are widely used for low bit rate representation of audio and video signals in multimedia and other applications, and was introduced in Chapter 10. While MPEG-1 described a two-channel format, MPEG-2 extended this to multichannel information. There are two versions of MPEG-2, one of which was developed to be backwards compatible with MPEG-1 decoders and the other of which is known as MPEG-2 AAC (for advanced audio coding) and is not backwards compatible. The MPEG-4 standards also include scalable options for multichannel coding. These standards are described in detail in ISSO 11172-3, 13818-3, 13818-7 and 14496 for those who want to understand how they work in detail.

The MPEG-2 BC (backwards compatible) version worked by encoding a matrixed downmix of the surround channels and the centre channel into the left and right channels of an MPEG-1 compatible frame structure. This could be decoded by conventional MPEG-1 decoders. A multichannel extension part was then added to the end of the frame, containing only the C, LS and RS signal channels, as shown in Figure 17.12. Upon decoding in an MPEG-2 surround decoder, the three additional surround components could be subtracted again from the L_0/R_0 signals to leave the original five channels. The main problems with MPEG-2 BC are that (a) the downmix is performed in the encoder so it cannot be changed at the decoder end, and (b) the data rate required to transfer the signal is considerably higher than it would be if backward compatibility were not an issue. Consequently the bit rate required for MPEG-2 BC to transfer 5.1- channel surround with reasonable quality is in the region of 600–900 kbit/s.

MPEG-2 AAC, on the other hand, is a more sophisticated algorithm that codes multichannel audio to create a single bit stream that represents all the channels, in a form that cannot be decoded by an MPEG-1 decoder. Having dropped the requirement for backward compatibility, the bit rate can now be optimised by coding the channels as a group and taking advantage of interchannel redundancy if required. The situation is now more akin to that with Dolby Digital, and the bit rates required for acceptable sound quality are also similar. The MPEG-2 AAC system contained contributions from a wide range of different manufacturers.

Figure 17.12 (a) MPEG-2-BC multichannel extension data appended to the MPEG-1-compatible two-channel frame. (b) Compatibility matrixing of surround information for MPEG-2-BC

The MPEG surround algorithms have not been widely implemented to date in broadcasting, film and consumer applications. Although MPEG-2 BC was originally intended for use with DVD releases in Region 2 countries (primarily Europe), this requirement appears to have been dropped in favour of Dolby Digital. MPEG two-channel standards, such as MPEG-1, Layer 3 (the well-known .MP3 format), have been widely adopted for consumer purposes, on the other hand.

17.7.5 MLP

Meridian Lossless Packing (MLP) is a lossless data reduction technique for multi-channel audio, licensed by Meridian Audio through Dolby Labs. It has been specified for the DVD-Audio format as a way of reducing the data rate required for high-quality recordings without any effect on sound quality (in other words, you get back exactly the same bits you put in, which is not the case with lossy processes like Dolby Digital and MPEG). Using this technique, a sufficient playing time can be obtained from the disk whilst still enabling high audio resolution (sample rate up to 192 kHz and resolution between 16 and 24 bits) and up to six channel surround sound.

MLP enables the mastering engineer to create a sophisticated downmix (for two-channel replay) of the multichannel material that occupies very little extra space on the disk, owing to the exploitation of similarities between this material and the multichannel version during lossless encoding. This downmix can have characteristics that vary during the programme and is entirely under the artistic control of the engineer.

There are also modes of MLP that have not really seen the light of day yet. For example, the system is extensible to considerable numbers of channels, and has an option to incorporate hierarchical encoding processes such as Ambisonics where sound field components rather than loudspeaker feeds are represented. This could be used in future as a means of overcoming the limitations of a loudspeaker-feed-based format for delivering surround sound to consumers.

17.8 Ambisonics

17.8.1 Principles

The Ambisonic system of directional sound pickup and reproduction is discussed here because of its relative thoroughness as a unified system, being based on some key principles of psychoacoustics. It has its theoretical basis in work by Gerzon, Barton and Fellgett in the 1970s, as well as work undertaken earlier by Cooper and Shiga.

Ambisonics aims to offer a complete hierarchical approach to directional sound pickup, storage or transmission and reproduction, which is equally applicable to mono, stereo, horizontal surround-sound, or full 'periphonic' reproduction including height information. Depending on the number of channels employed it is possible to represent a lesser or greater number of dimensions in the reproduced sound. A number of formats exist for signals in the Ambisonic system, and these are as follows: the A-format for microphone pickup, the B-format for studio equipment and processing, the C-format for transmission, and the D-format for decoding and reproduction. A format known as UHJ ('Universal HJ', 'HJ' simply being the letters denoting two earlier surround sound systems) is also used for encoding multichannel Ambisonic information into two or three channels whilst retaining good mono and stereo compatibility for 'non-surround' listeners.

Ambisonic sound should be distinguished from quadraphonic sound, since quadrophonics explicitly requires the use of four loudspeaker channels, and cannot be adapted to the wide variety of pickup and listening situations that may be encountered. Quadraphonics generally works by creating conventional stereo phantom images between each pair of speakers and, as Gerzon states, conventional stereo does not perform well when the listener is off-centre or when the loudspeakers subtend an angle larger than 60°. Since in quadraphonic reproduction the loudspeakers are angled at roughly 90° there is a tendency towards a hole-in-the-middle, as well as there being the problem that conventional stereo theories do not apply correctly for speaker pairs to the side of the listener. Ambisonics, however, encodes sounds from all directions in terms of pressure and velocity components, and decodes these signals to a number of loudspeakers, with psychoacoustically optimised shelf filtering above 700 Hz to correct for the shadowing effects of the head. It also incorporates an amplitude matrix that determines the correct levels for each speaker for the layout chosen. Ambisonics might thus be considered as the theoretical successor to coincident stereo on two loudspeakers, since it is the logical extension of Blumlein's principles to surround sound.

The source of an Ambisonic signal may be an Ambisonic microphone such as the Calrec Soundfield, or it may be an artificially panned mono signal, split into the correct B-format components (see below) and placed in a position around the listener by adjusting the ratios between the signals.

17.8.2 Signal formats

As indicated above, there are four basic signal formats for Ambisonic sound: A, B, C and D. The A-format consists of the four signals from a microphone with four sub-cardioid capsules orientated as shown in Figure 17.13 (or the pan-pot equivalent of such signals). These are capsules mounted on the four faces of a tetrahedron, and correspond to left-front (LF), right-front (RF), left-back (LB) and right-back (RB), although two of the capsules point upwards and two point downwards. Such signals should be equalised so as to represent the soundfield at the centre of the tetrahedron, since the capsules will not be perfectly coincident.

The B-format consists of four signals that between them represent the pressure and velocity components of the sound field in any direction, as shown in Figure 17.14. It can be seen that there is a similarity with the sum and difference format of

Figure 17.13 A-format capsule directions in an Ambisonic microphone

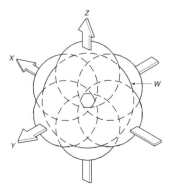

Figure 17.14 The B-format components W, X, Y and Z in Ambisonics represent an omnidirectional pressure component and three orthogonal velocity (figure-eight) components of the sound field respectively

two channel stereo, described in the previous chapter, since the B-format is made up of three orthogonal figure-eight components (X, Y and Z), and an omni component (W). All directions in the horizontal plane may be represented by scalar and vector combinations of W, X and Y, whilst Z is required for height information. X is equivalent to a forward-facing figure-eight (equivalent to M in MS stereo), Y is equivalent to a sideways-facing figure-eight (equivalent to S in MS stereo). The X, Y and Z components have a frontal, sideways or upwards gain of +3 dB or $\sqrt{2}$ with relation to the W signal (0 dB) in order to achieve roughly similar energy responses for sources in different positions.

A B-format signal may be derived from an A-format microphone. In order to derive B-format signals from these capsule-pair signals it is a simple matter of using sum and difference technique. Thus:

$$X = 0.5((LF{-}LB) + (RF{-}RB))$$
$$Y = 0.5((LF{-}RB) - (RF{-}LB))$$
$$Z = 0.5((LF{-}LB) + (RB))$$

W, being an omni pressure component, is simply derived by adding the outputs of the four capsules in phase, thus:

$$W = 0.5(LF + LB + RF + RB)$$

In a microphone W, X, Y and Z are corrected electrically for the differences in level between them, so as to compensate for the differences between pressure and velocity components. For example, W is boosted at very low frequencies since it is derived from velocity capsules that do not have the traditionally extended bass response of omnis.

B-format signals may also be created directly by arranging capsules or individual microphones in the B-format mode (two or three figure-eights at 90° plus an omni). The Z component is not necessary for horizontal information. If B-format signals are recorded instead of speaker feeds (D-format), subsequent manipulation of the soundfield is possible, and the signal will be somewhat more robust to interchannel errors.

Taking ϑ as the angle of the incidence in the horizontal plane (the azimuth), and η as the angle of elevation above the horizontal, then in the B-format the polar patterns of the different signals can be represented as follows:

$$W = 1$$
$$X = \sqrt{2}\cos\vartheta\cos\eta$$
$$Y = \sqrt{2}\sin\vartheta\cos\eta$$
$$Z = \sqrt{2}\sin\eta$$

The C-format consist of four signals L, R, T and Q, which conform to the UHJ hierarchy, and are the signals used for mono or stereo-compatible transmission or recording. The C-format is, in effect, a useful consumer matrix format. L is a two-channel-compatible left channel, R is the corresponding right channel, T is a third channel which allows more accurate horizontal decoding, and Q is a fourth channel containing height information. The proportions of B-format signals which are combined to make up a C-format signal have been carefully optimised for the best compatibility with conventional stereo and mono reproduction. If L + R is defined as Σ (similar to M in MS stereo) and L–R is defined as Δ (similar to S in MS stereo), then:

$$\Sigma = 0.9397W + 0.1856X$$

$$\Delta = j(-0.3420W + 0.5099X) + 0.6555Y$$

$$T = j(-0.1432W + 0.6512X) - 0.7071Y$$

$$Q = 0.9772Z$$

where j (or $\sqrt{-1}$) represents a phase advance of $90°$.

Two, three, or four channels of the C-format signal may be used depending on the degree of directional resolution required, with a two-and-a-half channel option available where the third channel (T) is of limited bandwidth. For stereo compatibility only L and R are used (L and R being respectively $0.5(\Sigma + \Delta)$ and $0.5(\Sigma - \Delta)$). The UHJ or C-format hierarchy is depicted graphically in Figure 17.15.

D-format signals are those distributed to loudspeakers for reproduction, and are adjusted depending on the selected loudspeaker layout. They may be derived from either B- or C-format signals using an appropriate decoder, and the number of speakers is not limited in theory, nor is the layout constrained to a square. Four speakers give adequate surround sound, whilst six provide better immunity against the drawing of transient and sibilant signals towards a particular speaker, and eight may be used for full periphony with height. The decoding of B- and C-format components into loudspeaker signals is too complicated and lengthy a matter to go into here, and is the subject of several patents that were granted to the NRDC (the UK National Research and Development Council, as was). It is sufficient to say

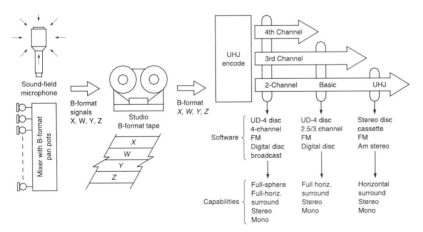

Figure 17.15 The C-format or UHJ hierarchy enables a variety of matrix encoding forms for stereo signals, depending on the amount of spatial information to be conveyed and the number of channels available

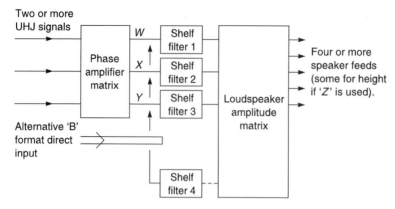

Figure 17.16 C-format signals are decoded to provide D-format signals for loudspeaker reproduction

that the principle of decoding involves the passing of two or more UHJ signals via a phase-amplitude matrix, resulting in B-format signals that are subjected to shelf filters (in order to correct the levels for head-related transfer functions such as shadowing and diffraction). These are passed through an amplitude matrix which feeds the loudspeakers (see Figure 17.16). A layout control is used to vary the level sent to each speaker depending on the physical arrangement of speakers. See also Fact File 17.4.

FACT FILE

17.4

Higher-order Ambisonics

The incorporation of additional directional components into the Ambisonic signal structure can give rise to improved directional encoding that covers a larger listening area than first-order Ambisonics. These second-order and higher components are part of a family of so-called 'spherical harmonics'. Horizontal Ambisonics can be enhanced by the addition of two further components, U and V, which have polar patterns described by:

$$U = 2\cos(2\vartheta)$$
$$V = 2\sin(2\vartheta)$$

provided that an appropriate decoder is implemented that can deal with the second-order components. Even higher-order components can be generated with the general form:

$$c_n \text{ (forwards)} = 2\cos(n\vartheta)$$
$$c_n \text{ (sideways)} = 2\sin(n\vartheta)$$

The problem with higher-order Ambisonics is that it is much more difficult to design microphones that produce the required polar patterns, although the signals can be synthesised artificially for sound modelling and rendering applications.

17.8.3 B-format-to-5.1 decoding

Although the original Ambisonic specifications assumed symmetrical rectangular or square loudspeaker layouts, Gerzon showed in 1992 how Ambisonic signals could be decoded with reasonable success to layouts such as the five-channel configuration described earlier. These are often referred to as 'Vienna decoders' after the location of the AES Convention at which these were first described. The sound image is in this case 'front biased', with better localisation characteristics in the frontal region than the rear, owing to the loudspeaker layout. This is an unavoidable feature of such a configuration in any case.

17.9 Surround sound monitoring

This section is principally concerned with monitoring environments and configurations for 5.1-channel surround sound, although many of the principles may be found to apply in other configurations. The Audio Engineering Society has published an information document on this topic, containing more detailed guidelines (see Recommended further reading at the end of this chapter).

17.9.1 Differences between two-channel and surround mixing rooms

There is a gradual consensus building around the view that rooms for surround monitoring should have an even distribution of absorbing and diffusing material. This is so that the rear loudspeakers function in a similar acoustic environment to the front loudspeakers. This is contrary to a number of popular two-channel control room designs that have one highly absorbtive end and the other end more reflective.

The effects of the acoustics of non-ideal control room acoustics on surround channels may be ameliorated if a distributed array of surround loudspeakers is used, preferably with some form of decorrelation between them to avoid strong comb filtering effects. (Appropriate gain/EQ modification should also be applied to compensate for the acoustic summing of their outputs.) This is more akin to the film sound situation, though, and may only be possible in larger dubbing stages. In smaller control rooms used for music and broadcasting mixing the space may not exist for such arrays. The ITU standard allows for more than one surround loudspeaker on either side and recommends that they are spaced equally on an arc from 60–150° from the front.

One of the difficulties of installing loudspeaker layouts according to the ITU standard, with equal spacing from the listening position and the surrounds at $110° ± 10°$, is the required width of the space. This arrangement often makes it appropriate for the room to be laid out 'wide' rather than 'long' (as it might be for two-channel setups). If the room is one that was previously designed for two-channel stereo the rotation of the axis of symmetry may result in the acoustic treatment being inappropriately distributed. Also the location of doors and windows may make the modification of existing rooms difficult. If building a new room for surround monitoring then it is obviously possible to start from scratch and make

---

FACT FILE

17.5   Loudspeaker mounting

In many studios it is traditional to mount the monitor loudspeakers flush with the front wall. This has the particular advantage of avoiding the reflection that occurs with free-standing loudspeakers from the wall behind the loudspeaker, causing a degree of cancellation at a frequency where the spacing is equal to one quarter of the radiated wavelength. It also improves the low-frequency radiation conditions if the front walls are hard. Nonetheless, it is hard to find places to mount five large loudspeakers in a flush-mounted configuration, and such mounting methods can be expensive. Furthermore the problems noted above, of detrimental reflections from rear loudspeakers off a hard front wall or speaker enclosure, can arise, depending on the angle of the rear loudspeakers. For such reasons, some sources recommend making the surfaces around the loudspeakers reflective at low frequencies and absorbent at mid and high frequencies.

The problem of low-frequency cancellation notches with free-standing loudspeakers can be alleviated but not completely removed. The perceived depth of the notch depends on the absorption of the surface and the directivity of the loudspeaker. By adjusting the spacing between the speaker and the wall, the frequency of the notch can be moved (downwards by making the distance greater), but the distance needed is often too great to be practical. If the speaker is moved close to the wall the notch position rises in frequency. This can be satisfactory for large loudspeakers whose directivity is high enough at middle frequencies to avoid too much rear radiation, but is a problem for smaller loudspeakers.

The use of a 5.1-channel monitoring arrangement (rather than five full-bandwidth loudspeakers), with proper bass management and crossovers, can in fact ameliorate the problems of free-standing loudspeakers considerably. This is because a subwoofer can be used to handle frequencies below 80–120 Hz and it can be placed in the corner or near a wall where the cancellation problem is minimised. Furthermore, the low-frequency range of the main loudspeakers can then be limited so that the cancellation notch mentioned above occurs below their cut-off frequency.

---

the room wide enough to accommodate the surround loudspeakers and absorption in more suitable places. See also Fact File 17.5.

17.9.2 Front loudspeakers in general

As a rule, front loudspeakers can be similar to those used for two-channel stereo, although noting the particular problems with the centre loudspeaker described in the next section. It has been suggested that low-directivity front loudspeakers may be desirable when trying to emulate the effect of a film mixing situation in a smaller surround control room. This is because in the large rooms typical of cinema listening the sound balancer is often well beyond the critical distance where direct and reflected sound are equal in level, and using speakers with low directivity helps to emulate this scenario in smaller rooms. Film mixers generally want to hear what the large auditorium audience member would hear, and this means being further from the loudspeakers than for small room domestic listening or conventional music mixing.

17.9.3 What to do with the centre loudspeaker

One of the main problems encountered with surround monitoring is that of where to put the centre loudspeaker in a mixing room. Ideally it should be of the same type or quality as the rest of the channels and this can make such speakers quite large. In 5.1 surround setups there is an increasing tendency to use somewhat smaller monitors for the five main channels than would be used for two-channel setups, handling the low bass by means of a subwoofer or two. This makes it more practical to mount a centre loudspeaker behind the mixing console, but its height will often be dictated by a control room window or video monitor (see below). The centre loudspeaker should be on the same arc as that bounding the other loudspeaker positions, as shown in the ITU layout above, otherwise the time delay of its direct sound at the listening position will be different from that of the other channels. If the centre speaker is closer than the left or right channels then it should be delayed slightly to put it back in the correct place acoustically.

The biggest problem with the centre loudspeaker arises when there is a video display present. A lot of 5.1 surround work is carried out in conjunction with pictures and clearly the display is likely to be in exactly the same place as one wants to put the centre speaker. In cinemas this is normally solved by making the screen acoustically 'transparent' and using front projection, although this transparency is never complete and usually requires some equalisation. In smaller mixing rooms the display is often a flat-screen plasma monitor or a CRT display and these do not allow the same arrangement.

With modestly sized solid displays for television purposes it can be possible to put the centre loudspeaker underneath the display, with the display raised slightly, or above the display angled down slightly. The presence of a mixing console may dictate which of these is possible, and care should be taken to avoid strong reflections from the centre loudspeaker off the console surface. Neither position is ideal and the problem may not be solved easily. Dolby suggests that if the centre loudspeaker has to be offset height-wise it could be turned upside down compared with the left and right channels to make the tweeters line up, as shown in Figure 17.17.

Interestingly, the flat-panel loudspeaker company, NXT, has shown large flat-panel loudspeakers that can double as projection display screens, which may be one way forward if the sound quality of the flat panels speakers can be made high enough.

17.9.4 Surround loudspeakers

The standard recommendations for professional setups suggest that the surround loudspeakers should be of the same quality as the front ones. This is partly to ensure a degree of inter-system compatibility. In consumer environments this can be diffi-

Figure 17.17 Possible arrangement of the centre loudspeaker in the presence of a TV screen, aligning HF units more closely

cult to achieve, and the systems sold at the lower end of the market often incorporate much smaller surround loudspeakers than front. As mentioned above, the use of a separate loudspeaker to handle the low bass (a so-called 'subwoofer') may help to ameliorate this situation, as it makes the required volume of all the main speakers quite a lot smaller. Indeed Bose has had considerable success with a consumer system involving extremely small satellite speakers for the mid–high frequency content of the replay system, mountable virtually anywhere in the room, coupled with a low-frequency driver that can be situated somewhere unobtrusive.

The directivity requirements of the surround loudspeakers have been the basis of some considerable disagreement in recent years. The debate centres around the use of the surround loudspeakers to create a diffuse, enveloping soundfield – a criterion that tends to favour either decorrelated arrays of direct radiators (speakers that produce their maximum output in the direction of the listener) or dipole surrounds (bi-directional speakers that are typically oriented so that their main axis does not point towards the listener). If the creation of a diffuse, enveloping rear and side soundfield is the only role for surround loudspeakers then dipoles can be quite suitable if only two loudspeaker positions are available. If, on the other hand, attempts are to be made at all-round source localisation (which, despite the evidence in some literature, is not entirely out of the question), direct radiators are considered more suitable. Given the physical restrictions in the majority of control rooms it is likely that conventional loudspeakers will be more practical to install than dipoles (for the reason that dipoles, by their nature, need to be free-standing, away from the walls) whereas conventional speakers can be mounted flush with surfaces.

A lot depends on the application, since film sound mixing has somewhat different requirements from some other forms of mixing, and is intended for large auditoria. Much music and television sound is intended for small-room listening and is mixed in small rooms. This was also the primary motivation behind the use of dipoles in consumer environments – that is, the translation of the large-room listening experience into the small room. In large rooms the listener is typically further into the diffuse field than in small rooms, so film mixes made in large dubbing stages tend not to sound right in smaller rooms with highly directional loudspeakers. Dipoles or arrays can help to translate the listening experience of large-room mixes into smaller rooms.

17.9.5 Subwoofers

Low-frequency interaction between loudspeakers and rooms has a substantial bearing on the placement of subwoofers or low-frequency loudspeakers. There appears to be little agreement about the optimum location for a single subwoofer in a listening room, although it has been suggested that a corner location for a single subwoofer provides the most extended, smoothest low-frequency response. In choosing the optimum locations for subwoofers one must remember the basic principle that loudspeakers placed in corners tend to give rise to a noticeable bass boost, and couple well to most room modes (because they have antinodes in the corners). Some subwoofers are designed specifically for placement in particular locations whereas others need to be moved around until the most subjectively satisfactory result is obtained. Some artificial equalisation may be required to obtain a reasonably flat overall frequency response at the listening position. Phase shifts or time-delay controls are sometimes provided to enable some correction of the time relationship of the subwoofer to other loudspeakers, but this will necessarily be a compromise with a single unit. A subwoofer phase shift can be used to optimise the sum of the subwoofer and main loudspeakers in the crossover region for a flat response.

There is some evidence to suggest that multiple low-frequency drivers generating decorrelated signals from the original recording create a more natural spatial reproduction than monaural low-frequency reproduction from a single driver. Griesinger proposes that if monaural LF content is reproduced it is better done through two units placed to the sides of the listener, driven $90°$ out of phase, to excite the asymmetrical lateral modes more successfully and improve LF spaciousness.

Others warn of the dangers of multiple low-frequency drivers, particularly the problem of mutual coupling between loudspeakers that takes place when the driver spacing is less than about half a wavelength. In such situations the outputs of the drivers couple to produce a level greater than would be predicted from simple summation of the powers. This is due to the way in which the drivers couple to the impedance of the air and the effect that one unit has on the radiation impedance of the other. The effect of this coupling will depend on the positions to which sources are panned between drivers, affecting the compatibility between the equalisation of mixes made for different numbers of loudspeakers.

17.10 Surround sound recording techniques

This section deals with the extension of conventional two-channel recording technique to multiple channels for surround sound applications, concentrating on standard 5(.1)-channel reproduction. Many of the concepts described here have at least some basis in conventional two-channel stereo, although analysis of the psychoacoustics of 5.1 surround has been nothing like as exhaustively investigated to date. Consequently a number of the techniques described below are at a relatively early stage of development and are still being evaluated.

The section begins with a review of microphone techniques that have been proposed for the pickup of natural acoustic sources in surround, followed by a discussion of multichannel panning and mixing techniques, mixing aesthetics and artificial reverberation, for use with more artificial forms of production such as pop music. Film sound approaches are not covered in any detail as they are well established and not the main theme of this book.

17.10.1 Principles of surround sound microphone technique

Surround sound microphone technique, as discussed here, is unashamedly biased towards the pickup of sound for 5.1 surround, although Ambisonic techniques are also covered because they are well documented and can be reproduced over five-channel loudspeaker systems if required, using suitable decoders. The techniques described in this section are most appropriate for use when the spatial acoustics of the environment are as important as those of the sources within, such as in classical music and other 'natural' recording. These microphone techniques tend to split into two main groups: those that are based on a single array of microphones in reasonably close proximity to each other, and those that treat the front and rear channels separately. The former are usually based on some theory that attempts to generate phantom images with different degrees of accuracy around the full 360° in the horizontal plane. (The problems of this are outlined in Fact File 17.6.) The latter usually have a front array providing reasonably accurate phantom images in the front, coupled with a separate means of capturing the ambient sound of the recording space (often for feeding to all channels in varying degrees). It is rare for such microphone techniques to provide a separate feed for the LFE channel, so they are really five-channel techniques not 5.1-channel techniques.

The concept of a 'main array' or 'main microphone configuration' for stereo sound recording is unusual to some recording engineers, possibly being a more European than American concept. The traditional European approach has tended to involve starting with a main microphone technique of some sort that provides a basic stereo image and captures the spatial effect of the recording environment in an aesthetically satisfactory way, and then supporting this subtly to varying degrees with spot mics as necessary. It has been suggested by some that many balances in fact end up with more sound coming from the spot mics than from the main array in practice, and that in this case it is the spatial treatment of the spot mics and any artificial reverberation that will have most effect on the perceived

FACT FILE 17.6   Surround imaging

It is difficult to create stable phantom images to the sides of a listener in a standard 5.1 surround configuration, using simple pairwise amplitude or time differences. If the listener turns to face the speaker pair then the situation may be improved somewhat, but the subtended angle of about 80° still results in something of a hole in the middle and the same problem as before then applies to the front and rear pairs. Phantom sources can be created between the rear speakers but the angle is again quite great (about 140°), leading to a potential hole in the middle for many techniques, with the sound pulling towards the loudspeakers. This suggests that those techniques attempting to provide 360° phantom

imaging may meet with only limited success over a limited range of listening positions, and might imply that one would be better off working with two- or three-channel stereo in the front and decorrelated ambient signals in the rear.

There is no escaping the fact that it is easiest to create images where there are loudspeakers, and that phantom images between loudspeakers subtending wide angles tend to be unstable or 'hole-in-the-middle'. Given this unavoidable aspect of surround sound psychoacoustics, one should always expect imaging in standard five-channel replay systems to be best between the front loudspeakers, only moderate to the rear, and highly variable to the sides, as shown below. Since the majority of material one listens to tends to conform to this paradigm in any case (primary sources in front, secondary content to the sides and rear), the problem is possibly not as serious as it might seem.

Good phantom images
between left, centre and right
loudspeakers

Typically poor and
unstable phantom images
between front and
surround loudspeakers

Typically poor and
unstable phantom images
between front and
surround loudspeakers

Only moderately satisfactory
phantom images between rear
loudspeakers, with a tendency
towards a 'hole in the middle'

result. This is covered in the next section and the issue is open to users for further experimentation.

One must accept also that the majority of consumer systems will have great variability in the location and nature of the surround loudspeakers, making it unwise to set too much store by the ability of such systems to enable accurate soundfield reconstruction in the home. Better, it seems, would be to acknowledge the limitations of such systems and to create recordings that work best on a properly configured reproduction arrangement but do not rely on 100 per cent adherence to a particular reproduction alignment and layout, or on a limited 'hot spot' listening position. Surround sound provides an opportunity to create something that works over a much wider range of listening positions than two-channel stereo, does not collapse rapidly into the nearest loudspeaker when one moves, and enhances the spatial listening experience.

17.10.2 Five-channel 'main microphone' arrays

Recent interest in five-channel recording has led to a number of variants on a common theme involving fairly closely spaced microphones (often cardioids) configured in a five-point array. The basis of most of these arrays is pair-wise time–intensity trading, usually treating adjacent microphones as pairs covering a particular sector of the recording angle around the array. The generic layout of such arrays is shown in Figure 17.18. Cardioids or even supercardioids tend to be favoured because of the increased direct-to-reverberant pickup they offer, and the interchannel level differences created for relatively modest spacings and angles, enabling the array to be mounted on a single piece of metalwork. The centre microphone is typically spaced slightly forward of the L and R microphones thereby introducing a useful time advance in the centre channel for centre-front sources.

The spacing and angles between the capsules are typically based on the so-called 'Williams curves', based on time and amplitude differences required between single pairs of microphones to create phantom sources in particular locations. (In fact the Williams curves were based on two-channel pairs and loudspeaker reproduction in

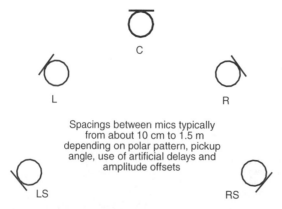

Spacings between mics typically
from about 10 cm to 1.5 m
depending on polar pattern, pickup
angle, use of artificial delays and
amplitude offsets

Figure 17.18 Generic layout of five-channel microphone arrays based on time–amplitude trading front of the listener. It is not necessarily the case that the same technique can be applied to create images between pairs at the sides of the listener, or that the same level and time differences will be suitable. There is some evidence that different delays are needed between side and rear pairs than those used between front pairs, and that inter-microphone crosstalk can affect the accuracy of stereo imaging to varying degrees depending on the array configuration and microphone type.) One possible configuration of many that satisfy Williams' psychoacoustic criteria is pictured in Figure 17.19. To satisfy the requirements for this particular array the front triplet is attenuated by 2.4 dB in relation to the back pair.

Some success has also been had by the author's colleagues using omni microphones instead of cardioids, with appropriate adjustments to the spacings according to 'Williams-style' time–amplitude trading curves (also with modifications to correct for different inter-loudspeaker angles and spacings to the sides and rear). These tend to give better overall sound quality but (possibly unsurprisingly) poorer front imaging. Side imaging has proved to be better than expected with omni arrays.

The closeness between the microphones in these arrays is likely to result in only modest low-frequency decorrelation between the channels. Good LF decorrelation is believed to be important for creating a sense of spaciousness, so these 'near-coincident' or 'semi-correlated' techniques will be less spacious than more widely spaced microphone arrays. Furthermore, the strong dependence of these arrays on precedence effect cues for localisation makes their performance quite dependent on listener position and front–rear balance.

The INA (Ideale Nieren Anordnung) or 'Ideal Cardioid Array' (devised by Hermann and Henkels) is a three-channel front array of cardioids (INA-3) coupled

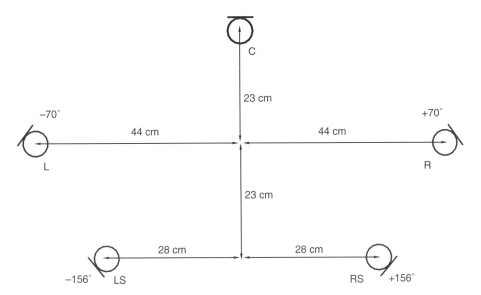

Figure 17.19 Five-channel microphone array using cardioids, one of a family of arrays designed by Williams and Le Dû. In this example the front triplet should be attenuated 2.4 dB with respect to the rear pair

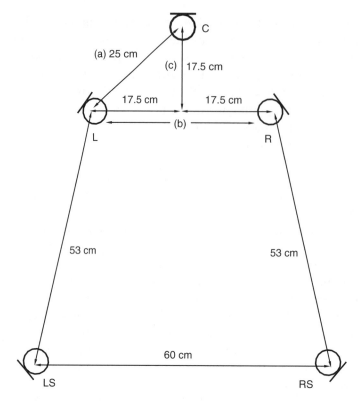

Figure 17.20 INA-5 cardioid array configuration (see Table 17.1)

with two surround microphones of the same polar pattern (making it into an INA-5 array). One configuration of this is shown in Figure 17.20, and a commercial implementation by Brauner is pictured in Figure 17.21. Table 17.1 shows some possible combinations of microphone spacing and recording angle for the front three microphones of this proposed array. In the commercial implementation the capsules can be moved and rotated and their polar patterns can be varied. The configuration shown in Figure 17.20 is termed an 'Atmokreuz' (atmosphere cross) by the authors. Its large front recording angle of 180° means that to use it as a main microphone it would have to be placed very close to the source unless all the sources were to appear to come from near the centre. This might make it less well placed for the surroundings. Such a configuration may be more suitable for general pickup slightly further back in the hall.

17.10.3 Separate treatment of front imaging and ambience

Many alternative approaches to basic microphone coverage for 5.1 surround treat the stereo imaging of front signals separately from the capture of a natural-sounding spatial reverberation and reflection component, and some are hybrid approaches without a clear theoretical basis. Most do this by adopting a three-channel variant on a conventional two-channel technique for the front channels,

Figure 17.21 SPL Atmos 5.1 Surround Recording System. (Courtesy of Sound Performance Lab)

as introduced in the previous chapter (sometimes optimised for more direct sound than in a two-channel array), coupled with a more or less decorrelated combination of microphones in a different location for capturing spatial ambience (sometimes fed just to the surrounds, other times to both front and surrounds). Sometimes the

Table 17.1 Dimensions and angles for the front three cardioid microphones of the INA array (see Figure 17.20). Note that the angle between the outer microphones should be the same as the recording angle

Recording angle $(\vartheta)^\circ$	Microphone spacing (a) cm	Microphone spacing (b) cm	Array depth (c) cm
100	69	126	29
120	53	92	27
140	41	68	24
160	32	49	21
180	25	35	17.5

front microphones also contribute to the capture of spatial ambience, depending on the proportion of direct to reflected sound picked up, but the essential point here is that the front and rear microphones are not intentionally configured as an attempt at a 360° imaging array.

The so-called 'Fukada Tree', shown in Figure 17.22, is based on a Decca Tree, but instead of using omni mics it mainly uses cardioids. The reason for this is to reduce the amount of reverberant sound pickup by the front mics. Omni outriggers are sometimes added as shown, typically panned between L–LS and R–RS, in an attempt to increase the breadth of orchestral pickup and to integrate front and rear elements. The rear mics are also cardioids and are typically located at approximately the critical distance of the space concerned (where the direct and reverberant components are equal). They are sometimes spaced further back than the front mics by nearly 2 metres, although the dimensions of the tree can be varied according to the situation, distance, etc. (Variants are known that have the rear mics quite close to the front ones, for example.) The spacing between the mics more closely fulfils requirements for the decorrelated microphone signals needed to create spaciousness, depending on the critical distance of the space in which they are used. (Mics should be separated by at least the room's critical distance for adequate decorrelation.) The front imaging of such an array would be similar to that of an ordinary Decca Tree (not bad, but not as precise as some other techniques).

The Dutch recording company, Polyhymnia International, has developed a variant on this approach that uses omnis instead of cardioids, to take advantage of their better sound quality. Using an array of omnis separated by about 3 metres between left–right and front–back they achieve a spacious result where the rear channels are well integrated with the front. The centre mic is placed slightly forward

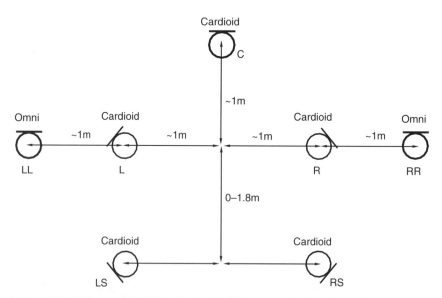

Figure 17.22 The so-called 'Fukada Tree' of five spaced microphones for surround recording

of left and right. It is claimed that placing the rear omnis too far away from the front tree makes the rear sound detached from the front image, so one gets a distinct echo or repeat of the front sound from the rear.

Hamasaki of NHK (the Japanese broadcasting company) has proposed an arrangement based on near-coincident cardioids (30 cm) separated by a baffle, as shown in Figure 17.23. Here the centre cardioid is placed slightly forward of left and right, and omni outriggers are spaced by about 3 metres. These omnis are low-pass filtered at 250 Hz and mixed with the left and right front signals to improve the LF sound quality. Left and right surround cardioids are spaced about 2–3 metres behind the front cardioids and 3 metres apart. An ambience array is used further back, consisting of four figure-eight mics facing sideways, spaced by about 1 metre, to capture lateral reflections, fed to the four outer channels. This is placed high in the recording space.

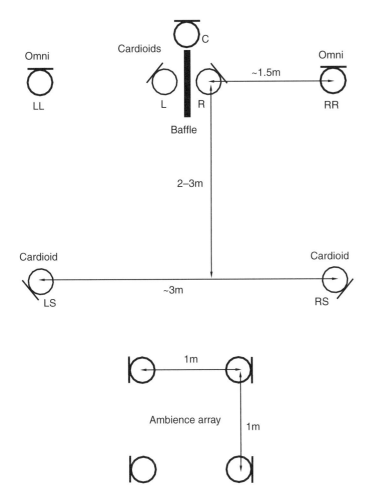

Figure 17.23 A surround technique proposed by Hamasaki (NHK) consisting of a cardioid array, omni outriggers and separate ambience matrix

Figure 17.24 Theile's proposed three-channel array for front pickup using supercardioids for the outer mics, crossed over to omni at LF. The spacing depends on the recording angle (C – R = 40 cm for 90° and 30 cm for 110°)

Theile proposes a front microphone arrangement shown in Figure 17.24. While superficially similar to the front arrays described in the previous section, he reduces crosstalk between the channels by the use of supercardioid microphones at ±90° for the left and right channels and a cardioid for the centre. (Supercardioids are more directional than cardioids and have the highest direct/reverberant pickup ratio of any first-order directional microphone. They have a smaller rear lobe than hypercardioids.) Theile's rationale behind this proposal is the avoidance of crosstalk between the front segments. He proposes to enhance the LF response of the array by using a hybrid microphone for left and right, that crosses over to omni below 100 Hz, thereby restoring the otherwise poor LF response. The centre channel is high-pass filtered above 100 Hz. Furthermore, the response of the supercardioids should be equalised to have a flat response to signals at about 30° to the front of the array (they would normally sound quite coloured at this angle). Schoeps has developed a prototype of this array, and it has been christened 'OCT' for 'Optimum Cardioid Triangle'.

For the ambient sound signal, Theile proposes the use of a crossed configuration of microphones, that has been christened the 'IRT cross' or 'atmo-cross'. This is shown in Figure 17.25. The microphones are either cardioids or omnis, and the spacing is chosen according to the degree of correlation desired between the channels. Theile suggests 25 cm for cardioids and about 40 cm for omnis, but says that this is open to experimentation. Small spacings are appropriate for more accurate

Figure 17.25 The IRT 'atmo-cross' designed for picking up ambient sound for routing to four loudspeaker channels (omitting the centre). Mics can be cardioids or omnis (wider spacing for omnis)

Front MS pair — Cardioid (forward), Figure-eight (sideways)

– – – – – – – – – – – – – – – – – – – Critical distance

Rear MS pair

Figure 17.26 Double MS pair arrangement with small spacing between front and rear pair

imaging of reflection sources at the hot spot, whereas larger spacings are appropriate for providing diffuse reverberation over a large listening area. The signals are mixed in to L, R, LS and RS channels, but not the centre.

A 'double MS' technique has been proposed by Curt Wittig and others, shown in Figure 17.26. Two MS pairs (see previous chapter) are used, one for the front channels and one for the rear. The centre channel can be fed from the front M microphone. The rear pair is placed at or just beyond the room's critical distance. S gain can be varied to alter the image width in either sector, and the M mic's polar pattern can be chosen for the desired directional response (it would typically be a cardioid). Others have suggested using a fifth microphone (a cardioid) in front of the forward MS pair, to feed the centre channel, delayed to time align it with the pair. If the front and rear MS pairs are co-located it may be necessary to delay the rear channels somewhat (10–30 ms) so as to reduce perceived spill from front sources into rear channels. In a co-located situation the same figure-eight microphone could be used as the S channel for both front and back pairs.

In general, the signals from separate ambience microphones fed to the rear loudspeakers may often be made less obtrusive and front–back 'spill' may be reduced by rolling off the high-frequency content of the rear channels. Some additional delay may also assist in the process of integrating the rear channel ambience. The precise values of delay and equalisation can only really be arrived at by experimentation in each situation.

17.10.4 Pseudo-binaural techniques

As with two-channel stereo, some engineers have experimented with pseudo-binaural recordings techniques intended for loudspeaker reproduction. Jerry Bruck adapted the Schoeps 'Sphere' microphone (described earlier) for surround sound purposes by adding bi-directional (figure-eight) microphones near to the 'ears' (omni mics) of the sphere, with their main axis front–back, as pictured in Figure 17.27. This microphone is now manufactured by Schoeps as the KFM360. The figure-eights are mounted just below the sphere transducers so as to affect their

(a)

(b)

Figure 17.27 (a) Schoeps KFM360 sphere microphone with additional figure-eights near the surface-mounted omnis. (b) KFM360 control box. (Courtesy of Schalltechnik Dr.-Ing. Schoeps GmbH)

frequency response in as benign a way as possible for horizontal sources. The outputs from the figure-eight and the omni at each side of the sphere are MS matrixed to create pairs of roughly back-to-back cardioids facing sideways. The size of the sphere creates an approximately ORTF spacing between the side-facing pairs. The matrixed output of this microphone can be used to feed four of the channels in a five-channel reproduction format (L, R, LS and RS). A Schoeps processing unit can be used to derive an equalised centre channel from the front two, and enables the patterns of front and rear coverage to be modified.

Michael Bishop of Telarc has reportedly adapted the 'double MS' technique described in the previous section by using MS pairs facing sideways, and a dummy head some 1–2.5 m in front, as shown in Figure 17.28. The MS pairs are used between side pairs of channels (L and LS, R and RS) and line-up is apparently

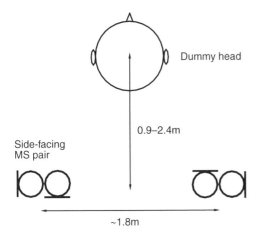

Dummy head

0.9–2.4m

Side-facing
MS pair

~1.8m

Figure 17.28 Double MS pairs facing sideways used to feed the side pairs of channels combined with a dummy head facing forwards to feed the front image

tricky. The dummy head is a model equalised for a natural response on loudspeakers (Neumann KU100) and is used for the front image.

17.10.5 Multimicrophone techniques

Most real recording involves the use of spot microphones in addition to a main microphone technique of some sort, indeed in many situations the spot microphones may end up at higher levels than the main microphone or there may be no main microphone. The principles outlined in the previous chapter still apply in surround mixing, but now one has the issue of surround panning to contend with. The principles of this are covered in more detail in section 17.11.

Some engineers report success with the use of multiple sphere microphones for surround balances, which is probably the result of the additional spatial cues generated by using a 'stereo' spot mic rather than a mono one, avoiding the flatness and lack of depth often associated with panned mono sources. Artificial reverberation of some sort is almost always helpful when trying to add spatial enhancement to panned mono sources, and some engineers prefer to use amplitude-panned signals to create a good balance in the front image, plus artificial reflections and reverberation to create a sense of spaciousness and depth.

17.10.6 Ambisonic or 'Soundfield' microphone principles

The so-called 'Soundfield' microphone, pictured in Figure 17.29, is designed for picking up full periphonic sound in the Ambisonic A-format (see section 17.8.2), and is coupled with a control box designed for converting the microphone output into both the B-format and the D-format. Decoders can be created for using the output of the Soundfield microphone with a 5.1-channel loudspeaker array, including that recently introduced by Soundfield Research. The full periphonic effect can only be obtained by reproduction through a suitable periphonic decoder and the use of a tetrahedral loudspeaker array with a height component, but the effect is quite stunning and worth the effort.

(a)

(b)

(c)

Figure 17.29 (a) The Soundfield microphone, (b) accompanying control box, (c) capsule arrangement. (Courtesy of SoundField Ltd)

The Soundfield microphone is capable of being steered electrically by using the control box, in terms of either azimuth, elevation, tilt or dominance, and as such it is also a particularly useful stereo microphone for two-channel work. The microphone encodes directional information in all planes, including the pressure and velocity components of indirect and reverberant sounds.

Figure 17.29(b) shows the physical capsule arrangement of the microphone, which was shown diagrammatically in Figure 17.13. Four capsules with sub-cardioid polar patterns (between cardioid and omni, with a response equal to $2 + \cos \vartheta$) are mounted so as to face in the A-format directions, with electronic equalisation to compensate for the inter-capsule spacing, such that the output of the microphone truly represents the soundfield at a point (true coincidence is maintained up to about 10 kHz). The capsules are matched very closely and each contributes an equal amount to the B-format signal, thus resulting in cancellation between variations in inherent capsule responses. The A-format signal from the microphone can be converted to B-format according to the equations given in section 17.8.2.

The combination of B-format signals in various proportions can be used to derive virtually any polar pattern in a coincident configuration, using a simple circuit as shown in Figure 17.30 (two-channel example). Crossed figure-eights are the most obvious and simple stereo pair to synthesise, since this requires the sum-and-difference of X and Y, whilst a pattern such as crossed cardioids requires that the omni component be used also, such that:

$$\text{Left} = W + (X/2) + (Y/2)$$
$$\text{Right} = W + (X/2) - (Y/2)$$

Figure 17.30 Circuit used for controlling stereo angle and polar pattern in Soundfield microphone. (Courtesy of Ken Farrar)

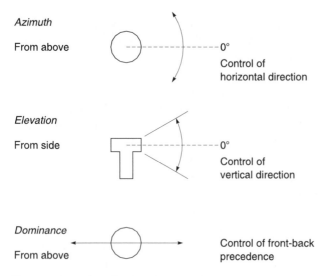

Figure 17.31 Azimuth, elevation and dominance in Soundfield microphone

From the circuit it will be seen that a control also exists for adjusting the effective angle between the synthesised pair of microphones, and that this works by varying the ratio between X and Y in a sine/cosine relationship.

The microphone may be controlled, without physical re-orientation, so as to 'point' in virtually any direction (see Figure 17.31). It may also be electrically inverted, so that it may be used upside-down. Inversion of the microphone is made possible by providing a switch which reverses the phase of Y and Z components. W and X may remain unchanged since their directions do not change if the microphone is used upside-down.

17.11 Multichannel panning techniques

The panning of signals between more than two loudspeakers presents a number of psychoacoustic problems, particularly with regard to appropriate energy distribution of signals, accuracy of phantom source localisation, off-centre listening and sound timbre. A number of different solutions have been proposed, in addition to the relatively crude pairwise approach used in much film sound, and some of these are outlined below. The issue of source distance simulation is also discussed.

Here are Michael Gerzon's criteria for a good panning law for surround sound:

The aim of a good panpot law is to take monophonic sounds, and to give each one amplitude gains, one for each loudspeaker, dependent on the intended illusory directional localisation of that sound, such that the resulting reproduced sound provides a convincing and sharp phantom illusory image. Such a good panpot law should provide a smoothly continuous range of image directions for any direction between those of the two outermost loudspeakers, with no 'bunching' of images close to any one direction or 'holes' in which the illusory imaging is very poor.

17.11.1 Pairwise amplitude panning

Pairwise amplitude panning is the type of pan control most recording engineers are familiar with, as it is the approach used on most two-channel mixers. As described in the previous chapter, it involves adjusting the relative amplitudes between a pair of adjacent loudspeakers so as to create a phantom image at some point between them. This has been extended to three front channels and is also sometimes used for panning between side loudspeakers (e.g. L and LS) and rear loudspeakers. The typical sine/cosine panning law devised by Blumlein for two-channel stereo is often simply extended to more loudspeakers. Most such panners are constructed so as to ensure constant power as sources are panned to different combinations of loudspeakers, so that the approximate loudness of signals remains constant.

Panning using amplitude or time differences between widely spaced side loudspeakers is not particularly successful at creating accurate phantom images. Side images tend not to move linearly as they are panned and tend to jump quickly from front to back. Spectral differences resulting from differing HRTFs of front and rear sound tend to result in sources appearing to be spectrally split or 'smeared' when panned to the sides.

In some mixers designed for five-channel surround work, particularly in the film domain, separate panners are provided for L–C–R, LS–RS, and front-surround. Combinations of positions of these amplitude panners enable sounds to be moved to various locations, but some more successfully than others. For example, sounds panned so that some energy is emanating from all loudspeakers (say, panned centrally on all three pots) tend to sound diffuse for centre listeners, and in the nearest loudspeaker for those sitting off-centre. Joystick panners combine these amplitude relationships under the control of a single lever that enables a sound to be 'placed' dynamically anywhere in the surround soundfield. Moving effects made possible by these joysticks are often unconvincing and need to be used with experience and care.

Research undertaken by Jim West at the University of Miami showed that, despite the limitations of constant power 'pairwise' panning, it proved to offer reasonably stable images for centre and off-centre listening positions, for moving and stationary sources, compared with some other more esoteric algorithms. Front–back confusion was noticed in some cases, for sources panned behind the listener.

17.11.2 'Ambisonic' panning laws

A number of variations of panning laws loosely based on Ambisonic principles have been attempted. These are primarily based on the need to optimise psychoacoustic localisation parameters according to low- and high-frequency models of human hearing. Gerzon proposed a variety of psychoacoustically optimal panning laws for multiple speakers that can theoretically be extended to any number of speakers. Some important features of these panning laws are:

- There is often output from multiple speakers in the array, rather than just two.
- They tend to exhibit negative gain components (out-of-phase signals) in some channels for some panning positions.
- The channel separation is quite poor.

A number of authors have shown how this type of panning could be extended to five-channel layouts according to the standards of interest in this book. McKinnie proposed a five-channel panning law based on similar principles, suitable for the standard loudspeaker angles. It is shown in Figure 17.32. Moorer also proposed some four and five channel panning laws, pictured in Figure 17.33 (only half the circle is shown because the other side is symmetrical). They differ because Moorer has chosen to constrain the solution to first order spatial harmonics (a topic beyond the scope of this book). He proposes that the standard ±30° angle for the front loudspeakers is too narrow for music, and that it gives rise to levels in the centre channel that are too high in many cases to obtain adequate L–R decorrelation, as well as giving rise to strong out-of-phase components. He suggests at least ±45° to avoid this problem. Furthermore, he states that the four-channel law is better behaved with these particular constraints and might be more appropriate for surround panning.

17.11.3 Head-related panning

Horbach of Studer has proposed alternative panning techniques based on Theile's 'association model' of stereo perception. This uses assumptions similar to those used for the Schoeps 'sphere' microphone, based on the idea that 'head-related' or pseudo-binaural signal differences should be created between the loudspeaker signals to create natural spatial images. It is proposed that this can work without crosstalk cancelling, but that crosstalk cancelling can be added to improve the full 3D effect for a limited range of listening positions.

Figure 17.32 Five channel panning law based on Gerzon's psychoacoustic principles. (Courtesy of Douglas McKinnie)

(a)

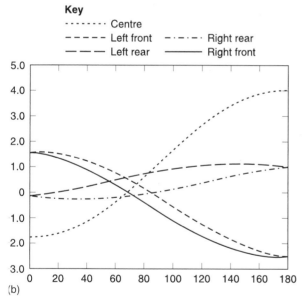

(b)

Figure 17.33 Two panning laws proposed by Moorer designed for optimum velocity and energy vector localisation with 2nd spatial harmonics constrained to zero. (a) Four-channel soundfield panning. The front speakers are placed at 30° angles left and right, and the rear speakers are at 110° left and right. (b) This shows an attempt to perform soundfield panning across five speakers where the front left and right are at 30° angles and the rear left and right are at 110° angles. Note that at zero degrees, the centre speaker is driven strongly out of phase. At 180°, the centre speaker is driven quite strongly, and the front left and right speakers are driven strongly out of phase. At low frequencies, the wavelengths are quite large and the adjacent positive and negative sound pressures will cancel out. At higher frequencies, their energies can be expected to sum in an RMS sense. (Courtesy of James A. Moorer)

In creating his panning laws, Horbach chooses to emulate the response of a simple spherical head model that does not give rise to the high-frequency peaks and troughs in response typical of heads with pinnae. This is claimed to create a natural frequency response for loudspeaker listening, very similar to that which would arise from a sphere microphone used to pick up the same source. Sources can be panned outside the normal loudspeaker angle at the front by introducing a basic crosstalk cancelling signal into the opposite front loudspeaker (e.g. into the right when a signal is panned left). Front–back and centre channel panning are incorporated by conventional amplitude control means. He also proposes using a digital mixer to generate artificial echoes or reflections of the individual sources, routed to appropriate output channels, to simulate the natural acoustics of sources in real spaces, and to provide distance cues.

Recommended further reading

AES (2001) *Proceedings of the 19th International Conference: Surround Sound – Techniques, Technology and Perception.* Audio Engineering Society

AES (2001) Technical document ESTD1001.0.01-05: *Multichannel surround sound systems and operations.* Available from web site: *http://www.aes.org*

Holman, T. (1999) *5.1 Surround Sound: Up and running.* Focal Press

ITU-R (1993) *Recommendation BS. 755: Multi-channel stereophonic sound system with or without accompanying picture.* International Telecommunications Union

Rumsey, F. (2001) *Spatial Audio.* Focal Press

Glossary of terms

AAC Advanced Audio Coding.

ABR Auxiliary Bass Radiator.

AC Alternating current.

A/D Analogue-to-digital conversion.

AES/EBU Audio Engineering Society/European Broadcasting Union: typically refers to a standard digital audio interface.

AF Audio Frequency.

AFL After Fade Listen.

AGC Automatic Gain Control.

Aliasing The generation of in-band spurious frequencies caused by using a sampling rate which is inadequate for the chosen frequency range, i.e. it is less than twice the highest frequency present in the signal.

AM Amplitude Modulation.

Amp The unit of electrical current, named in honour of André Marie Ampère (1775–1836).

Anechoic Chamber A highly absorbtive room; the walls, floor and ceiling of which are virtually non-reflective. Used for acoustical measurements of devices such as microphones and loudspeakers.

Antinode The part of a waveform where its propagating medium has maximum acceleration and minimum velocity.

ATRAC Adaptive Transform Acoustic Coding.

Bandwidth The range of frequencies over which a device will operate; formally, the measurement is taken from the points at which the response is 3 dB down at the frequency extremes compared with mid-frequencies.

BCD Binary Coded Decimal.

Bell Curve A narrow-band EQ curve shaped like a bell when displayed on a frequency response graph.

Bias An ultrasonic frequency added to the audio frequencies sent to an analogue tape recorder's record head to bias the tape into a more linear part of its operating range.

BPF Band Pass Filter

CD Compact Disc.

CMRR   Common Mode Rejection Ratio.

Compansion   The complete process of compression during the recording or transmission stages followed by reciprocal expansion during the replay or reception stages.

Compliance   'Springiness'.

CRC   Cyclic Redundancy Check.

Crossover   A device in a loudspeaker system for splitting the audio signal into frequency bands which can then be fed to appropriate speaker drive units (e.g. low frequency, mid range, and high frequency). A passive crossover is placed between the power amplifier and the speaker drivers. An active crossover splits the frequencies at line level ahead of the power amplifiers.

CU   Compliance unit.

D/A   Digital-to-Analogue conversion.

Damping factor   In power amplifiers, the impedance of the speaker it is driving divided by the amplifier's output impedance. It is an indication of the ability of an amplifier to control spurious movements of the speaker cones (particularly at low frequencies) which can be caused by resonances and energy storage in the drivers' suspension systems.

DASH   Digital Audio Stationary Head. An open-reel digital format.

DAT   Digital Audio Tape.

dBm   Signal level in decibels, referred to one milliwat (0 dBm = 775 mV across 600 ohms).

dBu   Signal level in decibels, referred to 0.775 V with an unspecified impedance value (775 mV = 0 dBu).

dBV   Signal level in decibels, referred to 1 volt (0 dBV = 1 volt).

dBv   The same as dBu, sometimes used in the USA.

DC   Direct current.

DCC   Digital Compact Cassette.

Decibel (dB)   In audio, the unit used to denote the logarithm of the ratio between two quantities, e.g. voltages or power levels. Also used to denote acoustical sound pressure level. Named in honour of Alexander Graham Bell (1847–1922).

De-emphasis   Reciprocal treble cut (see also **Pre-emphasis**) during a replay or reception process.

DI   Direct Inject.

Directivity   Defines the angle of coverage of a loudspeaker's output.

Directivity Factor   The number which denotes the ratio between a sound source's output on its axis of maximum radiation and its output if it were perfectly omni-directional and its total acoustical output were to be evenly spread all around it.

Directivity Index   Directivity Factor expressed in dB.

Dispersion   See **Directivity**.

Dither   A continuous low-level noise signal added to the programme prior to A/D conversion and quantisation or during signal processing.

DML   Distributed-mode Loudspeaker.

Drive unit   See **Driver**.

Driver   The component of a speaker system which actually vibrates or 'drives' the air, e.g. a speaker cone or tweeter.

DSD   Direct Stream Digital.

DSP   Digital Signal Processing.

DST   Direct Stream Transfer.

DVD   Digital Versatile Disc.

EBU   European Broadcasting Union.

Eigentone   A standing wave in a room which is set up when half the wavelength of a sound or a multiple of it is equal to one of the dimensions of the room (height, width, length).

EIN   Equivalent Input Noise.

EQ   Equalisation.

FET   Field Effect Transistor.

FIR   Finite Impulse Response.

FM   Frequency Modulation.

FX   Effects.

Haas Effect   If two sound sources emit similar sounds but one is delayed with respect to the other (up to about 50 mS) the ears perceive the non-delayed sound source to be the louder of the two, and the sound appears to come from a direction close to the non-delayed source, the exact location depending on the amount of delay between them. Beyond the 50 mS time difference, the ears tend to perceive the sounds as coming from two distinct sources.

Harmonics   (Also known as Overtones or Partials.) Components of a waveform which are multiples of the fundamental frequency; together with the starting transient they contribute much to the character or tone colour of a sound.

HDTV   High Definition Television.

Hertz (Hz)   The unit of vibration in cycles per second, named in honour of Heinrich Rudolf Hertz (1857–1894).

HPF   High Pass Filter.

HRTF   Head Related Transfer Function.

IIR   Infinite Impulse Response.

Impedance   Measured in ohms, it is a device's opposition to the flow of AC current. Reactive devices such as loudspeakers, capacitors and inductors exhibit impedances which vary with frequency.

LED   Light Emitting Diode.

Longitudinal wave   A wave in which the to and fro movement of the wave carrier is in the same plane as the wave's travel. A sound wave is an example of this, the source of sound pushing and pulling in the direction of the wave.

LP   Long Playing gramophone record.

LPF   Low Pass Filter.

MADI   Multichannel Audio Digital Interface.

Masking   A psychacoustic phenomenon, whereby quiet sounds in the presence of loud sounds and/or sounds with a similar frequency content will be rendered less audible.

MD   Minidisc.

MDCT   Modified Discrete Cosine Transform.

MFM   Miller Frequency Modulation.

MIDI   Musical Instrument Digital Interface.

MLP   Meridian Lossless Packing.

MMC   MIDI Machine Control.

MOL   Maximum Output Level.

MOSFET   Metal Oxide Semiconductor Field Effect Transistor.

MP3   Short for MPEG-1 Layer 3.

MPEG   Moving Pictures Expert Group. ('Empeg'.)

MPX   Multiplex.

MS   Main(or Middle)-Side.

MTC   MIDI Time Code.

NC   Noise Criterion.

Node   The part of a waveform where its propagating medium has maximum velocity and minimum acceleration.

Noise shaping   The technique of reducing noise in perceptually sensitive regions of the audio band at the expense of increasing it above the audio band, or in less sensitive regions, thereby improving the perceived signal-to-noise ratio. It is used in digital systems such as that used in SACD and over-sampling CD players.

NR   Noise Reduction. (Also Noise Rating.)

nWb/m   Nanowebers per metre. The unit of flux along a magnetic recording medium, named in honour of Wilhelm Eduard Weber (1804–1891).

Nyquist frequency   Half the sampling frequency in a digital system.

Ohm   The unit of resistance and impedance, named in honour of Georg Simon Ohm (1789–1854).

Overtones   See **Harmonics**.

PA   Public Address. (Also Power Amplifier.)

Pad   An input attenuator.

PAM   Pulse Amplitude Modulation.

Partials   See **Harmonics**.

PCM   Pulse Code Modulation.

PD   ProDigi. An open-reel digital format.

PDM   Pulse Density Modulation.

PFL   Pre-Fade Listen.

Phase   Two waves of the same frequency are 'in phase' when their positive and negative half-cycles coincide exactly in time. For example, two loudspeakers are in phase if, when fed by the same source signal, their cones move backwards and forwards in step with each other, and their acoustical outputs reinforce. If they are out of phase, cancellation of sound results. Electrical signals can similarly be in or out of phase, or in any intermediate relationship.

Phase response   A measure of phase lead or lag of signals across the frequency range as they pass through an electrical circuit or device.

Phon   A unit denoting the subjective loudness of a sound, the scale derived from research data.

Pink Noise   Noise which has equal energy per octave. Its frequency spectrum is therefore flat when the usual logarithmic horizontal frequency scale is used.

Power bandwidth   Superficially similar to **Bandwidth** (*qv*) for a power amplifier, but it is the range of frequencies over which the amplifier can deliver full power, with −3 dB being allowed at the frequency extremes. A power amplifier's frequency response is normally wider than its power bandwidth.

PPM   Peak Programme Meter.

Pre-emphasis   Treble boost applied during a recording or transmission process. See also **De-emphasis**.

PWM   Pulse Width Modulation.

PZM   Pressure Zone Microphone.

Q   Historically, the 'quality' of a tuned radio-frequency receiving circuit. A good sharp tuning which centred on the station of interest, greatly attenuating the unwanted frequencies to either side, was said to be of high quality or *Q*, and it could be quantified as set out below. In the audio industry it can denote a number of things, including:

(1) the bandwidth or 'sharpness' of an EQ curve. The *Q* is defined as the centre frequency divided by the bandwidth. See section 6.5.1.

(2) In a loudspeaker system: at its low-frequency resonant point, *Q* is the ratio between the speaker's output level here and its output level over the nominally flat part of its frequency range, expressed as a number. For example, if the output is 3 dB down at a speaker's LF resonant frequency (which is fairly typical), it has a *Q* of 0.707 ($20 \log Q = -3$ dB). If it is 6 dB down, it has a *Q* of 0.5 ($20 \log Q = -6$ dB).

Quantisation   After sampling a waveform, the quantisation process assigns a numerical value to each sample according to its amplitude. For example, a 16-bit digital system can assign one of a possible 65 536 values (2^{16}) to a particular sample, with no 'in-between' values being permitted.

RAM   Random Access Memory.

R-DAT   Rotary-head Digital Audio Tape (same as **DAT**).

Resonance   This takes place in a system at frequencies where the balance of its moving mass and its compliance gives rise to regions where a relatively small amount of input energy is required to produce vibrations of large amplitude compared with that required at most other frequencies.

RF   Radio frequency.

RIAA   Recording Industry Association of America.

RMS   Root-mean-square. The RMS heating power of a sine wave is 0.707× the value of its peak-to-peak measurement.

ROM   Read Only Memory.

SACD   Super Audio Compact Disc.

Sampling   (1) The process of encoding a signal digitally by registering it as discrete values of level at specified intervals of time (the sampling frequency), in contrast to analogue recording which registers the waveform continuously.

(2) The process of recording sounds into a 'sampler' which can then be edited and processed in various ways.

SCSI   Small Computer Systems Interface. ('Scuzzy'.)

S-DAT   Stationary-head Digital Audio Tape.

Sensitivity   For present purposes, sensitivity effectively denotes the efficiency with which a transducer converts electrical energy into acoustic energy (e.g. a loudspeaker) or vice versa (e.g. a microphone).

Shelving   Low- or high-frequency boost or cut with a gentle curve up to a 'shelf'.

Signal-to-noise ratio   The ratio in dB between the wanted signal and the unwanted noise in a system.

Sine wave   A wave which is made up of one single frequency.

Sinusoidal waveform   See **Sine wave**.

Slew rate   The maximum rate of change in volts/microsecond of which a circuit's output is capable.

SMART   System Managed Audio Resource Technique.

SMPTE   Society of Motion Pictures and Television Engineers. ('Simpty'.)

Solo   On a mixer, pressing 'solo' on a channel routes its post-fade output to the monitor output. It is the same as AFL.

SPDIF   Sony-Philips Digital Interface.

SPL   Sound pressure level.

SPP   Song Position Pointers.

SR   (1) Sound reinforcement. (2) Spectral recording (Dolby).

Standing wave   A standing wave is the result of reflections from room boundaries reinforcing each other at certain frequencies to create points where the sound pressure level is very high, and other points where it is very low.

SWR   Standing wave ratio.

Sync   Synchronisation.

THD   Total harmonic distortion.

TOA   Time of arrival.

Transducer   A device which converts one form of energy into another form of energy. For example, a loudspeaker converts electrical energy into acoustical energy.

Transverse wave   A wave in which the device or particles creating it move at right angles to the direction of the wave's travel, the carrying medium also oscillating at right angles to the wave's travel. An example is electromagnetic radiation, created by the electrons' up-and-down motion along the length of a transmitting aerial.

UHF   Ultra-high frequency.

UHJ   Universal HJ. The Ambisonics surround sound encoding and decoding system, the H and J denoting earlier systems.

VCA   Voltage controlled amplifier.

VCO   Voltage controlled oscillator.

VHF   Very high frequency.

VITC   ('vitcee') Vertical Interval Time Code.

Volt   The unit of electrical pressure, named in honour of Alessandro Volta (1745–1827).

VTR   Video tape recorder.

VU   Volume Unit.

Watt   The unit of electrical power, named in honour of James Watt (1736–1819).

White noise   Noise which has equal energy per Hz of frequency. Its frequency spectrum therefore rises by 3 dB per octave when the usual logarithmic horizontal frequency scale is used.

WORM   Write Once Read Many times.

XLR   Originally a part code for the ITT-Canon company's professional audio connector, the most familiar of which is the 3-pin microphone and balanced line XLR-3.

Appendix 1

Understanding basic equipment specifications

The performance of an audio system may be measured to determine its effect on a sound signal passed through it, and it may also be assessed subjectively (in other words, by listening to it). Theoretically, if a system introduces an audible modification to the sound signal then one should also be able to measure it, provided that the right test can be devised and suitable equipment is available, but the difficulty of achieving this ideal is always increasing as the audible differences between systems become ever smaller, and the absolute fidelity of recording and reproduction improves. Digital recording and processing has brought with it the possibility for recording and transmission with zero degradation, and there is still considerable debate about just what can be heard and what cannot. In the following sections it is not the intention to get involved in the debates which always rage in hi-fi circles concerning minute differences in sound quality between systems (which do of course exist, but are often explained badly), but rather that the reader should gain an insight into the most commonly encountered system specifications and what they mean, as well as describing the audible effects of different distortions on sound signals.

A1.1 Frequency response – technical

The most commonly quoted specification for a piece of audio equipment is its *frequency response*. It is a parameter which describes the frequency range covered by the device – that is, the range of frequencies which it can record or reproduce. To take a simple view, for high-quality reproduction the device would normally be expected to cover the whole audio-frequency range, which was defined earlier as being from 20 Hz to 20 kHz, although some have argued that a response which extends above the human hearing range has audible benefits. It is not enough, though, simply to consider the range of frequencies reproduced, since this says nothing about the relative levels of different frequencies or the amplitude of signals at the extremes of the range. If further qualification was not given then a frequency response specification of 20 Hz–20 kHz could mean virtually anything. It is important to compare devices' specifications on the same grounds, since otherwise little useful information can be gained.

a)

b)

- - - - -   Rising HF response

—·—·—·—   Uneven response

Figure A1.1 (a) Plot of a flat frequency response from 20 Hz to 20 kHz. (b) Examples of two non-flat frequency responses

The ideal frequency response is one which is 'flat' – that is, with all frequencies treated equally and none amplified more than others. Technically, this means that the gain of the system should be the same at all frequencies, and this could be verified by plotting the amplitude of the output signal on a graph, over the given frequency range, assuming a constant-level input signal. An example of this is shown in Figure A1.1(a), and it will be seen that the graph of output level versus frequency is a straight horizontal line between the limits of 20 Hz and 20 kHz – that is, a flat frequency response. Also shown in Figure A1.1 are examples of non-flat responses, and it will be seen that these boost some frequencies and cut others, affecting the balance between different parts of the sound spectrum (the audible effects of which are discussed in section A1.3). Typically, frequency response is quoted *with reference to the response at 1 kHz*. This means that the output level at 1 kHz is chosen as the level against which all other frequencies are compared, and would be given a relative level of 0 dB for this purpose. If the response at 5 kHz was said to be +3 dB ref. 1 kHz, this would mean that signals of 5 kHz would be amplified 3 dB more than signals at 1 kHz.

Although a graph gives the most detail about frequency response, since it shows what happens at every point in the range, often only figures are given in specifications. It is common to quote frequency response as the upper and lower limits of the frequency range handled by the device, giving two frequencies at which the response is '3 dB down' – in other words, where the response is 3 dB lower than the response at 1 kHz. It is implied in such a case that the response between these points is moderately flat, although this cannot be taken for granted in practice. Thus a response of 45 Hz–17 kHz (−3 dB) suggests that the device handles a frequency range between 45 and 17 000 Hz, at which extremes the response is 3 dB down. Below 45 Hz and above 17 000 Hz one would expect the response to fall off even further.

A more accurate way of specifying frequency response in figures, and one which leaves less room for misinterpretation, is to state a tolerance for allowed variations in level over the specified range. Thus a specification of 45 Hz – 17 kHz (±3 dB ref. 1 kHz) states that the response at any frequency between the limits will not deviate from the response at 1 kHz by more than 3 dB upwards or downwards.

A1.2 Frequency response – practical examples

Some practical examples may help to illustrate the above discussion of frequency response, and a table of typical specifications for a selection of devices is given in Table A1.1, so that they can be compared.

Firstly, purely electronic devices tend to have a flatter response than devices which involve a recording or reproduction process, since the latter usually incorporate mechanical, magnetic or optical processes which are more prone to distortions of all kinds. An amplifier is an example of the former case, and it is unusual to find a well-designed amplifier which does not have a flat frequency response these days – flat often to within a fraction of a decibel from 5 Hz up to perhaps 60 kHz. In the

Table A1.1 Examples of typical frequency responses of audio systems

Device	Typical frequency response
Telephone system	300 Hz–3 kHz
AM radio	50 Hz–6 kHz
Consumer cassette machine	40 Hz–15 kHz (±3 dB)
Professional analogue tape recorder	30 Hz–25 kHz (± 1 dB)
CD player	20 Hz–20 kHz (±0.5 dB)
Good-quality small loudspeaker	60 Hz–20 kHz (−6 dB)
Good-quality large loudspeaker	35 Hz–20 kHz (−6 dB)
Good-quality power amplifier	6 Hz–60 kHz (± 3 dB)
Good-quality omni microphone	20 Hz–20 kHz (±3 dB)

Frequency response – subjective

Subjectively, deviations from a flat frequency response will affect sound quality. If the aim is to carry through the original signal without modifying it, then a flat response will ensure that the original amplitude relationships between different parts of the frequency spectrum are not changed. If, say, low frequencies are boosted with respect to high frequencies, then the original sound will be modified, making it sound more bass heavy. It is important not to be side-tracked by the fact that the human ear's frequency response is not flat (see Chapter 2), since this fact has no bearing on the need for a flat response in audio equipment. In audio equipment the important factor is that sounds come out of a system as they went in.

Some forms of modification to the ideal flat response are more acceptable than others. For example, a gentle roll-off at the high-frequency (HF) end of the range often goes unnoticed, since there is not much sound energy at this point. Domestic cassette machines and FM radio receivers, for example, tend to have upper limits of around 15 kHz, but are relatively flat below this, and thus do not sound unpleasant. Frequency responses which deviate wildly from 'flat' over the audio-frequency range, on the other hand, sound much worse, even if the overall range of reproduction is wider than that of FM radio.

If the frequency response of a system rises at high frequencies then the sibilant components of the sound will be emphasised, music will sound very 'bright' and 'scratchy', and any background hiss will be emphasised. If the response is down at high frequencies then the sound will become dull and muffled, and any background hiss may appear to be reduced. If the frequency response rises at low frequencies then the sound will be more 'boomy', and bass notes will be emphasised. If low frequencies are missing, the sound will be very 'thin' and 'tinny'. A rise in the middle frequency range will result in a somewhat 'nasal' sound, perhaps having a rather harsh quality, depending on the exact frequency range concerned.

Concerning the effects of very low and very high frequencies, close to the limits of human hearing, it can be shown that the reproduction of sounds below 20 Hz does sometimes offer an improved listening experience, since it can cause realistic vibrations of the surroundings. Also, the ear's frequency response does not cut off suddenly at the extremes, but gradually decreases, and thus it is not true that one hears nothing below 20 Hz and above 20 kHz – one simply hears much less. Similarly, extended HF responses can sometimes help sound quality, and a gentle HF roll-off usually implies less steep filtering of the signal which in turn may result in improved quality.

---

other category, though, there are LP turntables, tape recorders, loudspeakers and microphones, to name but a few, and these are all much more difficult to design for a flat response.

Devices which convert sound into electricity or vice versa (in other words, transducers) are most prone to frequency response errors, and some loudspeakers have been known to exhibit deviations of some 10 dB or more from 'flat'. Since such devices are also affected by the acoustics of their surroundings it is difficult to divorce a discussion of their own response from a discussion of the way in which they interact with their surroundings. The room in which a loudspeaker is placed has a significant effect on the perceived response, since the room will resonate at certain frequencies, creating pressure peaks and troughs throughout the room. Depending on the location of the listener, some frequencies may be emphasised more than others, and this therefore makes it difficult to say what is the fault of the room and what is the fault of the speaker. A loudspeaker's response can be measured in so-called 'anechoic' conditions, where the room is totally absorbent and cannot produce significant effects of its own, although other methods now exist which do not require the use of such a room. A good loudspeaker will have a response which covers the majority of the audio-frequency range, with a tolerance of perhaps ±3 dB, but the LF end is less easy to extend than the HF end. Smaller loudspeakers will only go down to perhaps 50 or 60 Hz.

Analogue magnetic tape recorders use a number of equalisation processes to ensure that the frequency response is as flat as possible, but this can usually only be achieved with one of a few specified tape types and formulations. An unsuitable tape may result in a non-flat response, unless the machine can be realigned for the new tape. Cassette machines have a number of different tape-type settings to ensure that the frequency response and other parameters are optimum for each tape formulation. The frequency response of an analogue tape machine is likely to vary with recording level, since at higher recording levels the high frequencies become 'compressed' because the tape is incapable of retaining them. For this reason the response of such a machine is usually quoted at a relatively low recording level, perhaps 20 dB below reference level (see section 8.5).

LP records are equalised before they are cut, intentionally to give them a non-flat response, but this is re-equalised before reproduction to restore the correct frequency balance. The reason for this is explained in section A2.2. If the RIAA equaliser in the amplifier which reproduces the recording is not properly designed then it will not restore the correct frequency balance, and this can occur in cheap hi-fi equipment.

Microphones vary enormously in their characteristics, and their frequency response depends a lot on their polar pattern and design (see Chapter 4). Cheap consumer microphones may have a response which only extends up to 10 or 12 kHz, whereas professional mics may cover a range at least up to 20 kHz. The LF end of the spectrum is equally variable, with omnidirectional microphones having a much more extended LF response than other pickup patterns. A microphone's response often varies with the angle of incidence of the source.

A1.3 Harmonic distortion – technical

Harmonic distortion is another common parameter used in the specifications of audio systems. Such distortion is the result of so-called 'non-linearity' within a device – in other words, what comes out of the device is not exactly what went in. There are a number of types of non-linearity, but here it is the type which affects the shape of the waveform that is referred to. In Chapter 1 it was shown that only simple sinusoidal waveforms are completely 'pure', consisting only of one frequency without harmonics. More complex repetitive waveforms can be analysed into a set of harmonic components based on the fundamental frequency of the wave. Harmonic distortion in audio equipment arises when the shape of the sound waveform is changed slightly between input and output, such that harmonics are introduced into the signal which were not originally present, thus modifying the sound to some extent (see Figure A1.2). It is virtually impossible to avoid a small amount of harmonic distortion, since no device carries through a signal *entirely* unmodified, but it can be reduced to extremely low levels in amplifiers.

Harmonic distortion is normally quoted as a percentage of the signal which caused it (e.g.: THD 0.1 per cent @ 1 kHz), but, as with frequency response, it is important to be specific about what type of harmonic distortion is being quoted, and under what conditions. One should distinguish between *third*-harmonic distortion and *total* harmonic distortion, and unfortunately both can be abbreviated to 'THD' (although THD most often refers to total harmonic distortion). Total harmonic distortion is the sum of the contributions from all the harmonic components introduced by the device, assuming that the original wave has been filtered out, and is normally measured by introducing a 1 kHz sine wave into the device and measuring the resulting distortion at a recognised input level. The level and frequency of the sine wave used depends very much on the type of device and the test standard used. Third-harmonic distortion is a measurement of the amplitude of the third harmonic of the input frequency only, and is commonly found in tape recorder tests since the third harmonic is the most prominent in magnetic recording systems.

It may be important to be specific about the level and frequency at which the distortion specification is made, since in many audio devices distortion varies enormously with these parameters.

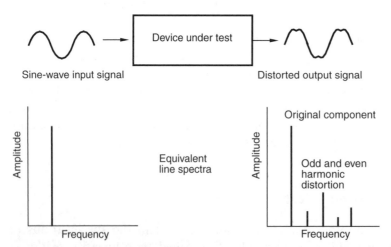

Figure A1.2 A sine wave input signal is subject to harmonic distortion in the device under test. The waveform at the output is a different shape to that at the input, and its equivalent line spectrum contains components at harmonics of the original sine-wave frequency

A1.4 Harmonic distortion – practical examples

In electrical devices such as amplifiers, distortion percentage does not usually change much with input level, but may vary slightly with frequency. With tape recorders distortion is very much a function of recording level and frequency, and can vary widely. In transducers, distortion usually remains at a moderately constant percentage with input level variation, but cheaper transducers can introduce fairly high levels of distortion.

Table A1.2 shows some typical quoted THD percentages for different audio devices, and it will be seen that they vary widely, with amplifiers and digital audio equipment having the lowest typical figures.

The distortion characteristics of digital audio systems are discussed in more detail in Chapter 10, and will not be covered further here. Analogue tape machine distortion performance is often quoted in the form of a *maximum output level* or *MOL* figure, which is the recording level at which third-harmonic distortion reaches a certain percentage, considered to be the maximum sensible recording level. In professional recorders this is 3 per cent at 1 kHz, and in consumer cassette machines it is 5 per cent at 315 Hz. Typically, one might expect this distortion percentage to be reached at a recording level of around 10–12 dB above reference level in professional recorders with a good modern tape, and around 4–8 dB above reference level

FACT FILE
A1.2

Harmonic distortion – subjective

Harmonic distortion is not always unpleasant, indeed many people find it quite satisfying and link it with such subjective parameters as 'warmth' and 'fullness' in reproduced sound, calling sound which has less distortion 'clinical' and 'cold'. Since the distortion is harmonically related to the signal which caused it, the effect may not be unmusical and may serve to reinforce the pitch of the fundamental in the case of even-harmonic distortion.

The sound of third-harmonic distortion is easy to detect on pure tones, but less easy on music, and can be heard when recording a tone on to a tape recorder at high level whilst comparing the sound 'off tape' with the input signal. The tone no longer sounds 'pure', but

has an edge to it. It contains a component one octave and a fifth above the fundamental tone.

Because distortion tends to increase gradually with increasing recording level in tape recorders, the onset of distortion is less noticeable than it is when an amplifier 'clips', for example, and many analogue tape recordings contain large percentages of harmonic distortion which have been deemed acceptable. Amplifier clipping, on the other hand, is very sudden and results in a 'squaring-off' of the audio waveform when it exceeds a certain level, at which point the distortion becomes severe. This effect can be heard when the batteries are going flat on a transistor radio, or when a hi-fi loudspeaker is driven exceedingly hard from a low-powered amplifier, and sounds like a serious breaking up of the sound on peaks of the signal. If tested with sine-wave sources, the result is as shown in Fact File 7.2.

446   Appendix 1

Table A1.2 Typical THD percentages

Device	% THD
Good power amplifier @ rated power	< 0.05% (20 Hz–20 kHz)
16 bit digital recorder (via own convertors)	< 0.05% (−15 dB input level)
Loudspeaker	< 1% (25 W, 200 Hz)
Professional analogue tape recorder	< 1% (ref. level, 1 kHz)
Professional capacitor microphone	< 0.5% (1 kHz, 94 dB SPL)

in cassette machines. Analogue tape machines and reference levels are discussed in more detail in section 8.5.

A1.5 Dynamic range and signal-to-noise ratio

Dynamic range and signal-to-noise (S/N) ratio are often considered to be interchangeable terms for the same thing. This may be true, but depends on how the figures are arrived at. S/N ratio is normally considered to be the number of decibels between the 'reference level' and the noise floor of the system (see Figure A1.3). The noise floor may be weighted according to one of the standard curves which attempts to account for the potential 'annoyance' of the noise by amplifying some parts of the frequency spectrum and attenuating others (see Fact Files 1.4 and A1.3). Dynamic range may be the same thing, or it may be the number of decibels between the *peak* level and the noise floor, indicating the 'maximum-to-minimum' range of signal levels which may be handled by the system. Either parameter quoted without qualification is difficult to interpret.

For example, the specification 'Dynamic range = 68 dB' for a tape recorder means very little, since there is no indication of the reference points or weightings, whereas 'S/N ratio, CCIR 468-3 (ref. 1 kHz, 320 nWb m^{-1}) = 68 dB' tells the reader virtually all that is required. It says that the noise has been measured to the CCIR 468-3 weighting standard, and that it measures at 68 dB below the level of a 1 kHz

Figure A1.3 Signal-to-noise ratio is often quoted as the number of decibels between the reference level and the noise floor. Available dynamic range may be greater than this, and is often quoted as the difference between the peak level and the noise floor

---

FACT FILE A1.3 — Noise weighting curves

As discussed in Fact File 1.4, weighting filters are used when measuring noise to produce a figure which more closely represents the subjective annoyance value of the noise.

Some examples of regularly used weighting curves are shown in the diagram, and it will be seen that they are similar but not the same. Here 0 dB on the vertical axis represents the point at which the gain of the filter is 'unity', that is where it neither attenuates or amplifies the signal. The 'A' curve is not normally used for measuring audio equipment noise, since it was designed for measuring acoustic background noise in buildings. The various DIN and CCIR curves are more commonly used in audio equipment specifications.

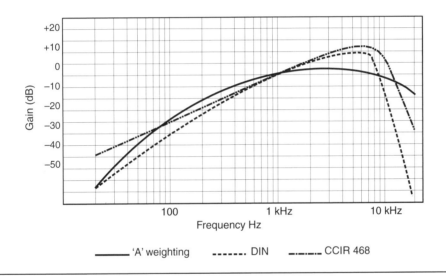

—— 'A' weighting   ------ DIN   –·–·– CCIR 468

tone recorded at a magnetic level of 320 nWb m^{-1}. This could at least be compared directly with another machine measured in the same way, even if the reference level was different, although the difference between the two reference levels would have to be taken into account. It is difficult, though, to compare S/N ratios between devices measured using different weighting curves.

In analogue tape recorders, dynamic range is sometimes quoted as the number of decibels between the 3 per cent MOL (see section A1.4) and the weighted noise floor. This gives an idea of the available recording 'window', since the MOL is often well above the reference level. In digital recorders, the peak recording level is really also the reference level, since there is no point in recording above this point due to the sudden clipping of the signal, and thus dynamic range and S/N ratio are often referred to this point, although some manufacturers have chosen to refer to a level 15 dB below it.

Table A1.3 Typical values for CCIR weighted S/N ratio

Device	S/N ratio
Consumer cassette machine (without noise reduction)	50 dB (ref. 315 Hz, 200 nWb m^{-1})
Professional analogue tape machine (without noise reduction) @ 15 ips (38 cm s^{-1})	65 dB (ref. 1 kHz, 320 nWb m^{-1})
16 bit digital audio recorder	94 dB (ref. peak level)
Professional power amplifier	108 dB (ref. max. output)

Table A1.3 gives some typical values for S/N ratio found in audio equipment.

A1.6 Wow and flutter

Wow and flutter are names used to describe speed (pitch) variations of a tape machine or turntable. Wow is applied to slow variations in speed and flutter is applied to faster variations in speed. The figures depend very much on the mechanical quality of the device and its state of wear and cleanliness. Again a weighting filter (usually to the DIN standard) is used when measuring to produce a figure which closely correlates with one's perception of the annoyance of speed variations. Specifications are usually quoted as WRMS (Weighted Root-Mean Square, a form of average), but occasionally peak figures may be used which will be worse than the RMS figures. Long-term speed accuracy is also quoted in many cases – this being the anticipated overall drift in speed of the machine over a reel of tape. These days speed drift is less of a problem than it used to be, with machines remaining stable to within hundredths of a per cent over the length of a reel. Drift is only really a problem if two machines are to be synchronised.

A typical figure for a good analogue tape machine would be better than 0.02 per cent WRMS, and good cassette machines can approach this figure also. Cheap cassette machines coupled with poor tapes can cause the figure to rise considerably, with some examples approaching 0.5 per cent or more. A good LP turntable may achieve 0.02 per cent results, but again cheaper models will be worse. Digital audio recorders and CD players do not suffer from wow and flutter in the same way as analogue transports, since the audio data from the tape or disk is first passed through a so-called *timebase corrector*, prior to conversion, which removes any speed variations resulting from mechanical instability in the transport.

A machine with poor W&F results will sound most unpleasant, with either uncomfortable 'wowing' in the pitch of notes, or fast flutter which gives rise to a 'roughness' in the sound aptly described by the word 'flutter', and possibly some intermodulation distortion (see section A1.7).

A1.7 Intermodulation (IM) distortion

IM distortion results when two or more signals are passed through a non-linear device. Since all audio equipment has some non-linearity there will always be small amounts of IM distortion, but these can be very low.

Unlike harmonic distortion, IM distortion may not be harmonically related to the frequency of the signals causing the distortion, and thus it is audibly more unpleasant. If two sine-wave tones are passed through a non-linear device, sum and difference tones may arise between them (see Figure A1.4). For example, a tone at $f_1 = 1000$ Hz and a tone at $f_2 = 1100$ Hz might give rise to IM products at $f_1 - f_2 = 100$ Hz, and also at $f_1 + f_2 = 2100$ Hz, as well as subsidiary products at $2f_1 - f_2$ and so on. The dominant components will depend on the nature of the non-linearity.

IM distortion can also arise when speed variations of a tape transport or LP turntable modulate the signals reproduced from them. For example, a tape transport with speed variations at 25 Hz modulating a reproduced signal at 1000 Hz could give rise to IM products at 975 Hz and 1025 Hz.

Low IM distortion figures are an important mark of a high-quality system, since such distortion is a major contributor to poor sound quality, but it is less often quoted than THD (see section A1.4).

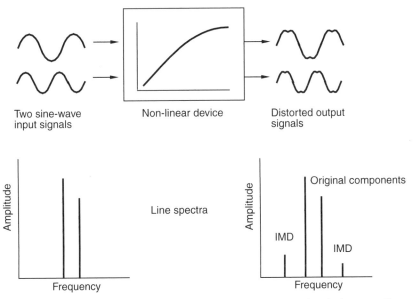

Figure A1.4 Intermodulation distortion between two input signals in a non-linear device results in low level sum-and-difference components in the output signal

A1.8 Crosstalk

Crosstalk figures describe the expected amount of break-through from one channel of a device to another. For example, in a stereo tape recorder crosstalk may arise between the left and right channels, or in a multitrack recorder between adjacent tracks. In general, crosstalk is undesirable. It is usually quoted either as negative decibels relative to the causatory signal (e.g.: −53 dB), or as decibels channel separation (e.g.: 53 dB).

Crosstalk may arise within the electronics of a device (such as by electromagnetic induction between tracks on a printed-circuit board), magnetically (by induction within the heads of a tape machine), or externally (say between cables in a duct). For good performance in a multitrack tape machine and mixer it is essential that crosstalk is very low, since the operator will not want components of one channel being audible on another. For stereo recorders and reproducers the requirement is not so stringent.

A typical figure for a good multitrack analogue recorder in reproduce mode is between 40 and 50 dB, but this is usually much worse between tracks in record and adjacent tracks in sync reproduce (see Fact File 8.3). In digital equipment the crosstalk between channels is exceptionally low (around −90 dB), since crosstalk is rejected as part of the replay decoding process. In analogue LP cartridges separation is quite poor, as it is also in analogue stereo FM radio and TV broadcasting (around 25–30 dB), but is normally adequate for maintaining stereo separation.

Record players

Record players are obsolete technology, but it is felt that information should remain in the literature describing how to get the best out of the format, not least because inadequate or inept setting up of equipment can cause permanent damage to records, and because it seems likely that the LP will persist for some time yet. Also, there is still valuable material to be found here which is not always available on CD.

A2.1 Pickup mechanics

The replay stylus motion should describe an arc offset from the vertical by 20°, as shown in Figure A2.1. This will be achieved if the arm height at the pivot is adjusted such that the arm tube is parallel to the surface of the record when the stylus is resting in the groove. The stylus tip should have a cone angle of 55°, as shown in Figure A2.2. The point is rounded such that the tip makes no contact with the bottom of the groove. Stylus geometry is discussed further in Fact File A2.1.

The arm geometry is arranged so that a line drawn through the cartridge body, front to back, forms a tangent to the record groove at a point where the stylus rests in the groove, at two points across the surface of the record: the outer groove and the inner position just before the lead-out groove begins. Figure A2.3 illustrates this. Note that the arm tube is bent to achieve the correct geometry. Alternatively, the arm tube can be straight with the cartridge headshell set at an offset angle which achieves the same result. The arc drawn between the two stylus positions shows the horizontal path of the stylus as it plays the record. Due to the fact that the arm has a fixed pivot, it is not possible for the stylus to be exactly tangential to the groove throughout its entire travel across the record's surface, but setting up the arm to meet this ideal at the two positions shown gives a good compromise, and a correctly designed and installed arm can give less than ±1° tracking error throughout the whole of the playing surface of the disc.

Alignment protractors are available which facilitate the correct setting up of the arm. These take the form of a rectangular piece of card with a hole towards one end which fits over the centre spindle of the turntable when it is stationary. It has

FACT FILE

A2.1

Stylus profile

Two basic cross-sectional shapes exist for a replay stylus – conical and elliptical, as shown in the diagram. The elliptical profile can be seen to have a smaller contact area with the

Conical   Elliptical

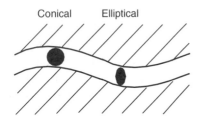

wall of the groove, and this means that for a given tracking weight (the downforce exerted by the arm on to the record surface) the elliptical profile exerts more force per unit area than does the conical tip. To compensate, elliptical styli have a specified tracking force which is less than that for a conical tip. The smaller contact area of the elliptical tip enables it to track the small, high-frequency components of the signal in the groove walls, which have short wavelengths, more faithfully. This is particularly advantageous towards the end of the side of the record where the groove length per revolution is shorter and therefore the recorded wavelength is shorter for a given frequency. Virtually all high-quality styli have an elliptical profile or esoteric variation of it, although there are still one or two high-quality conical designs around. The cutting stylus is, however, always conical.

a series of parallel lines marked on it (tangential to the record grooves) and two points corresponding to the outer and inner groove extremes. The stylus is lowered on to these two points in turn and the cartridge and arm are set up so that a line drawn through the cartridge from front to back is parallel to the lines on the protractor.

The original cutting stylus is driven across the acetate in a straight line towards the centre, using a carriage which does not have a single pivot point like the replay arm, and it can therefore be exactly tangential to the groove all the way across the disk. The cutting lathe is massively engineered to provide an inert, stable platform. There are some designs of record player which mimic this action so that truly zero tracking error is achieved on replay. The engineering difficulties involved in implementing such a technique are probably not justified since a well-designed and well-set-up arm can achieve excellent results using just a single conventional pivot.

A consequence of the pivoted arm is that a side thrust is exerted on the stylus during play which tends to cause it to skate across the surface of the record. This is a simple consequence of the necessary stylus overhang in achieving low tracking error from an arm which is pivoted at one end. Consider Figure A2.4. Initially it can be considered that the record is not rotating and the stylus simply rests in the groove. Consider now what happens when the record rotates in its clockwise direction. The stylus has an immediate tendency to drag across the surface of the record in the arrowed direction towards the pivot rather than along the record groove. The net effect is that the stylus feels a force in a direction towards the centre of the record

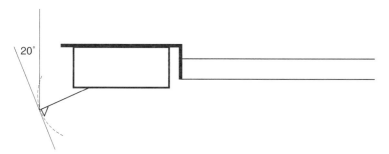

Figure A2.1 Stylus vertical tracking geometry

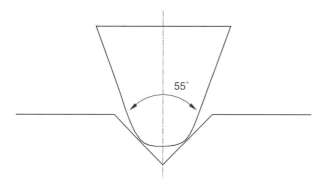

Figure A2.2 Stylus cone angle

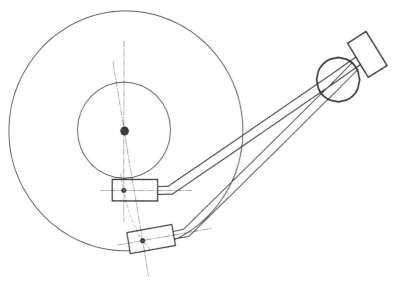

Figure A2.3 Ideal lateral tracking is achieved when a line through the head-shell forms a tangent to the groove

Figure A2.4 The rotation of the disk can create a force which pulls the arm towards the centre of the disc

causing it to bear harder on the inner wall of the groove than on the outer wall. One stereo channel will therefore be tracked more securely than the other, and uneven wear of the groove and stylus will also result. To overcome this, a system of bias compensation or 'anti-skating' is employed at the pivot end of the arm which is arranged so that a small outward force is exerted on the arm to counteract its natural inward tendency. This can be implemented in a variety of ways, including a system of magnets; or a small weight and thread led over a pulley which is contrived so as to pull the arm outwards away from the centre of the record; or a system of very light springs. The degree of force which is needed for this bias compensation varies with different stylus tracking forces but it is in the order of one-tenth of that value (see Fact File A2.2).

Although every cartridge will fit into every headshell apart from one or two special types, one must be aware of certain specifications of both the arm and the cartridge in order to determine whether the two are compatible. In order for the stylus to move about in the groove, the cantilever must be mounted in a suitable suspension system so that it can move to and fro with respect to the stationary cartridge body. This suspension has compliance or springiness and is traditionally specified in (cm/dyne) $\times$ 10^{-6}, abbreviated to cu ('compliance units'). This is a measure of how many centimetres (in practice, fractions of a centimetre!) the stylus will deflect when a force of 1 dyne is exerted on it. A low-compliance cartridge will have a compliance of, say, 8 cu. Highest compliances reach as much as 45 cu. Generally, values of 10–30 cu are encountered, the value being given in the maker's specification.

FACT FILE A2.2   Tracking weight

The required weight varies from cartridge to cartridge. A small range of values will be quoted by the manufacturer such as '1 gram ±0.25 grams' or '1–2 grams' and the exact force must be determined by experiment in conjunction with a test record. Firstly, the arm and cartridge must be exactly balanced out so that the arm floats in free air without the stylus moving either down towards the record surface or upwards away from it, i.e.: zero tracking force. This is generally achieved by moving the counterweight on the end of the arm opposite to the cartridge either closer to or further away from the pivot until an exact balance point is found. The counterweight is usually moved by rotating it along a thread about the arm, or alternatively a separate secondary weight is moved. This should be carried out with bias compensation off.

When a balance has been achieved, a tracking weight should be set to a value in the middle of the cartridge manufacturer's values. Either the arm itself will have a calibrated tracking force scale, or a separate stylus balance must be used. The bias compensation should then be set at the appropriate value, which again will have either a scaling on the arm itself or an indication in the setting up instructions. A good way to set the bias initially is to lower the stylus towards the play-in groove of a rotating record such that the stylus initially lands mid-way between these widely spaced grooves on an unused part of the surface of the record. Too much bias will cause the arm to move outwards before dropping into the groove. Too little bias will cause the arm to move towards the centre of the record before dropping into the groove. Just the right amount will leave the arm stationary until the relative movement of the groove itself eventually engages the stylus. From there, the optimum tracking and bias forces can then be determined using the test record according to the instructions given. In general, a higher tracking force gives more secure tracking but increases record wear. Too light a tracking force, though, will cause mistracking and damage to the record grooves.

A2.2 RIAA equalisation

The record groove is an analogue of the sound waves generated by the original sources, and this in itself caused early pioneers serious problems. In early electrical cutting equipment the cutter stylus *velocity* remained roughly constant with frequency, for a constant input voltage (corresponding to a falling amplitude response with frequency) except at extreme LF where it became of more constant *amplitude*. Thus, unequalised, low frequencies would cause stylus movements of considerably greater excursion per cycle for a given stylus velocity than at high frequencies. It would be difficult for a pickup stylus and its suspension system inside the cartridge body to handle these relatively large movements, and additionally low frequencies would take up relatively more playing surface or 'land' on the record curtailing the maximum playing time. Low-frequency attenuation was therefore used during cutting to restrict stylus excursions.

Figure A2.5 RIAA recording and reproducing characteristics

In modern record cutting, a standard known as RIAA equalisation has been adopted which dictates a recorded *velocity* response, no matter what the characteristics of the individual cutting head. Electrical equalisation is used to ensure that the recorded velocity corresponds to the curve shown in Figure A2.5(a). A magnetic replay cartridge will have an output voltage proportional to stylus *velocity* (its unequalised output would rise with frequency for a constant amplitude groove) and thus its output must be electrically equalised according to the curve shown in Figure A2.5(b) in order to obtain a flat voltage–frequency response.

The treble pre- and de-emphasis of the RIAA replay curve have the effect of reducing HF surface noise. An additional recommendation is very low 20 Hz bass cut on replay (time constant 7960 μs) to filter out subsonic rumble and non-programme-related LF disturbance. The cartridge needs to be plugged into an input designed for this specific purpose; the circuitry will perform the above discussed replay equalisation as well as amplification.

A2.3 Cartridge types

The vast majority of cartridges in use are of the moving-magnet type, meaning that the cantilever has small powerful magnets attached which are in close proximity to the output coils. When the stylus moves the cantilever to and fro, the moving magnets induce current in the coils to generate the output. The DC resistance of the coils tends to be several hundred ohms, and the inductance several hundred millihenries (mH). The output impedance is therefore 'medium', and rises with frequency due to the inductance. The electrical output level depends upon the velocity with which the stylus moves, and thus for a groove cut with constant deviation the output of the cartridge would rise with frequency at 6 dB per octave.

The velocity of the stylus movements relative to an unmodulated groove is conveniently measured in cm s^{-1}, and typical output levels of moving magnet cartridges are in the order of 1 mV cm^{-1} s^{-1}.

The average music programme produces cartridge outputs of several millivolts, and an upper limit of 40 or 50 mV will occasionally be encountered at mid frequencies. Due to the RIAA recording curve, the output will be less at low frequencies but not necessarily all that much more at high frequencies owing to the falling power content of music with rising frequency. A standard input impedance of an RIAA input of 47 k has been adopted, and around 40 dB of gain ($\times 100$) is needed at mid frequencies to bring the signal up to line level.

Another type of cartridge which is much less often encountered but has a strong presence in high-quality audio circles is the moving-coil cartridge. Here, the cantilever is attached to the coils rather than the magnets, the latter being stationary inside the cartridge body. These cartridges tend to give much lower outputs than their moving-magnet counterparts because of the need to keep coil mass low by using a small number of turns, and they have a very low output impedance (a few ohms up to around a hundred) and negligible inductance. They require 20–30 dB more gain than moving magnets do, and this is often provided by a separate head amplifier or step-up transformer, although many high-quality hi-fi amplifiers provide a moving-coil input facility. The impedance of such inputs is around 100 ohms or so.

A2.4 Connecting leads

Owing to the inductive nature of the output impedance of a moving-magnet cartridge, it is sensitive to the capacitance present in the connecting leads and also that present in the amplifier input itself. This total capacitance appears effectively in parallel with the cartridge output, and thus forms a resonant circuit with the cartridge's inductance. It is the high-frequency performance of the cartridge which is affected by this mechanism, and the total capacitance must be adjusted so as to give the best performance. Too little capacitance causes a frequency response which tends to droop several decibels above about 5 kHz, sharply rising again to a 2–3 dB peak with respect to the mid band at around 18–20 kHz, which is the result of the resonant frequency of the stylus/record interface, the exact value depending upon the stylus tip mass. Adding some more capacitance lifts the 5–10 kHz trough and also curtails the tip mass resonant peak to smooth out the frequency response. Too much capacitance causes attenuation of the highest frequencies giving a dull sound.

Around 300–400 pF total is the usual range of capacitance to be tried. Adding capacitance is most conveniently carried out using special in-line plugs containing small capacitors for this purpose. Assume that 200 pF is present already, and try an extra 100 pF, then 200 pF. Alternatively, small polystyrene capacitors can be purchased and soldered between signal and earth wires of the leads inside the plugs or sockets. Sometimes solder tags are present in the base of the record player itself which are convenient. NEVER solder anything to the cartridge pins. The cartridge can very easily be damaged by doing this.

Moving-coil cartridges have a very low output impedance and negligible inductance, and their frequency response is therefore not affected by capacitance.

A2.5 Arm considerations

The effective mass of the pickup arm (Figure A2.6), which is the inertial mass of the arm felt by the stylus, coupled with the cartridge's suspension compliance, together form a resonant system, the frequency of which must be contrived such that it is low enough not to fall within the audio band but high enough to avoid coinciding with record warp frequencies and other LF disturbances which would continually excite the resonance causing insecure tracking and even groove jumping. Occasionally, large, slow excursions of the cones of the speaker woofers can be observed when the record is being played, which is the result of non-programme-related, very low-frequency output which results from an ill-matched arm/cartridge combination.

A value of 10–12 Hz is suitable, and a simple formula exists which enables the frequency to be calculated for a given combination of arm and cartridge:

$$f = 1000/(2\pi\sqrt{(MC)})$$

where f = resonant frequency in hertz, M = effective mass of the arm + mass of the cartridge + mass of hardware (nuts, bolts, washers) in grams, C = compliance of the cartridge in compliance units.

For example, consider a cartridge weighing 6 g, having a compliance of 25 cu; and an arm of effective mass 20 g, additional hardware a further 1 g. The resonant frequency will therefore be 6.2 Hz. This value is below the optimum, and such a

Figure A2.6 The SME Series V pickup arm. (Reprinted with permission of the Society of Manufacturing Engineers, USA)

combination could give an unsuitable performance due to this resonance being excited by mechanical vibrations such as people walking across the floor, record warps, and vibrations emanating from the turntable main bearing. Additionally, the 'soft' compliance of the cartridge will have difficulty in coping with the high effective mass of the arm, and the stylus will be continually changing its position in the groove somewhat as the arm's high inertia tends to flex the cartridge's suspension and dominate its performance.

If the same cartridge in an arm having an effective mass of 8 g is considered, then $f = 8.4$ Hz. This is quite close to the ideal, and would be acceptable. It illustrates well the need for low-mass arms when high compliances are encountered. The resonance tends to be high Q, and this sharp resonance underlines the need to get the frequency into the optimum range. Several arms provide damping in the form of a paddle, attached to the arm, which moves in a viscous fluid, or some alternative arrangement. This tends to reduce the amplitude of the resonance somewhat, which helps to stabilise the performance. Damping cannot, however, be used to overcome the effects of a non-optimum resonant frequency, which must still be carefully chosen.

A2.6 Laser pickups

The idea of reading a record groove with a laser beam rather than a stylus has been mooted for quite some time, and in 1990 a player using such a technique finally appeared. It is a very attractive proposition for record libraries due to the fact that record wear becomes a thing of the past. However, a laser beam does not push particles of dust aside as does a stylus, and the commercial system needs to be fed with discs which are almost surgically clean, otherwise signal drop-outs occur. Two entirely separate laser beams are in fact used, one reading the information from each wall of the groove. Error concealment circuitry is built in, which suppresses the effects of scratches. Towards the centre of the record, short wavelengths occupy a proportionately smaller area of the groove than at the perimeter of the disc, and the width of the laser beam means that difficulties in reading these high-amplitude HF signals occur. The frequency response of the player therefore droops by around 10 dB if the highest frequencies towards the end of the side are of a high amplitude.

CD-type features are offered such as pause, track repeat and track search, which are very useful. The player does not suffer from traditional record player ills such as LF arm/cartridge resonance, rumble, and wow and flutter. Costing not much under £10 000, the player obviously has a limited appeal, the professional user being the main potential customer.

Recommended further reading

AES (1981) *Disk Recording – An Anthology*, Vols 1 and 2. Audio Engineering Society
BS 7063. British Standards Office

Earl, J. (1973) *Pickups and Loudspeakers*. Fountain Press
Roys, H. E. (1978) ed. *Disk Recording and Reproduction*. Dowden, Hutchinson and
    Ross

See also *General further reading* at the end of this book.

General further reading

Alkin. G. (1989) *Sound Techniques for Video and TV*, 2nd edition. Focal Press

Alkin, G. (1996) *Sound Recording and Reproduction*, 3rd edition. Focal Press

Ballou, G. (2001) ed. *Handbook for Sound Engineers – The New Audio Cyclopedia*, 3rd edition. Focal Press

Borwick, J. (1995) ed. *Sound Recording Practice*. Oxford University Press

Capel, V. (1994) *Newnes Audio and Hi-Fi Engineer's Pocket Book*, 3rd edition. Butterworth-Heinemann

Eargle, J. (1992) *Handbook of Recording Engineering*. Van Nostrand Rheinhold

Eargle, J. (1990) *Music, Sound, Technology*. Van Nostrand Rheinhold

Huber, D. M. (2001) *Modern Recording Techniques*, 5th edition. Focal Press

Leonard, J. A. (2001) *Theatre Sound*. A. and C. Black

Nisbett, A. (1994) *The Sound Studio*, 6th edition. Focal Press

Roberts, R. S. (1981) *Dictionary of Audio, Radio and Video*. Butterworths

Talbot-Smith, M. (1995) *Broadcast Sound Technology*, 2nd edition. Focal Press

Talbot-Smith, M., ed. (1999) *Audio Engineer's Reference Book*, 2nd edition. Focal Press

Woram, J. (1989) *Sound Recording Handbook*. Howard W. Sams and Co.

Index

Italic page numbers refer to illustrations

CHESTER COLLEGE LIBRARY